T0327627

Fire Risk Management

Fire Risk Management

Principles and Strategies for Buildings and Industrial Assets

Luca Fiorentini
Fabio Dattilo

This edition first published 2023
© 2023 John Wiley & Sons Ltd

The right of Luca Fiorentini and Fabio Dattilo to be identified as the authors of this work has been asserted in accordance with law.

Registered Offices
John Wiley & Sons, Inc., 111 River Street, Hoboken, NJ 07030, USA
John Wiley & Sons Ltd, The Atrium, Southern Gate, Chichester, West Sussex, PO19 8SQ, UK

For details of our global editorial offices, customer services, and more information about Wiley products visit us at www.wiley.com.

Wiley also publishes its books in a variety of electronic formats and by print-on-demand. Some content that appears in standard print versions of this book may not be available in other formats.

Library of Congress Cataloging-in-Publication Data:

Names: Fiorentini, Luca, 1976- author. | Dattilo, Fabio, author. | John Wiley & Sons, publisher.
Title: Fire risk management : Principles and Strategies for Buildings and Industrial Assets / Luca Fiorentini, Fabio Dattilo.
Description: Hoboken, New Jersey : JW-Wiley, [2023] | Includes bibliographical references and index.
Identifiers: LCCN 2023021900 (print) | LCCN 2023021901 (ebook) | ISBN 9781119827436 (hardback) | ISBN 9781119827443 (pdf) | ISBN 9781119827450 (epub) | ISBN 9781119827467 (ebook)
Subjects: LCSH: Fire protection engineering. | Fire risk assessment. | Fire prevention. | Risk management.
Classification: LCC TH9145 .F49 2023 (print) | LCC TH9145 (ebook) | DDC 628.9/22--dc23/eng/20230622
LC record available at https://lccn.loc.gov/2023021900
LC ebook record available at https://lccn.loc.gov/2023021901

Cover Image(s): © Possawat/Getty Images, Keith Lance/Getty Images, rocketegg/Getty Images
Cover design: Wiley

Set in 9.5/12.5pt STIXTwoText by Integra Software Services Pvt. Ltd., Pondicherry, India

To my father, Carlo Fiorentini.

The memory of his passion for fire safety, his immeasurable expertise and above all his example at work with TECSA and with fire safety professional associations and also in our family accompanies me in my professional life every day, with the hope that I can always do my best and also leave a small contribution of my own to the world of fire-safety engineering and industrial risk assessment, which he made known to me and which I have always been close to, appreciating this whole world and developing a passion to be part of it.

Luca Fiorentini

To Carlo Fiorentini, father of Luca, pioneer and master in risk assessment and fire safety. His passion for his work, in-depth knowledge and love for his family mark our path like milestones.

Fabio Dattilo

Contents

Foreword

"Fire safety between prescription and performance".

Fire safety, in deliberately general terms, is a discipline of extreme complex application. This is primarily because, although it presents itself as a specialised sector, it affects almost the entirety of the profiles in which the design of an activity is declined; if we think that the fire-safety strategy developed for a given commercial activity conditions the choice of furnishings and fittings, we immediately realise the breadth of the profiles and perspectives impacted by the discipline. Second, because fire safety runs through and affects all phases of the development of an activity, starting from design and ending with daily management, once implemented, the fire-safety strategy must be applied in the operation of the activity and cyclically measured in expected performances.

Considering therefore the breadth of the regulated profiles and the immanence of fire safety in management processes, it is easy to understand how the discipline cannot be relegated among the recurring fulfilments to be carried out once and for all, but must find an integrated place in the production cycle and constitute an opportunity to improve the overall management process of the activity.

A little further elaboration is needed on this point.

In fact, fire safety – particularly in its prevention portion – was perceived as a separate process that had to accompany, through fire-safety design, the technical development of a given project, and that ended with obtaining a favourable opinion from the competent authorities and obtaining certification once the project was implemented (where applicable).

Such an approach was undoubtedly favoured by a prescriptive regulatory approach that, by providing for predetermined standards, allowed for the certainty of compliance once they had been integrated.

In fact, this approach can certainly not be considered the most efficient; the inherent limitation of the prescriptive approach precisely lies in the general and abstract nature of the standardisation and thus in the rigid application of standards:

- on the one hand, they condition the possibility of developing innovative solutions in the case of new works;
- on the other hand, they do not allow the utilisation in the fire strategy of strengths that may be available in existing works and activities through compensation with other requirements that are not fully sufficient.

Granted, with all its limitations, but the prescriptive approach is somewhat reminiscent of the Platonic view of reality in which the project constitutes the ideality and its application represents its imperfect mirror.

The fact is that, probably also due to instances of severe and continuous innovation in architectural, engineering and supporting technological development, a more performance-oriented approach to fire-safety management has progressively established itself, the development of which has gone hand in hand with the affirmation of the centrality of risk assessment and the empowerment of the activity owner in this regard.

In short, the fire protection designer has been allowed to play a central role in constructing the fire protection strategy – as if he or she were the 'demiurge' that connects reality to bring it closer to its ideality – with the owner's guarantee and commitment to ensure that the assumptions underlying the design and expected performance are maintained over time.

Having said this, in such a largely established context, it would make no sense – in addition to being inconsistent with the obligations assumed by the owner with respect to the service – to manage fire prevention 'fulfilments' in a minimal and fractional manner in the context of the entire business cycle.

On the contrary, also thanks to the technological development of support tools, the integration of fire safety into the broader system of business process management constitutes an opportunity for overall improvement, both to strengthen safety and performance monitoring and to extend a participative and conscious approach of all actors involved.

This book, applicable to civil buildings as well as to industrial assets, enforces a holistic view of fire strategy design to be coupled with a conscious management of assets overtime to ensure the maintenance of the performances identified to achieve an acceptable level of fire risk from the earliest design stages and from the risk-based identification of fire scenarios by the competent application of sound approaches and methods.

Damiano Tranquilli
Head of Safety, Environment and Quality of Rete
Ferroviaria Italiana, Direzione operativa stazioni. Head of Safety,
GS Rail, Operations. Italian Ferrovie dello Stato Group

Foreword

According to its current technical definition, *risk* is the potential for realisation of unwanted, adverse consequences to human life, health, property or to the environment. Estimation of risk (for an event) is usually based on the expected value of the conditional probability of the event occurring times the consequence of the event, given that it has occurred. In this context, fire risk management can be considered as the process of firstly understanding and characterising fire hazard in a building, unwanted outcomes that may result from a fire, and secondly developing optimal and robust fire strategies to reduce risk or, at least, control its occurrence.

Recent tragic fire events such as the fire of the Grenfell Tower in London (2017) and of the Torre dei Moro in Milano (2021) have shown the importance of integrating the fire risk analysis from the beginning of the building design process, in order to identify the best fire strategy to be implemented in the construction. In both cases, the composite facades heated up rapidly and allowed the fire to spread faster, pass through windows and advance from floor to floor up and down the building's facade.

In this book, the authors, thanks to their personal experience in fire-safety design and accident analysis, provide a comprehensive treatise of fire risk management. First, they describe recent fires, failed strategies and lessons learned. As a second step, they define the appropriate measures for fire risk assessment and the acceptable fire risk levels (according to national and international rules and performance-based codes) representing the first step in fire risk management. Then, the authors explain the state-of-the-art fire risk assessment and the fire-safety design leading to risk mitigation.

All the aspects of fire risk management are considered, including, for example, fault tree analysis, barrier performance, fire growth, fire spread and smoke movement, compartments, occupant response and evacuation models. Critical aspects of risk, such as the correct analysis of event consequences on people, environment, property and business continuity, are included. Finally, a note on explosions and appendices dedicated to railway stations, process industries and warehouse storage buildings are included.

The wide experience of the authors, both on civil buildings and industrial assets, along with their clarity and scientific rigor, make the book a unique and comprehensive essay on fire risk management.

Prof. Dr. Eng. Bernardino Chiaia
Head of the Center SISCON 'Safety of Infrastructures
and Constructions', Politecnico di Torino (Italy)

Foreword

Fire risk management in contexts where the magnitude of damage is potentially very high is a particularly complex business. The history of major accidents teaches that they are typically determined by a variety of logically connected and antecedent causes to the facts, revealing that prevention is a multidisciplinary and multi-level theme, which is constituted on a stratification of decisions and controls, to be planned and supervised with the highest time priority.

Largest industrial organisations have long time ago understood that serious risks like this – which shake the foundations of entrepreneurial certainties linked to human, industrial, economic and reputational heritage – need to be matched, even before an adequately articulated architecture of measures, an iterative and very robust assessment system in order to properly understand accident phenomena in their possibilities, create organisational awareness and management competence among the professional figures involved, and reach a risk management plan capable of providing adequate strategies and responses.

Process control measures, as well as prevention and protection measures, while qualifying the organisation in terms of performance, activate investment procedures that are sometimes very demanding; therefore, the decisions connected to this must be carefully weighed, making use of all the available technologies and specific competencies to define the best actions to protect safety.

In this perspective, this editorial work is precious because, starting from a very broad and usable explication of the fundamental notions, it allows us to understand the importance of conducting weighted and customer-specific analyses and decisions. In fact, there are different methodologies and approaches for risk assessment, and it is now clear that the same performance result – in the design phase – can be achieved with a different dosage of technical-plant engineering solutions, organisational-managerial solutions and/or behavioural solutions, which turns into different costs and sustainability of the results for the operators or users of the assets. What are the most appropriate choices? What implications and charges do these choices entail on the operational management of processes? Since safety is the ultimate and common goal of all the involved actors, fire risk management is obviously not a theme that is affirmed only when the analysis is carried out, nor it is resolved in the effective completion of an authorisation process: the assessment process must accompany a project from its birth and continue throughout its life, consolidating its being as plural process in terms of ownership, temporal development and a variety of analytical and methodological focuses.

Risk assessment becomes a mindset to be used regularly. Appropriately fast and accurate methodologies must correspond to this; the use of resources must in fact be modular so that the efforts of calculation, representation, discussion and investment are diversified and concentrate where needed. Conversely, adopting inadequate methodologies necessarily involves a high risk on detriment of the asset under consideration, for the simple fact that some risk scenarios may be unknown and therefore not well controlled.

Finally, a good risk assessment provides clear and accurate outputs. Based on this, an effective competence network can be established for the benefit of all components of the organisation concerned. It is no longer just a matter of fostering the ability to react at zero time; rather, the foundations are laid for a widespread governance culture causal elements as well as elements not directly conducive to, obtainable only through an adequate study that moves the centre of the time axis away from the moment of the accident.

Vito Carbonara
Sabo S.p.A. (Italy) –Technical Procurement and Logistics Director

Preface

Heraclitus, an ancient Greek philosopher, asserted that everything in the world flows ('Panta Rei') and that fire represents universal becoming better than anything else because fire itself is the 'arché', the principle from which all things are generated.

For the philosopher, this is becoming not random and chaotic but is regular and orderly, provided one knows the rules.

In this volume, we have tried to explain the complex rules governing fire in a simple way, using methods, from the simplest to the most refined, such as the engineering approach.

Studying the development of smoke and heat in fires, knowing the effects they have on people and buildings, helps a great deal in adopting the right strategies for preventing and containing fires.

But the approach taken in the book is deliberately holistic in the sense that each individual strategy can have a great influence on the others, and therefore fire prevention must be seen as a whole.

And as a whole, the success (or failure) of the strategies implemented also depends on the behaviour of the people involved, behaviour that must be framed within a safety management perspective.

A volume that purports to present the historical discipline of fire prevention but with a new methodological approach based on the performance to be achieved rather than on strongly prescriptive but often uncritical methods and requirements.

Happy reading.

Luca Fiorentini and Fabio Dattilo

Acknowledgments

First of all, we would like to express our sincere thanks to Riccardo Di Camillo (P.Eng.).

Riccardo Di Camillo is Head of Fire Safety and Emergency Planning at Grandi Stazioni Rail S.p.A. – Operations, where he deals with all safety issues including permitting activities for the major Italian railway stations. Given his expertise in dealing with very large and complex railway infrastructures as well as with their renovation and modification plans, Riccardo gave us an important and fundamental support in developing all the fire strategy elements in the chapter with the title 'fire strategies'. Fire risk mitigation should be based on a fire strategy conceived to be reliable over time, focused, auditable, and Riccardo, being a professional engineer specialised in fire-safety engineering, also offered us the practical experience in managing fire strategy elements on a daily basis in complex railway stations and infrastructure. This allowed us to highlight how the link between risk analysis, the basis of a performance approach, must necessarily find fulfilment in the implementation of an effective strategy over time as a commitment by organisations to ensure that an acceptable level of fire risk is maintained over time. Riccardo showed how the effective maintenance of the basic elements of the strategy must take into account the complications associated with the normal day-to-day management of the infrastructure for which he works, posed by the constant transformations during the necessary operational continuity, the presence of the public, the intersection with other infrastructure, and nonetheless the architectural complexity, the extension and the use of historical assets. By masterfully managing these aspects within the scope of his work, relating to all stakeholders, he enabled us to describe in a simple, clear and effective manner the problems and methods to seek their solution in the combination of actions aimed on the one hand at identifying and measuring the fire risk and on the other hand at managing the risk over time.

A heartfelt thank you people at TECSA S.r.l. (www.tecsasrl.it) who deal every day in fire risk assessment and industrial risk assessment consulting activities, overcoming the challenges posed by complexity and sharing the professional growth of the entire organisation that complexity itself poses to all those who are called upon to ensure safety over time. Through TECSA activities it is possible every day to measure oneself against important and unique experiences that impose the need to disseminate and share the lessons learnt so that we can increasingly not only speak a common language but also acquire a common understanding. TECSA gave us the material to prepare the case studies in this book summarising some experiences.

Finally, considering the fact that fire safety is an achievement of the organisations for themselves to protect their people, their contractors and third parties working there, the environment, their assets and their business continuity, it is most important to thank Dr. Germano Peverelli, President and CEO of Sabo S.p.A. (www.sabo.com), a fine chemical company operating for more than 80 years and under the requirements of the Seveso major accident EU Directive. We appreciated the proactive attitude of the company in dealing with fire and industrial risks as issues to be conjugated with the business. We should thank those guys firstly not only for having allowed high-level risk identification and management activities to be carried out using modern methodologies, but also for having established a relationship over the years characterised by seriousness and a common will to assess and manage fire and industrial risks in the best way and without compromise as a fundamental value for the organisation and all the involved stakeholders including the authorities having jurisdiction. Some of their continuous investments for safety and their commitment widely transpire from several summarised case studies presented in this book for which we thank them.

List of Acronyms

AHJ	Authority Having Jurisdiction
AIChE	American Institute of Chemical Engineers
AIIA	Associazione Italiana di Ingegneria Antincendio (SFPE Italy)
ALARP	As Low as Reasonably Practicable
ANSI	American National Standards Institute
API	American Petroleum Institute
ASET	Available Safe Egress Time
ATEX	Explosive Atmosphere
BFA	Barrier Failure Analysis
BIA	Business Impact Analysis
BLEVE	Boiling Liquid Expanding Vapour Explosion
BS	British Standard
BSI	British Standard Institute
CEI	Comitato Elettrotecnico Italiano
CEN	European Committee for Standardisation
CENELEC	European Committee for Electrotechnical Standardisation
CFD	Computational Fluid Dynamics
CLP	Classification, Labelling and Packaging (EU Regulation)
COMAH	Control of Major Accident Hazards (Regulation)
DCS	Distributed Control System
DNV	Det Norske Veritas (now DNV-GL)
DOWF&EI	Dow Fire and Explosion Index
EIV	Emergency Isolation Valve
ETA	Event Tree Analysis
EVAC	Evacuation
EWS	Early Warning System
F&EI	Fire and Explosion Index
F&G	Fire and Gas
FARSI	Functionality, Availability, Reliability, Survivability and Interaction
FDS	Fire Dynamics Simulator
FEM	Finite-Element Method

FERA	Fire and Explosion Risk Assessment
FMEA	Failure Modes and Effects Analysis
FMECA	Failure Modes and Effects Criticality Analysis
FMEDA	Failure Modes and Effects Diagnostics Analysis
F–N	Frequency–Number (of fatalities)
FPSO	Floating Production and Offloading
FRA	Fire Risk Assessment
FRM	Fire Risk Management
FSE	Fire-Safety Engineering
FSM	Fire-Safety Management
FSMS	Fire-Safety Management System
FTA	Fault Tree Analysis
GSA	Gestione Sicurezza Antincendio (Fire-Safety Management)
HAC	Hazardous Area Classification
HAZAN	Hazards Analysis
HAZID	Hazards Identification
HAZOP	Hazard and Operability
HEP	Human Error Probability
HMI	Human–Machine Interface
HRA	Human Reliability Analysis
HRR	Heat Release Rate
HS	Health and Safety
HSMS	Health and Safety Management System
HSE	Health, Safety, Environment
HVAC	Heating, Ventilation and Air Conditioning
ICI	Imperial Chemical Industries
IEC	International Electrotechnical Commission
IMO	International Maritime Organisation
IPL	Independent Protection Layer
ISO	International Standard Organisation
ISO-TR	ISO-Technical Report
ISO-TS	ISO-Technical Specification
LFL	Low Flammability Level
LGN	Liquid Natural Gas
LOC	Loss of Containment
LOPA	Layer of Protection Analysis
LPG	Liquified Petroleum Gas
MARS	Major Accidents Reporting System
MIL-STD	Military Standard (US)
MOC	Management of Change
NFPA	National Fire Protection Association (USA)
NIST	National Institute for Standards and Technology (USA)
OHSAS	Occupational Health and Safety Assessment Series
P&IDs	Process and Instrumentation Diagrams
PDCA	Plan Do Check Act
PED	Pressure Equipment Directive (EU)

PFD	Probability of Failure on Demand
PHA	Preliminary (Process) Hazards Analysis
PSV	Pressure Safety Valve
QRA	Quantitative Risk Assessment
RAGAGEPs	Recognised and General Accepted Good Engineering Practices
RAM	Reliability Availability Maintainability
RAMS	Reliability Availability Maintainability Safety
RBD	Reliability Block Diagram
RHR	Heat Release Rate
R_{env}	Risk for Environment
R_{life}	Risk for Occupants
R_{pro}	Risk for Assets and Business Continuity
RM	Risk Management
RSET	Required Safe Egress Time
SFPE	Society of Fire Protection Engineers
SIF	Safety Instrumented Function
SIL	Safety Integrity Level
SIS	Safety Instrumented System
SMS	Safety Management System
TNO	Nederlandse Organisatie voor Toegepast Natuurwetenschappelijk Onderzo
TOR	Terms of Reference
UNI-VVF	Italian Specific Technical Regulation
UVCE	Unconfined Vapour Cloud Explosion
VCE	Vapour Cloud Explosion

About the Companion Website

This book is accompanied by a companion website:
www.wiley.com/go/Fiorentini/FireRiskManagement

This website includes a presentation of the main principles expressed in the book along with a number of pictures and diagrams that can be used during training sessions, as well as a number of fully developed case studies from several domains to illustrate different fire risk assessment and management methods from real experiences:

1) Fire risk assessment of a production plant for a Fire & Explosion Risk Assessment (FERA)
2) Fire risk assessment of a production plant for a Quantitative Risk Assessment (QRA)
3) Fire risk assessment of listed historical building
4) Fire risk assessment of listed Industrial production and warehouse site
5) Semi quantitative fire and explosion risk assessment of major accidents with LOPA and physical and effects modelling

1

Introduction

Building and industrial asset complexity is increasing, and new fire threats are emerging. Risk-based approach, instead of prescriptive rules, can give a better perspective to various stakeholders (not only in the design phase but also in the operation phase), but an effective fire risk assessment should be based on sound foundations around fire characteristics, building/industrial asset characteristics and people characteristics as well as the interactions among these elements. Fire-safety level should be managed and maintained during the life cycle of the asset and, in particular, during design and operation phases, including any emergency situation that may arise. Fire strategy should be defined, shared and communicated among stakeholders that often have different knowledge and feeling about fire protection measures, assessment methods, codes and standards. This book allows the readers achieving a common and intuitive overview of the process to select, design and operate a fire strategy in a risk-based framework, in which the strategy, as a pool of different measures, is not unique.

Given a fire scenario, the proper fire strategy should be defined given the risk (magnitude and probability of occurrence), the risk-reduction factor, the cost to implement and to maintain the measures, the vulnerabilities, etc. Resilience is achieved when fire risk assessment allows the consideration of the relevant fire scenarios, and their mitigation in frequency and magnitude to an acceptable level, given a defined risk criterion, is put in place and maintained over the time with a sound fire-safety strategy, known and shared among the stakeholders.

Stakeholders should be aware of considering the fire strategy as a common and shared holistic approach that goes beyond the differences among the parties (*in primis* the 'famous' gap among architects or civil engineers and fire protection engineers) to a specific additional and inalienable objective for the building performance: fire safety.

This approach would benefit from the increasingly common collaborative and working environments (even digital) that could solicit a common discussion around fire-safety issues.

According to the complexity of the building/asset under consideration, the readers will gain an overview of the general approach to achieve a structured fire-safety strategy in the design phase to be maintained over time, based on fire risk assessment results and eventually coupled with performance-based approaches for alternative solutions.

This workflow, based on fire-safety principles and from the examples gained, in terms of lessons learnt from real and severe fire events, is regulation-free and codes-neutral. This framework may become the basis of a common fire-safety culture among professionals with different expertise and from different environments, including the people who should manage fire safety during the operation phase under the use cases defined during earlier design.

Fire Risk Management: Principles and Strategies for Buildings and Industrial Assets, First Edition.
Luca Fiorentini and Fabio Dattilo.
© 2023 John Wiley & Sons, Inc. Published 2023 by John Wiley & Sons, Inc.
Companion Website: www.wiley.com/go/Fiorentini/FireRiskManagement

These use cases, built around fire-safety objectives, should nowadays face complexity of socio-technical organisations using basic fire-safety principles. For professional engineers who want to adopt performance-based approaches (often built around the use of sophisticated tools), this book serves as a reminder of the objectives to be achieved considering the fire-safety fundamentals. For all the involved stakeholders, the content discusses the fire-risk-based workflow to be followed to verify and document the achievement of the performance as well as the requirements for the building/asset owners to maintain over time the required performance levels for each preventive/mitigative safety measure in the selected fire strategy as defined by a consistent fire risk assessment activity that becomes, together with the fire engineer, the main element of the entire process, while increasing the responsibilities of the expert himself:

> *"The fire engineer needs a certain toughness – and I am referring to intellectual toughness. The engineer must be able to be tested, challenged and deal with matters in a rigorous, analytical and, above all, honest way".*
>
> Margaret Law, OBE, 1990

2

Recent Fires and Failed Strategies

Over the past few decades, fire safety has taken some significant steps forward which have made it possible to achieve high safety standards today in a variety of application areas, from civil to industrial to maritime. The merit of this advance in different sectors can be found in the overlapping of multiple factors that, layering one on top of the other, have considerably thickened the level of fire safety potentially achievable today. A fundamental role is undoubtedly played by the progress made in the field of materials technology, fire protection systems and local regulatory requirements, which, in each country, has imposed increasingly restrictive measures that are primarily oriented towards expected performance rather than the prescription of already 'pre-packaged' design parameters that showed great limitations in applicability given the current complexity in each sector and daily emerging threats.

Despite this, the more or less recent news headlines continue to be populated by incidents involving fires and/or explosions that attract media attention due to their severe consequences in terms of fatalities, injuries, damage to the artistic-historical heritage, environmental pollution and so on. Excluding wildfires and fires on aircraft from the list, a simple search reveals the following fires to be identified among the major news stories of recent years:

- Oil depot explosion in Cuba – 7 August 2022 – 1 fatality, 121 injured and 16 missing;
- 'Moro' Tower fire in Milan (IT) – 29 August 2021 – no fatalities;
- Beirut port explosion – 4 August 2020 – 218 fatalities;
- ICS plant fire in Avellino (IT) – 13 September 2019 – complete destruction of the plant;
- Notre Dame fire in Paris – 15 April 2019 – damage to historical and artistic heritage;
- Grenfell Tower fire in London (UK) – 14 June 2017 – 72 fatalities;
- Norman Atlantic ferry fire – 28 December 2014 – 9 fatalities and 20 missing;
- ThyssenKrupp fire in Turin (IT) – 6 December 2007 – 7 fatalities;
- Deep Water Horizon drilling offshore platform explosion and fire in Gulf of Mexico – 22 April 2010 – 11 fatalities and complete loss and severe environmental damage;
- Lac – Mégantic crude oil train rail disaster in Quebec (CA) – 6 July 2013 – 47 fatalities (42 confirmed and 5 presumed victims).

It is striking to observe how, despite the long period of time that separates us from the event chronologically most distant to us, incidents of these kinds are still terribly topical and equally nefarious because of the high degree of exposure that characterises them.

In order to understand the reasons that led the authors to write this book, a number of incidents are illustrated in the following sections, with the aim of discussing, at this stage albeit briefly, the reasons that led to the failure of the planned fire-safety strategies.

Fire Risk Management: Principles and Strategies for Buildings and Industrial Assets, First Edition.
Luca Fiorentini and Fabio Dattilo.
© 2023 John Wiley & Sons, Inc. Published 2023 by John Wiley & Sons, Inc.
Companion Website: www.wiley.com/go/Fiorentini/FireRiskManagement

All the incidents are simple examples, among the most well known at international level, of failed fire strategies where the severities have been determined and/or escalated by a number of multiple failures, often occurred in different stages, including the design phase and the operation phase.

2.1 Torre dei Moro

2.1.1 How It Happened (Incident Dynamics)

High-rise building known as 'Torre dei Moro' is composed of

- an approximately 60 m high tower consisting of 18 above-ground floors for exclusively residential use;
- 2 underground floors;
- some lower bodies for commercial and residential use, for a total of 77 residential building units, out of a total of 84 building portions.

It follows that the building has a mixed-type configuration and is consisted of (i) 3100 m^2 of production area, (ii) 2300 m^2 of commercial area, (iii) 420 m^2 of tertiary area, and (iv) 3700 m^2 of residential area.

The tower is composed of a reinforced concrete frame: the first five horizons of the complex are composed of prefabricated trussed slabs, lightened with polystyrene blocks and cast-in-place completion castings, while the remaining levels are composed of prefabricated latero-concrete panels with cast-in-place completion castings. The staircase ramp is made of a reinforced concrete mix with solid slabs, the roofing of the units above the business premises is flat, while that of the tower is made of prefabricated joists with brick-lightening blocks.

Visually, the tower appears as a parallelepiped with balconies jutting out on the largest sides with a curved profile, while on the two largest sides of the parallelepiped there are two curved sails that cover the balcony parapets and continue beyond the building outline (Figure 2.1).

Fire started on the balcony of flat C, exposed on one of the two side sails, located on the fifteenth floor of the building. Initial fire has been recorded as very severe and fully developed since the beginning.

Fire first started visibly with the presence of abundant smoke, which was followed by the rapid spread of flames over the building via the mast external façade and also involving the insulation panels applied as an external cladding to the main structure of the building. It then rapidly spread to the panels located on the side end of the balcony, further feeding and, above all, favouring the downward spread of the flames, made possible by the phenomenon of the dripping of the polymer constituting the panels and the fall of the same (Table 2.1).

Table 2.1 General information about 'Torre dei Moro' fire.

Who	High-rise residential building
What	Façade fire
When	29 August 2021
Where	Milan (Italy)
Consequences	Severe damages to the building and adjacent residential premises; no fatalities
Credits	Luca Fiorentini (TECSA S.r.l.)

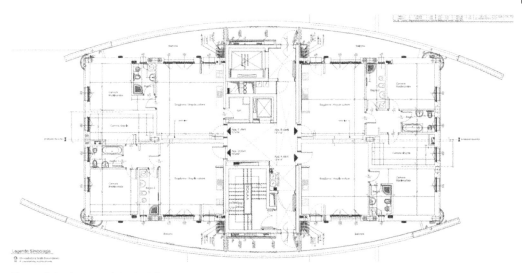

Figure 2.1 Layout of the building.

The spread of the fire was rapid and followed a totally atypical evolution, extending not only towards the upper floors favoured by the upwards development of the flames, but also laterally and, above all, towards the lower floors, involving the entire building, up to the shops on the ground floor, including the car park.

Not only were the sail panels unable to contain and/or at least contain the combustion phenomenon, but, on the contrary, they also caused the fire to develop rapidly, creating new hotspots (located in the lower part of the building) completely disconnected from the first one (located on the fifteenth floor of the tower), thus affecting parts of the building that would not normally have been involved in the fire in any way.

Although the flames spread outside the building, they then penetrated inside the flats through the windows and doors.

The flames that spread on the west façade of the tower passed through the outer structure known as the 'sail', affecting the materials on the balconies. These acted as vectors for the introduction of the flames inside the flats at some locations contributing to the resulting damages.

The evidence found on site highlights the differentiated damage suffered by the flats. By virtue of the materials on the balconies, these were attacked by the flames, carrying the fire from the outside to the inside. The wind, which was almost always present, especially on the upper floors, helped channel hot smoke and flames into the houses, first affecting the window frames and then penetrating into the flats.

On one and the same landing, for example, there are houses where only one is affected by partial smoke damage, while the one next door suffered a heat stroke that destroyed the rooms. The dwellings face opposite façades.

An analysis of the angle of the flames shows that the material from which the fire originated was located in the niche in the end section inside the balcony, close to the closing part of the sail.

The action of the flames therefore caused the metal end wall of the balcony to collapse, effectively creating a direct communication with the cavity where the PVC rainwater drainage pipes are located.

This service cavity is a single duct extending from the ceiling down to the ground, and in fact creates a high 'chimney'-type effect, in which the flames also develop quickly thanks to the turbulence created inside.

Figure 2.2 Building during the fire.

Photographic evidence shows the presence of flames at the top of the façade, in correspondence with the cavity, an indication that they are fully spread inside this duct (cavity) where they gain speed precisely because of the aforementioned 'chimney effect'.

The vigorous dripping of the incandescent insulation triggers a fire at the base of the cavity, causing the rapid extension of the flames that enveloped the entire building.

The fire was somehow brought under control after many hours of extinguishing activities by the fire brigade department, who used ladder trucks to reach (even if only partially) the upper floors of the building, and highly qualified personnel.

It is quite evident that (i) the speed of flame propagation, (ii) its continuous feeding, (iii) its spreading not only upwards but also downwards and sideways, and (iv) its ability to involve even a completely detached façade oriented in the opposite direction show that the initial ignition found its source of combustion in the materials that are used to compose and clad the building.

Significant damage to the attics of many floors above ground level was immediately reported, without the involvement of people.

After the critical phase of the fire (Figure 2.2), in constant liaison with the fire brigade department continuously present at the site of the damaging event, work was started to save the unsafe parts still present on the façades.

2.2 Norman Atlantic

Regardless of the wide prescriptions of the maritime regulations, fire on board is still a significant cause of losses for human beings. This case study underlines how the structural weaknesses of the management system may affect the phase of incident management, which is amplified by the assumed condition of being at sea (longer time required for rescuers to arrive at the incident site). The outcomes of this fragility can be devastating, heavily affecting first the abandon of the ship and then the rescue activities. This case study is explicitly focused on the fire investigation, neglecting all the other aspects related to the same incident, emerging from the complex world of maritime transportation (Table 2.2).

Table 2.2 General information about 'Norman Atlantic' fire.

Who	Ro-Ro Pax ferryboat
What	Fire on board
When	2014
Where	Adriatic Sea (Italy)
Consequences	9 fatalities, 14 lost
Credits	Rosario Sicari (TECSA S.r.l. - forensic division)

2.2.1 How It Happened (Incident Dynamics)

The ferryboat Norman Atlantic was an Italian ship rented by a Greek company for ferry crossing between the two countries. The night of the incident, the route Patras–Igoumenitsa–Ancona was planned, and 55 crew members were on board. When leaving from Igoumenitsa to Ancona, at 23.28 on 27 December 2014, the cargo consisted of about 130 heavy vehicles, 417 passengers and 88 cars. The navigation was regular until 03.23 (UTC), when the fire alarm sprang into action on deck 4, near the frame no. 156. Because of a smoke sighting coming out from the lateral openings of the deck, a sailor was appointed to carry out an inspection on deck 4. He referred that the alarm was attributable to the smoke coming from the auxiliary diesel engine belonging to a reefer truck, which was not connected to the electrical supply of the ship. After a few minutes, at 03.27, the master brought himself to the flying bridge deck on the starboard side and observed the flames coming out from the openings of deck 4. The first Engineer Officer activated the manual deluge system (known as 'drencher'), following the master's order. Meantime, the Chief Engineer Officer and his personnel abandoned the engine room because of the excessive smoke, while the two engines of the ship stopped definitively. The ship went in a blackout, and the emergency generator, placed on deck 8, was incapable of providing energy to the emergency utilities, including the emergency pump. At the same time, the cooling team uselessly tried to cool deck 5, but steam came from the fire hoses instead of liquid water. The emergency management, especially during its first stages, revealed as chaotic. During the rescue operations, some passengers fell into the sea, while others threw the remaining life rafts in the sea, with no possibility to properly use them. At the end of the search and rescue operations, 452 people were rescued, including 3 illegal immigrants; 9 victims and 14 lost in the sea were also counted. The ship was then tugged to Bari, where it has been under the lens of the investigators.

The longitudinal section of the ship is shown in Figure 2.3.

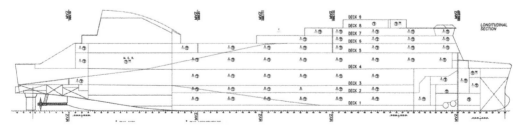

Figure 2.3 Longitudinal section of the ship, with fire compartments.

2.3 Storage Building on Fire

The fire that occurred on 22 April 2011 had a large spread on the roof and, in particular, on water-proofing and insulating layers, and a thin-film photovoltaic (PV) system was installed on the roof and on some portions of the interior of the building. Later, another fire occurred on 26 March 2012 on another building of the same storage centre, and it involved the roof layers (waterproofing and insulating) and some portions of the building interiors (Table 2.3).

The activities usually carried out in the storage centre were in the list of those checked out by the Ministry of Home Affairs because of fire-safety concerns but, at the time of main event (April 2011), these were operating without formal authorisation of the Ministry.

The main event (22 April 2011) occurred when the work to install the waterproofing and insulating layers and the thin-film photovoltaic system was being carried out on the building roof (a temporary consortium of companies was in charge of it).

2.3.1 How It Happened (Incident Dynamics)

On 22 April 2011 (sunny and windy day), at 15h00′, probably, the present staff noticed the fire, at 15h10′ they alerted the company emergency squad and at 15h20′ they alerted firefighters (Figure 2.4).

Table 2.3 General information about the 'Storage building' fire.

Who	Storage building inside a storage building area
What	Fire on roof and internal compartment of the building
When	April 2011
Where	Nola, Italy
Consequences	Building damages, no fatalities
Credits	Giovanni Manzini

Figure 2.4 Photo of the burned roof and the installed PV system.

2.4 ThyssenKrupp Fire

The accident occurred at a pickling and annealing line, two operations that are usually conducted in the same plant. The main technical challenges of the process derive from the need to run both the thermal and electrochemical processes continuously, even when the coils have finished. In order to comply with continuous process constraints, the subsequent coils must be welded, and this introduces a discontinuous process. As a consequence, some complications arise in the architecture of the lines. The coil is handled via hydraulic systems that use mineral oil. This oil is not usually flammable, but it is of course combustible. Hydraulic circuits are fed with high-pressure oil. In this case, the pressure ranged between 70 and 140 bar. Under these conditions, highly flammable spray/mists can originate from small leaks (Table 2.4).

2.4.1 How It Happened (Incident Dynamics)

On 6 December 2007, five workers were on the night shift (22:00 p.m. to 6:00 a.m.) on the annealing and pickling line. Another three workers were on the line either to substitute or train other workers. At 00.35, the line was restarted after a 84 min stop to remove some paper lost from a previously treated coil. As there was no automatic control system on the inlet section for the axial coil position, the coil, after some time, started to rub against the line structure, which was made of iron carpentry. The location of this scraping was identified just above flattener no. 2, while scraping occurred to coil no. 1. The rubbing lasted for several minutes, and, as a consequence, produced sparks and local overheating. A local fire, which involved paper and the hydraulic oil released from previous spills, started from these circumstances. A small pool fire started in the flattener area, which is depicted in Figures 2.5 and 2.6, and subsequently spread to roughly 5 m^2, involving the flattener and its hydraulic circuits. At roughly 00.45, the workers realised there was a fire and took some measures to control it. First, they stopped the entry section of the line, reduced the line velocity, seized some fire extinguishers and went close to the fire to attack it from at least two directions. After some seconds, they decided to also use a fire hydrant, so one of them walked to the fire hydrant and a second one handled the fire hose. At that moment, one of the several pipes of the hydraulic circuits involved in the fire (roughly 10 mm inner diameter) collapsed and released a jet of high-pressure oil from the pipe fitting. The pipe, which is depicted in Figure 2.7, was fed at a pressure of 70 bar from the main pump station, which was still running. As a consequence, a spray of hydraulic oil was released into the already existing fire.

Table 2.4 General information about the 'ThyssenKrupp' fire.

Who	Steel plant
What	Jet fire
When	2007
Where	Turin (Italy)
Consequences	7 fatalities
Credits	Luca Marmo (Politecnico di Torino)
	Norberto Piccinini (Politecnico di Torino)
	Luca Fiorentini (TECSA S.r.l.)

Figure 2.5 Area involved in the accident. Right: unwinding section of the line; left: the front wall impinged by flames.

Figure 2.6 The flattener and the area involved in the accident. Details of the area struck by the jet fire, view from the front wall.

The ignition of the spray was immediate, due to the contemporary presence of the pool fire near the release point, and this resulted in a huge jet fire that struck the eight workers. Figure 2.8 shows a map of the site with the presumed positions of the workers (no reliable witnesses could be found concerning this topic) and the extension of the area in which the jet fire took place. The length of the jet fire has not been determined precisely since it hit the front wall that was located at a distance of more than 10 m from the broken hosepipe. The footprint of the jet fire is clearly visible on the wall in Figure 2.9. The spread angle of the jet fire was roughly 30° since some scattering occurred against various equipment. The fire also spread backwards with respect to the hose direction, and in such a manner, it involved the large area indicated in Figure 2.8. This spread of the fire was determined by the interaction of the fluid released at very high velocity with a number of fixed structures located in the vicinity of the release point.

Figure 2.7 Details of the hydraulic pipe that provoked the flash fire.

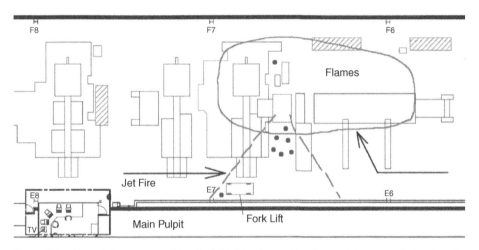

Figure 2.8 Map of the area struck by the initial jet fire and by the resulting extended and prolonged fire. The dots represent the presumed position of the workers at the moment the jet was released.

Then, a total of 13 pipes were collapsed in a few minutes. Many of these pipes were under pressure and continuously fed by the pump station, hence provoking a huge spread of oil and of the flames. The pressure in the hydraulic circuit dropped due to the huge oil leak; thus, the intensity of the jet fire diminished very quickly. At the same time, the fire reached its maximum size (see Figure 2.8) as it was being fed by the released hydraulic oil that burned in a pool. The first pipe collapsed in a time interval of between 00.45′49″ and 00.48′24″; the pumps were stopped by the automatic control system at 0.53′10″ due to the low-level switch system on the basis of the oil level in the main reservoir. In this time interval, at least 400 l of oil escaped.

The jet fire struck the eight workers who were fighting the first fire. The worker who was close to the fire hydrant (see Figure 2.8) was sheltered by a forklift and only suffered minor burns. Six workers were struck by the jet fire and suffered third-degree burns covering from 60 to 90% of their bodies. They died over the following months. One, who went to the back of the plant to fight the fire, was trapped and died immediately. The fire spread to the machines and lasted for approximately two hours before the fire brigade from the National Fire Corps could extinguish it.

2.5 Refinery's Pipeway Fire

Since the years of construction of the petrochemical pole of Mellili-Priolo-Augusta (near the 1950s), there was not a significant sensibility towards the urbanistic and territorial compatibility issues that were regulated, for those aspects of industrial risk, only many years later in 1999 (D.Lgs 334/99) and 2001 (D.M. 9/5/2001), with a ministerial decree aimed at the regulation, according to the Seveso legislation, of these aspects since the phase of creation of the local urbanistic rules. In this area, a rapid urban development was observed, without any control for the areas next to the establishments, while the main terrestrial communication infrastructures (rail and road networks) divided the industrial establishments for long distances, creating high-vulnerability points in case of incident but, at the same time, being too much difficult to be modified. The incident here described happened on 30 April 2006 when a mixture of hydrocarbons caught fire because of a leakage of crude oil from 1 of about 100 pipes crossing a subway (Table 2.5).

Table 2.5 General information about the 'industrial pipeway' fire.

Who	Oil refinery
What	Pool fire and BLEVE (boiling liquid expanding vapour explosion)
When	30 April 2006
Where	Syracuse (Italy)
Consequences	Highway closed for 53 days, 14 injured people and more than €25 million of total incident costs
Credits	Salvatore Tafaro (Italian National Fire Brigade)

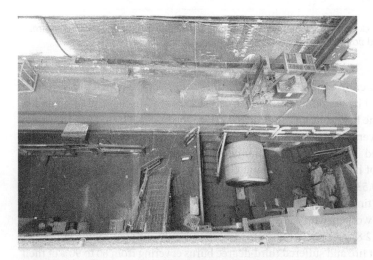

Figure 2.9 Footprint of the jet fire on the front wall.

The authors highlighted how it is important to compare the facility's incident records with data from similar industries in order to also learn from the experience of others. Several databases are available to meet this objective, both national and international. For example, in Europe the Major Accident Reporting System (MARS) is a shared industrial accident notification scheme that was established by the Seveso Directive in 1982. The database allows examining the historical data related to the industrial accidents in order to draw the lessons learnt and prevent the occurrence of a future unwanted event or, at least, to mitigate their consequences.

Some of the contents of this example are taken from the MARS report. The MARS database can be consulted online for free at https://emars.jrc.ec.europa.eu.

2.5.1 How It Happened (Incident Dynamics)

On 30 April 2006, at 3:40 p.m., in one of the three refineries of the industrial pole near Syracuse, near the subway of ex SS114 (highway Siracusa-Catania), a leakage of crude oil occurred from an insulated pipe with a significant diameter (DN 500), which was then ignited, setting it on fire and extending flames both the upstream and downstream of the subway itself, also involving a part of the pipeway near some liquified petroleum gas tanks.

Approximately three hours after the leak detection, the onsite emergency response plan was triggered, performing the following:

- mobilisation of the onsite fire brigade;
- isolation of the pipelines starting with the leaking one, as a precautionary measure;
- closure of road ex SS114 passing through the establishment.

The onsite fire brigade mobilised initially two emergency response vehicles, spraying foam on the area where the smoke was generated. At 5:42 p.m., the shift commander of the onsite fire brigade alerted the public fire department in Siracusa once he understood that the fire was out of control. After a few minutes, another fire front developed from the entry of the subway in road Nr 9/0 (uphill), and this fire was also controlled initially by the onsite emergency response services with other two firefighting vehicles.

At 6:10 p.m., the public fire brigade arrived from Siracusa and took the command of the emergency response management. At 6:50 p.m., a first BLEVE explosion occurred in another pipeline containing probably light hydrocarbons (Figure 2.10) and was followed by successive explosions of pipelines in the same trench, all due to the overheating of the products inside the pipeline consequent to the heat irradiation of the fire in the trench (Figure 2.11).

The situation inside the subway at the moment of the fire (as shown in Figure 2.12) highlights a not-orderly arrangement of the piping, not taking into account different typologies of the substances inside the pipes (e.g. flammable substances near toxic ones or near the fire system pipe). The arrangement left no possibility to operate in case of verification and/or maintenance; moreover, a proper active fire protection system was absent, excluding the general firefighting system made of hydrants.

The subway and its trench contained about 100 pipes crossed by oil products and other hazardous substances, as well as fluids of utilities (including steam, nitrogen, water, air and so on). The pipes were not owned by the same company; indeed, in the pipeway there were the pipes of four different companies.

Since the very beginning of the fire, the seriousness of the incident was clear, and therefore the fire brigade from different districts of Sicily was asked to intervene, working for more than 80 hours, in addition to the local district fire brigade and the company fire brigade.

Figure 2.10 Damages of the piping uphill the road. Gash and ruptures caused by BLEVE.

Figure 2.11 Some damaged pipes downwards the road. There is also the pipe of the fire system.

SOTTOPASSO SS 114

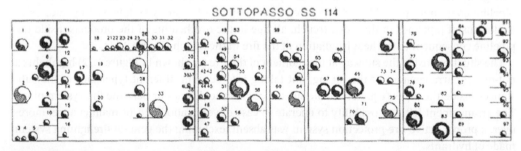

Figure 2.12 Transversal section of the subway before the incident.

During the firefighting activities (Figure 2.13), the fire brigade personnel found difficult to control the fire for two main reasons: the lack of water resources, for the unavailability of the fire water system in the establishment that was damaged by the fire, and the fact that many pipes engulfed by the fire, especially the ones having major diameters, had a huge hold-up between two consecutive blocking valves, between 100 and 350 m^3.

Figure 2.13 Photos of the extinguishment operation.

Some of the main effects of the fire were the following:

- Dense and extremely pollutant smoke; fortunately, because of the favourable weather conditions, there were no bad consequences for the surrounding residential areas.
- The runoff of the product dispersed from the piping along the pipeway's trench for a significant length such that the fire extended downhill the trench, with no possibility to contrast this effect with appropriate containment systems.
- Unexpected collapse of some pipes, especially those containing light hydrocarbons, causing not-serious injuries to some firefighters.
- Escape of flammable gas/vapours in correspondence of the flanges because of the lacking seal of the gaskets due to the overpressure reached.

At 6:55 p.m., the decision was taken to shut down all installations inside the establishment as a precautionary measure. The Prefect ordered the closure of all gates of the establishment in accordance with the Offsite Emergency Plan, which caused the interruption of all vehicle circulation in the area around the accident place for several days. The circulation on the Siracusa-Catania railway was interrupted for 48 hours, and the ship circulation in the port of Augusta was interrupted for approximately 36 hours. The road ex-SS114 was closed (Figure 2.14) for 53 days, waiting for the results of the static structural tests and the substitution of the damaged asphalt covering.

Fire extinguishing operations lasted 48 hours: the fire seriously damaged the pipelines uphill and downhill the subway, as well as aerial pipelines passing parallel to the trench or crossing it. Further limited damage occurred to the subway. During the accident, four members of the onsite fire brigade, two plant operators and eight members of the public Fire brigade were hospitalised for more than 10 days due to burns, contusions and/or intoxication. The offsite emergency has not caused health consequences to the population living in the neighbouring communes. The major of Priolo ordered the closure of the public schools for one day and invited the population to stay indoors immediately after the accident as a precautionary measure. The onsite costs of the incident are both related to material losses (€22 million) and response, clean-up and restoration activities (€5.65 million). The offsite costs are only related to material losses, for a total amount of €50 000. 'Rebuilding' refers to the repair of the subway, the substitution of the pavement, and the road signs and the guardrails of ex SS114 damaged by the fire on an approximately 100 m long road section.

Figure 2.14 A helicopter view of the area.

The fire extinguishing operations have been managed to confine the fire in the area corresponding to the subway without extinguishing the fire completely, in order to completely eliminate the presence of hydrocarbons from the hardly accessible trench, where hydrocarbon vapours could have ignited again at any point. The incident attracted the national media interest.

In order to restart the activity, it was necessary, for the Authorities Having Jurisdiction (AHJ), to find the technical corrective measures that the management of the society would have to adopt within a defined target date, for avoiding the recurrence of the event, finding the direct and indirect causes, together with the factors that have negatively influenced the dynamics of the incident (for instance, amplifying its effect).

2.6 Refinery Process Unit Fire

Some of the contents of this example are taken from the MARS report. The EU official MARS database on major accidents (Seveso Directive related) can be consulted online for free at https://emars.jrc.ec.europa.eu (Table 2.6).

Table 2.6 General information about the 'refinery process unit' fire.

Who	Crude oil refinery
What	Fire
When	2 September 2005
Where	Province of Genova (Italy)
Consequences	No safety consequences; economic losses = €7 million
Credits	Gianfranco Peiretti (HSE Manager, IPLOM S.p.A. Refinery)

2.6.1 How It Happened (Incident Dynamics)

On 1 September 2005, at 21:40, a fire broke out affecting the heavy fuel purification plant of the refinery.

The company revealed the following statements concerning the accident.

The shift personnel present testified that a strong hiss was heard lasting a few seconds, which was followed with the ignition of the released product and that there had not been any noticeable pressure changes (like pressure waves, etc.). The shift personnel present onsite and, in particular, the shift foreman who was staying on the main access ramp to the atmospheric distillation plants confirmed that they saw the jet fire ignited from the top and propagated down (from approximately 18 m above ground down to approximately 14 m in the area covered between the exchangers E1718 A/B/C and the reactor R1702). Immediately, the shift foreman activated the onsite emergency plan and informed the gate guard in order to alert the fire brigade. At 21:40, the fire brigade was alerted. The jet fire affected the quench line with the 3″ hydrogen pipe, which ruptured after six-minute exposure with the consequent ignition of the hydrogen. The fire took a cylindrical form from the bottom to the top starting at 14 m height affecting the above-located plant parts, including diathermic oil preheating section. Approximately 30 minutes after the fire ignited, an 8″ fuel pipe of the diathermic oil system ruptured and subsequently ignited the product. The fire was kept under control and evolved without noticeable changes until consumption of the fuel once the pipes were shut off according to the emergency response plan that was activated.

The fire was extinguished at 1:20 h on 2 September 2005 (3 hours and 40 minutes after the fire initiated), and the state of emergency was called off by the fire brigade at 1:45 h.

No damages to people were reported due to the accident.

According to evaluations of damages, there was not any environmental damage, and this evaluation was also confirmed by an environmental indicator assessment performed by the Agenzia Regionale per la Protezione dell'Ambiente (ARPA) of Genoa (competent authority that investigated the initial environmental effects of the fire).

The accident affected the catalytic hydro-treatment plant named Unit 1700. The main characteristics of the plant are as follows:

- Capacity: 1650 t/d light fuel treatment section.
- Capacity: 145o t/d heavy fuel treatment section; design and construction Tecnimont Construction – January–August 1997; operation started in September 1997.
- Unit 1700 was designed to improve the characteristics of light and heavy fuel oil produced in the refinery by treating the fuel fractions with high-pressure hydrogen.

The technology employed was essentially applied in treating the fuel oil with high-pressure hydrogen, on a specific catalyst, such as to eliminate the sulphur in the fuel oil, produce hydrogen sulphide, hydrogenate the hydrocarbons and improve the other characteristics.

The plant was designed with two heating–reaction–fractionating sections considering different characteristics of the charges to be treated: one for light fuel-oil mixtures (Unit 1700) from the topping and the other for the heavy fuel-oil mixtures from vacuum, respectively, while foreseeing one single gas purification and compression section for the recycled gas to be reintegrated in the circuit.

The charge, made of a heavy fuel-oil mixtures coming from the vacuum distillation plant, is sent to Unit 1700b through three pipelines equipped with flow control systems. Once the water contained is eliminated, the charge is pre-heated and then transferred to the reaction section working at 60 bars of pressure.

Before the charge enters the reactor, the charge is mixed with the hot recycled hydrogen and then heated up to optimal temperature for the catalytic reaction (on average 360 °C) through a closed circuit of heating oil. An appropriate catalyst inside the reactor facilitates the desulphuration reactions inside the reactor (hydrogenation of sulphur in H_2S).

The reactor effluents, constituted by desulphurised fuel oil and a gas mixture, is cooled down and sent to a liquid–gas separator. The gas principally constituted of hydrogen is washed and purified from the hydrogen sulphide before being recycled, while the desulphurised heavy fuel oil feeds a pre-stripping column (T-1707), and the head flow of this column is then sent to the stripping column of the light fuel-oil section (T-1701).

The pre-stripped heavy fuel fraction (bottom T-1707) is heated and sent to the main fractionating column (T-1703) in which following products are separated:

- Uncondensable gas;
- Virgin nafta;
- Light desulphurised fuel oil and heavy desulphurised fuel oil.

The light fuel oil is sent from the fractionating column to a stripper (T-1704) in which hydrocarbon tails and water are eliminated. The heavy fuel oil is cooled down and transferred to storage.

The area affected by the accident corresponds to the heavy fuel-oil purification circuit comprising reactor R1702, the charge/hot oil circuit heat exchange train, the hydrogen injection circuit (quench) and reactor R1702 comprising the control instrumentation.

The incident put at risk about 20 onsite people and 6 emergency personnel (offsite). The material losses have been quantified as €5 million, while the response, clean-up and restoration costs as €7.6 million. The severity of the incident required the interruption of the road and railway crossing nearby. The incident had large media coverage. Figure 2.15 shows some photos of the incident.

Figure 2.15 Photos of the incident. *Source:* Courtesy of IPLOM S.p.A.

Figure 2.15 (Cont'd)

Proposed events are just examples of severe fires and/or explosions in different domains. All domains are vulnerable to these threats. It appears very clear to all involved stakeholders that compliance-based fire safety is no more sufficient in our times due to a number of factors, all of them governed by the complexity associated with all the assets we daily design and operate and should be considered socio-technical systems.

For all of them, we should know the answers, from a fire-safety perspective having the correct level of details to some important questions about the fire scenarios:

- What can go wrong?
- How likely is it?
- What are the impacts?

This is the aim of fire risk assessment to be fed by a number of different approaches such as historical experience (starting from real events), analytical methods, intuition coupled with expert judgement, strict compliance with requirements from regulation and technical standards and or the demonstration of alternative fulfilment of performances connected with the requirements. If a comprehensive, not selective, approach is recommended, it can be demonstrated how many lessons learnt can be derived from some of the presented cases.

3

Fundamentals of Risk Management

Fire safety management (FSM) can be defined as a measure aimed at the management of an activity under safe conditions, both during operations and in emergency conditions, through the adoption of an organisational structure that provides for roles, tasks, responsibilities and procedures.

Fire safety is just an integral part of the activities to preserve health and safety in all sectors.

Just like management of health and safety, an FSM plays a fundamental role in organisations to implement, maintain, monitor, measure and eventually correct or improve the level of fire safety given the organisational context (people, processes and goals). Having a health and safety management system (HSMS) allows to improve health and safety (HS) culture across the organisation (both by imposing accountability on top management and stakeholders and boosting motivation, empowerment and confidence/trust of workers), reduce near misses and incidents (and associated costs including business disruption), achieve and demonstrate compliance (to both legal and voluntary requirements), define the hazards and risks associated with the organisation's activities, make efforts to eliminate them or set up controls (monitored over time) to minimise effects, periodically assess the HS performance via specific KPIs and take actions to improve it (identifying opportunities) and increase recognition of the stakeholders and relevant parties, including authorities having jurisdiction. But this is also typical of an FSM: the two shares many common features. Indeed, HSMS, eventually integrated with other management systems or in the general organisation system, guarantees a consistent, structured and coherent approach to manage HS aspects, given the context, over time (a sort of systematic approach to safety case maintenance) by defining an organisational structure, systematic processes and their associated resources, an effective HS risk assessment methodology and a review process to identify opportunities, given the goal to prevent incidents systematically and continuously, and retaining documented evidences. Summarising it can be seen as the organisational tool, split into pillars, enabling four main activities: commitment to HS, understanding of hazards and associated risks, risk management and learning from experience from the design phase to the decommissioning phase.

Among several requirements for implementing an effective management system that covers the main pillars (such as safety culture, safety competency and training, contractors' management, incident investigation, auditing activities, etc.), it is possible to mention the following three requirements:

1) Choose a consistent methodological framework and adapt it to the specific context of the organisation considering the risks, the complexity, the organisational structure and the available/needed resources and combining the management system pillar in a coherent, solid and consistent way.

Fire Risk Management: Principles and Strategies for Buildings and Industrial Assets, First Edition.
Luca Fiorentini and Fabio Dattilo.
© 2023 John Wiley & Sons, Inc. Published 2023 by John Wiley & Sons, Inc.
Companion Website: www.wiley.com/go/Fiorentini/FireRiskManagement

2) Define a policy, agreed among the stakeholders (including the top management), describing the intention and commitment of the organisation, the objectives and scope, the approach to the management system implementation and maintenance over time (including its review activities), any eventual legal requirements (e.g. Seveso Safety Management System as per Directive 2012/18/EU of the European Parliament and of the Council of 4 July 2012 on the control of major-accident hazards involving dangerous substances, amending and subsequently repealing Council Directive 96/82/EC) and/or best practices (e.g. ISO 2018a) to be followed.

3) Identify the context and key roles and responsibilities, processes/activities and assets to determine the current status and select the proper methodological framework to conduct the risk assessment process in all its components (risk identification, risk analysis and risk evaluation as defined in ISO 31000) and consider management of change over time.

Having clarified that the management of fire risk is not dissimilar to the management of any other types of risks, an excellent reference to address this topic is the ISO 31000 Standard – Risk Management – Principles and Guidelines.

Given this important assumption, it is fundamental to explain to the stakeholders of fire safety some important concepts about risk and risk management activities.

3.1 Introduction to Risk and Risk Management

'Risk management' (RM) is undoubtedly one of the most frequently used phrases in the current social and economic scenario, the importance of which has progressively expanded in line with the regulatory evolution which, in a consolidated manner, has taken on risk and its management as a criterion for the responsibility of the individual and the organisation.

Risk management is now an integral part of company processes, not simply as a legal necessity but increasingly as a factor and opportunity for consolidation and development of the organisation and production processes.

Based on this assumption, the need emerges to focus on 'risk management' methods, not so much to find a definition – which, although effective, would not offer great applicative utility – but to substantiate its implementation with methodologies able to offer logical coherence to all phases of the process.

Completeness and effectiveness of risk management are in fact directly correlated to the capacity of the process to develop in phases and levels directly related to the articulation of the organisation and the processes analysed.

Considering the complexity and operational breadth of the company organisations, and therefore the network of processes, both internal and external, that condition the pursuit of results, it is quite clear that the risk management process is not immune from the risk of losing coherence of the parameters adopted in the analysis, altering the outcome of the assessment and jeopardising the possibility of consistent revision in the updating and comparison phases.

Of course, it is not a question of 'harnessing' risk management through the search for aprioristic rules that limit its potential in terms of flexibility and adherence to the specific realities analysed, which is – a bit simplistically but effectively – the added value of the process.

It is only a matter of identifying methodologies that assist the development of risk management in a flexible but coherent manner in all phases and levels, guaranteeing its analytical and evaluative rigour, and its applicative and comparative replicability, ensuring its effectiveness with respect to the complexity of modern organisations and the network of relevant processes, in an organic vision of organisational, technical and management factors.

Risk management can have variously articulated perimeters and purposes of analysis.

Consider, for example, the risk linked to the safety management of a production unit located in a single location, on the outskirts of an average provincial town, characterised by a single operational process and in a context in which no interference between production processes belonging to different employers takes place.

Imagine instead the same production unit that is part of a wider production organisation, belonging to a holding company that splits the operating processes between several subsidiaries, relocating them to several plants where different employers operate with interference between their respective suppliers.

Again, consider the hypothesis that the corporate development indicated above leads to the location of the production unit in the context of complex infrastructures in which several owners/managers/employers subject to different reference regulations operate, and not always coordinate from a technical and managerial point of view, such as, for example, in airports and railway stations where, moreover, the presence of third parties (public) is predominant; to be practical, think of the employer of a catering activity within an airport.

The perimeter of the risk analysis would seem to be limited in the first case, i.e. that of the single production unit; moreover, the identification of processes relevant to risk management would be reduced to the only production process located in the production facility.

The perimeter of the analysis is more articulated in the second case and enormously more complex in the third.

But what would happen if in the single production unit of the first case radiographic materials or highly harmful chemical products were treated?

The perimeter of the analysis would be unchanged from a strictly technical point of view, although considerably more complex; the management profile would instead be enormously more complex and open to external factors where it is generally considered, in force of the law, that the employer must take appropriate measures to prevent the technical measures adopted from causing risks to the population or deteriorating the external environment by periodically checking the continued absence of risk.

Finally, it is clear that the complexity of the risk management process is accentuated when our catering production unit moves into the station (equally, albeit on a smaller perimeter, where it moves into a multiplex cinema or a shopping centre).

The technical and management process, and the corresponding risk factors, is significantly extended in correspondence with the management of the station premises, where other commercial, service and railway production units operate alongside the catering unit, without prejudice to the presence of the public.

The catering unit, like the others, is the bearer of the management of its own risks, which must be coordinated with those of the other production units.

But the same production unit is the client of the initial set-up of the space in use and the service and maintenance activities of the same, in which, however, it interferes with the supply of condominium services provided through other contractors on behalf of the station manager.

Likewise, again with regard to safety and fire prevention, the fitting out and operation of the unit must remain consistent and coordinated with the fire prevention design of the station, which in turn is developed in primary and secondary activities included in the building complex.

It is also necessary to consider the hypothesis in which the controlling company of our catering production unit arranges for the sale of products and the size of the space in use to make the point of sale subject to fire prevention controls with the need to coordinate, also in terms of time, the relevant documentation of the single unit with that of the station.

All of this is articulated in an extensive activity of cooperation and collaboration that sees the relationship between the station manager and the manager of the production unit installed, with charges of promotion of coordination on the part of the subject from time to time client or promoter of the modification, management and coordination of the risks introduced and relative diffusion to the other parties for the corresponding evaluation from the point of view of their respective interests.

Furthermore, the development of the same process over time in relation to the changes made from time to time by each of the parties in question (set-up, entrusting of services and works, etc.) and the repetition and updating of the cooperation and collaboration process are described above.

To further accentuate the complexity of the process, consider the need for the station manager to guarantee adequate safety conditions with respect to third parties present in the station for whom there is no margin for process management other than those limited to signage in the areas.

Irrespective of the 'colour' with which the three conditions represented above and the deliberately simplistic nature of the relative characterisation have been represented, it undoubtedly emerges that risk management, already with respect to the technical and managerial safety profile alone, is a multi-stage process (work safety, fire prevention, administrative requirements, etc.) and is carried out on several levels of interaction between different but mutually relevant and interfering subjectivities.

In this sense, the need for a methodology that organically covers different phases and connects different levels with logical consistency and applicative replicability, also for updating and comparing the results in relation to the changes that have occurred, appears evident and of pressing necessity in order to guarantee the complete mapping of technical and management activities in the reciprocal interferences.

In the three cases represented above, with the development of our catering unit, no reference was made by chance to the decisions of the controlling holding company.

Paradoxically, and in purely theoretical terms, the scope of risk management analysis could be extended to infinity or almost infinity; to confirm this, we reiterate the need for a methodology that assumes consistent parameters and replicates them in all phases of the process.

Some correlations are, however, directly formulated by the legislator, where it provides (e.g. some specific regulations on industrial risk management) that the risk of the technical and management process goes back to the body in terms of responsibility, or when the body itself is presumed to be exempted from responsibility where it has adopted and effectively implemented an organisation and management model ensuring a corporate management system with respect to specific obligations.

In addition, perhaps even more explicitly and with reference to security processes, the importance of organisational factors is laid down in the text of other Italian laws, which states that 'the guarantee positions [...] also apply to a person who, though not having a regular office, actually exercises the legal powers relating to each of the persons defined therein'.

In other words, and with reference to corporate risk management, organisational factors cannot be included in the risk management process, even if from time to time aimed at more specific profiles such as safety or environmental risk assessment or in terms of sustainability.

Just think, for example, of the case in which the company organisation divides management and commercial competencies and the top management entrusts the safety function delegations to the manager in charge of management.

Any misalignment of processes shall not correspond to the same responsibilities with respect to any non-compliance.

In other words, with the risk management processes, the possibility of instrumental use of function delegations is in some way undermined, and the assessment of risk as a criterion for making the organisation responsible refers to a more detailed analysis of the consistency between organisation and responsibility.

The above considerations, albeit in a concise and in many ways summary manner, are intended solely to highlight the assumption that risk management, due to the importance assumed in the current regulatory framework and the complexity of the reference scenarios, requires to be 'based' on methodologies capable of replicating logical and evaluative consistency in all phases and levels of the process regardless of the scope of the analysis.

With these premises, this book is intended to be a valid reference, for professionals and technicians in the risk management sector, to the rigorous approach of the barrier-based risk management. The basic idea behind this approach is to integrate the traditional and well-defined risk management processes using a barrier-based perspective, basically the one suggested by James Reason with his well-known Swiss cheese model (Figure 3.1).

Throughout this book, a consistent vocabulary is used to avoid any misunderstanding in using extensive and complex terminology. The main reference for this is the ISO Guide 73, Risk Management – Vocabulary.

For the intent of this book, it is supposed that the reader knows that risk is different from a hazard (the risk is the future impact of an uncontrolled hazard or, better, the future uncertainty created by the hazard). This is a fundamental difference since all methods are intended to assess the risks associated with inherent methods. This is outmost valid for fire and explosion risks too, including, in particular, industrial domains for the risk for major accidents.

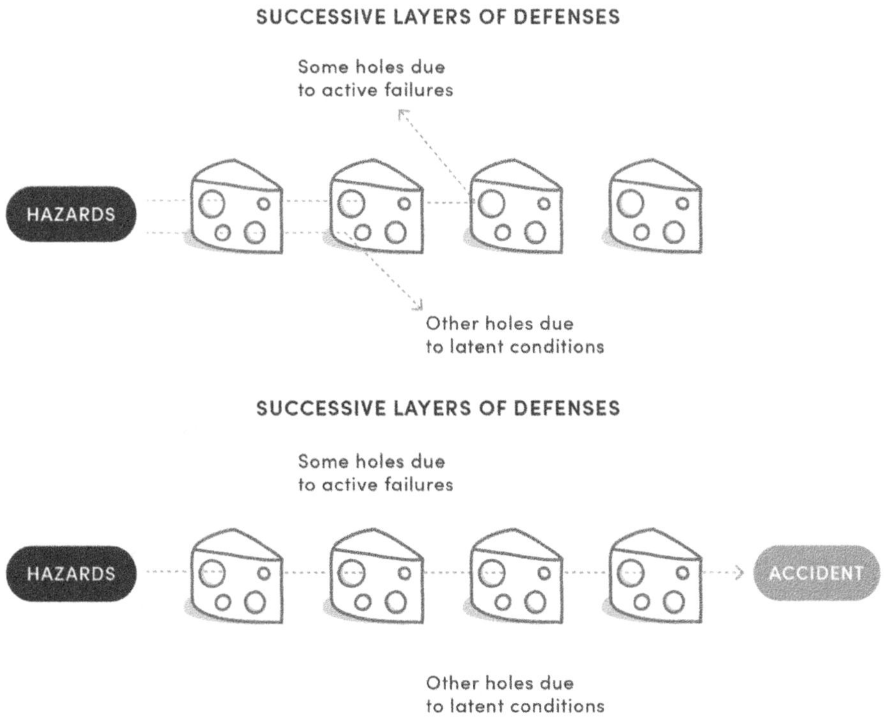

Figure 3.1 Swiss cheese model.

3.2   ISO 31000 Standard

Regardless of the type, entity, size and complexity of an organisation, many regulations and laws (both national and international) increasingly require the adoption of management systems that cover risk management. The adoption of a risk management model that over time complies with the international technical standard ISO 31000 can increase the effectiveness of this action: the organisation's efforts are made consistent with a general model that is already consolidated, widely tested and used.

The ISO 31000 technical standard aims to provide principles and guidelines for risk management. The standard, of voluntary adoption, preserves a universal conception such as to make it applicable to any company context, regardless of the nature of the risks associated with the organisation's activities, adapting itself in a systematic, transparent and credible way, with the possibility of a progressive approach.

The ISO 31000 technical standard provides guidance to ensure adequate risk management in organisations. The content of the standard universally applies to any organisation regardless of entity, type, business sector and size. For this reason, the indications of the technical standard must then find appropriate customisation depending on the specific context of application. Moreover, the document is valid throughout the entire life of the organisation, depicting the whole life cycle for risk management. The standard contains proper references that can be applied to any activity, within the same organisation, and embraces decision-making at all levels.

But first, what is risk? According to the vocabulary worldwide used in the sector, as stated in the international standards, the risk is the 'effect of uncertainty on objectives'. This definition recalls the necessity to explain what an objective is. An objective, from an organisational point of view, can be defined as the business goals, thus including not only the need to maximise profit but also the safety goals, reputational goals, environmental goals and so on. Having clarified what risk is, as stated in the previous paragraph, it is clear that risk can be seen from three different points of view (Figure 3.2).

These three different perspectives reflect different perceptions of the risk. All of them share the occurrence of an event, whose effect can be positive, negative or both. Indeed, an effect can be seen as a deviation from the expected. This deviation might result in opportunities or threats. In the end, the risk is usually described in terms of hazardous sources, potential events (scenarios), and their consequences and likelihood. It is clear that when talking of fire incidents, a negative perspective is adopted.

Other definitions are summarised in Table 3.1.

Having defined what the risk is, the next question is 'What risk management is?'. According to clause 3.2 of ISO 31000, it can be described as the 'coordinated activities to direct and control an organization with regard to risk'. The expression, widely used both in ISO 31000 and in ISO/TR 31004, refers to the principles, framework and process that should be set up by every organisation in order to manage risk effectively.

Risk management is carried out in the identification, analysis and evaluation phases with

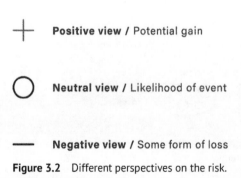

Positive view / Potential gain

Neutral view / Likelihood of event

Negative view / Some form of loss

Figure 3.2   Different perspectives on the risk.

Table 3.1 Definitions of 'risk'.

Definition	Source	Pros	Cons
1) Risk is the potential that a chosen action or activity (including the choice of inaction) will lead to a loss (an undesirable outcome).	Wikipedia	Definition takes into account the fact that also inactivity and not only activity may rise risks. This is important in a systematic approach to risk management in process-oriented organisations. Also, risk is seen in a general way with the term 'outcome' that is 'undesirable'.	No specific cons identified.
2) The effect of uncertainty on objectives.	ISO 31000	This definition is synthetic and general. It covers all the domains and sectors, and it is not specific for any kind of risk. 'Uncertainty' is introduced as the element that increases risks, leading to an undesired event that coincides with the partial or total failure in reaching objectives that have been previously defined. This is a high-level definition on the basis of the modern HLS ISO management systems. Risk, as defined in the standard, could also be seen as an opportunity.	This definition can be mastered by risk management professionals: it is undoubtedly not so usual among general people to understand the link among 'uncertainty' and 'risk'. Also, in most of the cases, the risk effect can be connected to harm, damages, etc. rather that the potential to fail in reaching objectives. Those events, especially those related with HSE aspects, can be linked to the failure of an organisation in reaching a proper safety level. Definition requests for the users to have a high level and systemic vision over organisation and its processes.
3) Combination of the probability of an event and the consequence of the event.	ISO 17776	Very useful when dealing with quantitative risk assessments as in the offshore oil and gas field to which the ISO 17776 relates with.	No specific cons identified.

the aim of identifying any corrective actions that may be necessary to meet the risk acceptability criterion that the organisation has set for itself, also in relation to any legal requirements. Through this process, the organisation communicates with stakeholders, monitors and reviews the risks and control measures put in place to ensure that no further action is needed to reduce the risk levels achieved. This logical process is described in detail by ISO 31000: the interested reader will find more detailed information in the standard as this book is not intended to cover these topics so in-depth.

In a few words, organisations embracing a structured approach to risk management have the possibility to improve their resilience to complexity and have a higher chance to reach their strategic objectives. And that is why fire risks must also be managed in this way.

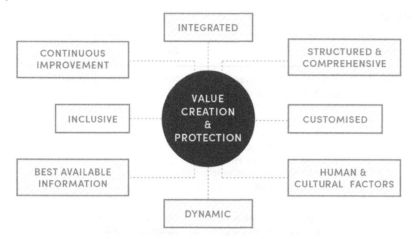

Figure 3.3 The principles of RM according to ISO 31000.

3.2.1 The Principles of RM

When managing risk, it is fundamental to consider the principles that enable the creation of solid RM framework and processes, allowing an organisation to dominate the effects of uncertainty on its objectives. In particular, the ISO 31000 defines the principles that an organisation should satisfy, describing the logic behind an effective and efficient RM. Indeed, their satisfaction allows to create and protect value, as shown in Figure 3.3.

To manage risk effectively, each organisation should adhere to the principles shown in Figure 3.3 at all levels.

Each organisation should therefore adopt a structured and complete approach to risk management: this is what ISO 31000 calls the framework (Figure 3.4). It ensures that the information derived from the risk management process is properly used as a basis for decision-making and the definition of responsibilities at all levels of the organisation.

This framework needs to be designed using the components shown in Figure 3.5.

3.3 ISO 31000 Risk Management Workflow

As reminded in the ISO Guide 73, the RM framework is a 'set of components that provide the foundations (i.e. the policy, objectives, mandate and commitment) and organizational arrangements (i.e. the plans, relationships, accountabilities, resources, processes and activities) for designing, implementing, monitoring, reviewing and continually improving risk management throughout the organization' (Figure 3.6).

A short description of these components is discussed in the following section.

3.3.1 Leadership and Commitment

Leadership and commitment ensure the integration of the RM into all the organisational activities. Being an RM leader requires solid skills in predicting and accepting changes that may be involved in the behaviour, culture, processes and should be reflected within the policy, always monitoring the

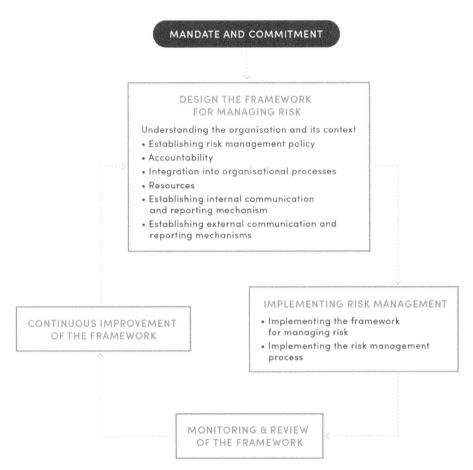

Figure 3.4 The RM framework.

Figure 3.5 Components of a risk management framework.

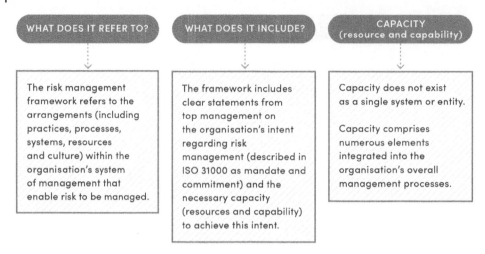

Figure 3.6 Risk management framework.

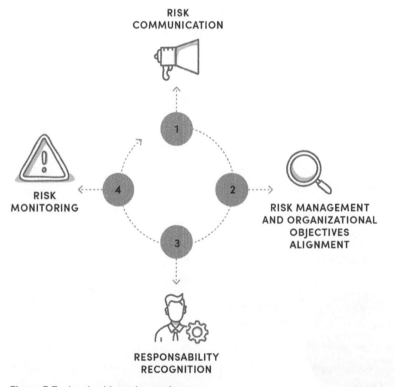

Figure 3.7 Leadership and commitment.

expected performance while managing risks, and in the RM framework. Of course, meeting the principles of ISO 31000 is among the best attempts for excellence in risk management (Figure 3.7).

3.3.2 Understanding the Organisation and Its Contexts

In building the risk management process, it is also of paramount importance to prioritise the context factors, both external and internal (Figure 3.8).

Its definition allows the organisation to articulate its objectives, define the internal and external parameters to be taken into account in operational risk management, and define the scope of the process and the criteria of risk acceptability.

The requirements (both mandatory/voluntary or internal/external) related to RM are shown in Figure 3.9.

3.3.3 Implementation of the RM Framework

In order to implement the RM framework, an organisation should perform the four steps shown in Figure 3.10.

Figure 3.8 Internal and external context factors.

Figure 3.9 Identify the requirements related to risk management.

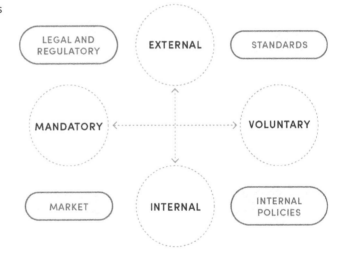

Figure 3.10 Implementing the risk management framework.

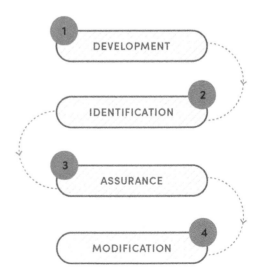

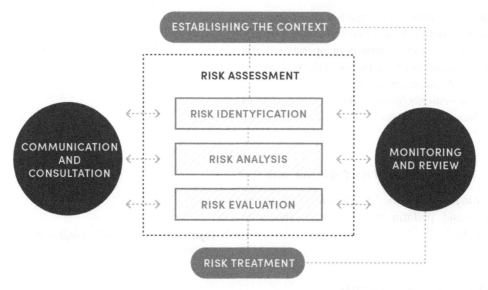

Figure 3.11 Scheme of the risk management process according to ISO 31000.

3.3.4 The Risk Management Process

This process includes the activities described in Figure 3.11.

3.4 The Risk Assessment Phase

Risk assessment is at the heart of the entire management process as well represented in Figure 3.12. In particular, once the objectives, criteria, scope and degree of depth have been established, it constitutes the fundamental element for the construction, implementation and periodic assessment of the corporate risk management system. The risk assessment process consists of three phases: the identification of the risk, its analysis and its evaluation. Some techniques to implement this phase are discussed in the next chapters of this book.

In particular, given a brief description of some commonly used risk identification methods (FTA, ETA, structured brainstorming, HAZOP/HAZID and matrices) that are among those strongly applicable to that initial step of the risk assessment, the focus will be on the barrier-based approach to the risk analysis phase, discussing both Bow-Tie and LOPA.

The risk assessment provides an understanding of the risks, their causes and consequences, and their probabilities (intended as the probability of the hazard as modified by present and planned measures). This provides important input to decide the following factors:

- If an activity must be performed.
- How to maximise opportunities for improvement.
- Which risks need to be treated in order to reduce (possibly further) their level.
- How to prioritise risk treatment options.
- How to select the most appropriate risk treatment strategies capable of bringing high levels of risk to tolerable or acceptable levels, including in relation to the resources required (e.g. associated costs and time).

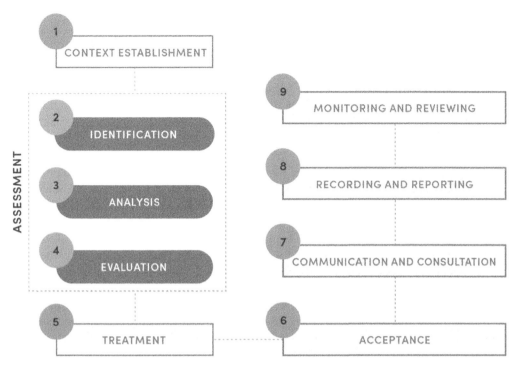

Figure 3.12 The risk assessment phase in the context of the RM process.

Risk assessment may often require a multidisciplinary approach as the organisation's risks can cover a wide range of causes and consequences, and, as anticipated, combinations of risks and control measures often need to be considered.

3.5 Risk Identification

Each organisation should identify the hazards associated with its processes, products and services, sources of risk, areas of impact, events (including context changes), their causes and potential consequences. The objective of this phase is to generate an exhaustive list of risks based on those events that could create, increase, prevent, accelerate or delay the achievement of the objectives. A comprehensive identification of risks is essential as a risk not identified at this stage will not be considered in the subsequent analysis.

The identification should include those risks whose source is under or not under the control of the organisation, even when it is not immediate to identify their source or causes. Risk identification should include an examination of domino effects, identifying cascading and cumulative effects.

Each organisation should apply the risk identification tools and techniques that best suit its objectives and capabilities, as well as the risks it faces. Relevant up-to-date information is important in identifying risks; therefore, it is important to know the background information, where available. In the following chapters of this book, a description of the most used methods is provided.

3.6 Risk Analysis

Risk analysis is the phase in which the understanding of risk is developed. It provides input to the following phase of risk evaluation and enables a decision to be made as to which risks need further treatment in relation to the established acceptance criteria, as well as identifying the most appropriate strategies and methods for doing so. This phase can also provide input to the decision-making process of choosing between several treatment options covering different types and levels of risk.

Risk analysis consists of determining, by combining them, the consequences and associated probabilities of risk events identified in the previous phase of hazard identification, taking into account the presence – or absence – and the effectiveness of existing control measures. The level of consequences and their likelihood of occurrence are then combined to determine the level of risk (Figure 3.13).

The risk analysis is based on the causes and sources of risk, their consequences and the likelihood of these consequences occurring. Therefore, the factors influencing the level of severity of the consequences and their probability must necessarily be identified. It should be borne in mind that a single event can have multiple consequences and have an impact on several targets.

For example, a fire scenario can impact the occupants of a building, its structural capacity, its strategic role (if an hospital or a city town), the reputation of a company and the surrounding environment.

This phase normally includes an estimate of the severity of the consequences (effects) that may arise from an event, situation or circumstance and their associated probability of occurrence in order to measure the level of risk. However, sometimes in simple cases where the expected consequences are not significant and the associated probability of occurrence is estimated to be extremely low, a single parameter may be sufficient to make an estimate.

The methods used for risk analysis may be qualitative, semi-quantitative or quantitative. The degree of detail required will depend on the particular application, the availability of reliable data, and the decision-making needs of the organisation and of course the type of risk (and its potential effects). Some methods and the degree of detail of the analysis may be prescribed by a specific law with which the organisation is obliged to comply.

Qualitative assessments define the consequences, the probability and the level of risk through levels of significance, such as 'high', 'medium' and 'low', thus assessing the resulting level of risk from the combination of consequences and probability through a qualitative criterion.

The semi-quantitative methods use numerical scales for consequences and probabilities and combine these levels to produce a level of risk by adopting a specific formula. The scales can be linear or logarithmic or have other types of relationships; the formulas used are the most varied.

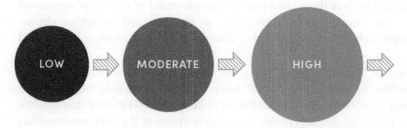

Figure 3.13 Level of risk.

The quantitative risk analysis (QRA – 'Quantitative Risk Assessment') estimates, through numerical values often expressed according to codified parameters, including risk indices, the level of consequences and their probability, and produces values of the risk level in specific units defined in the previous phase of context identification. A QRA may not always be possible or desirable due to insufficient information on the system or activities analysed, lack of data, complex influence of human factors, complexity and severity in terms of the extent of the complete survey, etc. In fact, sometimes quantitative analysis is not applied because such an effort is not required and it is sufficient to adopt a semi-quantitative or qualitative analysis that is not only synthetic but also immediately expendable and clearly effective.

In cases where qualitative analysis is used, it is good to provide a clear explanation of the terminology adopted and the reasoning behind the definition of the criteria used.

Even in cases where a QRA is adopted, it is good to keep in mind that calculated risk levels are always estimates. It must therefore be ensured that these risks are not attributed a level of accuracy and precision inconsistent with the accuracy of the data and methods used.

The results of this phase must be expressed in clear terms depending on the type of risk and in a form that provides easy input for the next phase of risk assessment.

3.6.1 Analysis of Controls and Barriers

Risk levels will depend on the adequacy and effectiveness of existing control measures. The analysis of control barriers should highlight the following questions:

- What are the existing control measures for a particular risk?
- Whether these barriers are capable of dealing adequately with the risk so that it is managed at an acceptable level.
- Whether the control measures are operating as intended (as designed) and whether their effectiveness can be demonstrated (even over time).

The answers to these questions can only be provided with certainty if adequate documentation is available and if the way in which the processes under analysis are managed is well known, as well as the conditions under which the barriers can correctly reduce the risk assigned to them.

The level of effectiveness (or, using a synonym, of maturity) of a particular control measure can be expressed qualitatively, semi-quantitatively or quantitatively, according to the type of analysis adopted. Although it is generally difficult to express a measure of effectiveness in a highly accurate way, it is nevertheless useful to express and record a measure of the effectiveness of the control barrier in such a way that it is possible to make an informed judgement as to whether it is convenient to improve the performance of the same barrier or to adopt a different risk control measure.

3.6.2 Consequence Analysis

The consequence analysis, to be combined for the risk estimation with the assessment of the probability of occurrence, allows to determine the nature and type of impact that could occur assuming that a particular event could happen. The same event may have different impacts of different magnitude, affect the fulfilment of different objectives of the organisation and influence different stakeholders. The types of consequences to be analysed and the targets that are affected must be established at the context definition stage.

Consequence analysis can range from a simple description of the findings to detailed quantitative modelling.

   EXPERT   PROBABILITY
HISTORICAL DATA   JUDGEMENT   ESTIMATION

Figure 3.14 Frequency analysis and probability estimation.

Consequences may have a low level in terms of impact but have a high probability of occurrence, or vice versa a high level of magnitude and a low probability of occurrence, or intermediate values. In some cases, it may be appropriate to focus on risks with very severe consequences regardless of their probability of occurrence. In other cases, it may be equally important to analyse risks with low-level consequences but whose impact is frequent or even chronic with cumulative or not negligible long-term effects.

The analysis of consequences requires that the following factors should be considered:

- Existing control barriers are taken into account together with all relevant factors that have an effect on the magnitude of the consequences (including, of course, any 'escalation factors').
- The consequences of the risk are always related to the objectives of the risk management system originally defined.
- Consequences relevant to the scope of the risk assessment are considered, i.e. those in line with the defined scope and work perimeter.
- Secondary consequences (so-called domino effects) are also considered, such as those affecting systems, activities, equipment or organisations connected to the risk assessment scope, also taking into account the common causes of failure and the impact on critical systems and processes for the organisation.

3.6.3   Frequency Analysis and Probability Estimation

Probability of occurrence estimation is generally performed according to three approaches that can be used individually or jointly (Figure 3.14).

3.7   Risk Evaluation

The objective of the risk evaluation is to assist the organisation in the decision-making process, based on the results of the risk analysis, about which risks need to be treated, also identifying the priorities for the implementation of such treatment.

The risk evaluation is carried out by comparing the risk levels estimated in the previous phase of the analysis with thresholds of acceptability (or tolerability) defined by a pre-established criterion. Based on this comparison, the need for further action to reduce the level of risk is analysed. These decisions should take into account the assumptions and models on the basis of which the previous phase of analysis was carried out, in order to properly consider the tolerances and sensitivities of the data obtained. It is understood that such decisions should also be made in accordance with legal requirements where defined.

Possible decisions include the following:

- If a risk must be subjected to the next stage of treatment.
- Set priorities for treatment.
- Which of the possible alternative solutions to achieve the desired result should be followed, also in relation to the resources that need to be employed.

Sometimes the risk evaluation may lead to the decision to undertake further in-depth risk analysis studies; other times it may lead to the decision to maintain existing control barriers thus accepting the pre-existing level of risk. These decisions are influenced by the company's risk appetite and the criteria of acceptability and tolerability that the company has set when setting its objectives, in accordance with the risk management policy.

3.7.1 Acceptability and Tolerability Criteria of the Risk

The simplest approach to defining a risk acceptability and tolerability criterion is to divide risks into two main categories: those requiring treatment and those not requiring treatment. This approach leads to seemingly simple results, but it does not reflect the uncertainties inherent in estimating risks and defining the boundary between risks that will need to be treated and those that will not. Once it is clarified that the risk acceptability and tolerability criterion must be defined by the organisation on the basis of its own principles and risk sensitivity, the decision whether or not to treat a risk may depend on the costs and benefits of implementing any additional control measures. For this reason, a cross-sectoral approach across various industries is to divide risks into the following three bands (Figure 3.15):

- A higher band, where the level of risk is considered intolerable whatever the benefit of the risky activity. In this band, the treatment of the risk is essential whatever the cost.
- An intermediate band, where the relationship between the costs and benefits expected from the implementation of additional measures is considered, comparing opportunities and potential consequences.
- A lower range, where the level of risk is considered acceptable or such that no further treatment is required.

Figure 3.15 Risk acceptability and tolerability thresholds.

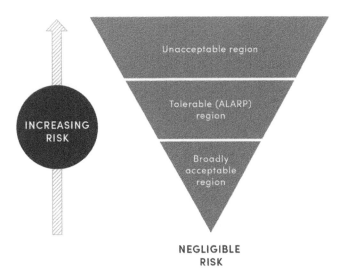

One of the most common methods for the representation of the risk acceptability and tolerability ranges is the risk matrix, within which reference ranges based on effect categories and/or probability classes can be identified.

The risk matrix is a useful tool for the graphical visualisation of risks and the combination of their magnitude and frequency levels (Figure 3.16). It is particularly used when a semi-quantitative analysis is carried out. This type of analysis, as already discussed, uses a numerical approach, typical of a quantitative analysis, together with simplifying and conservative assumptions about the evaluation of the level of severity of consequences, the evaluation of the frequency of occurrence of the initiating causes of an event and the efficiency of control measures. Generally, the results of a semi-quantitative analysis are expressed in orders of magnitude. However, a risk matrix is also generally used for qualitative analysis, such as that shown in Figure 3.16. In this figure, both probability and severity are expressed in qualitative terms, which must be evaluated by an experienced team to assign the appropriate level of risk, given by the combination of a given severity class with a specific probability class. On the other hand, in a semi-quantitative analysis, the frequency of occurrence is generally expressed on occasions per year (occ/year), whereas the consequences are identified through a progressive level from 1 (the least severe) to 5 (the most severe), in relation to the severity of the

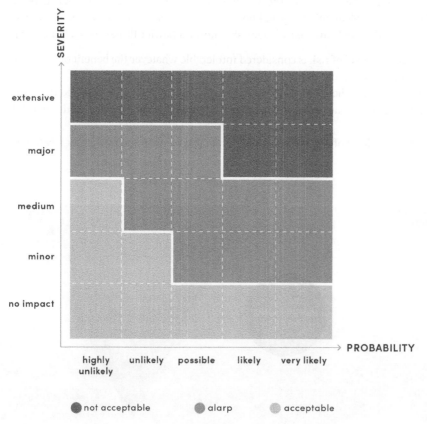

Figure 3.16 Example of a risk matrix with level of acceptability regions.

expected consequence. In the example, the grey region defines the most severe risks. A risk falling in this region often requires the immediate stop of the organisation's activity in this area, being absolutely unacceptable. The blue area of the matrix identifies a particular region of the matrix where the risk could be accepted (tolerable risk). The risks in this region require an ALARP (As Low As Reasonably Practicable) study. Briefly, this is a cost–benefit analysis of the potential intervention required to further mitigate the risk in order to target the region of acceptability. Since mitigation may require an economic effort that is not justified by the reduction of the level of risk, a risk falling within the ALARP region could be accepted as such: the managers (or, in general, who is responsible for the risk) will take responsibility for this choice based on a cost–benefit analysis.

Finally, the green region is about acceptable risks: no further ALARP mitigation or study is required for them.

ALARP is not the only risk acceptance criterion available. There are also other criteria, such as GAMAB (globally at least as good), MEM (minimum endogenous mortality), MGS (at least the same level of safety) and NMAU (not more than unavoidable), but it is not the subject of this book to address them, so the ALARP study, one of the most widespread in the world of risk management, is discussed below.

The graphical representation of the risk level through the matrix allows the introduction of the risk prioritisation (Figure 3.17), i.e. the possibility of assigning a degree of priority to the various risk treatment options that will be offered, as better explained below.

The prioritisation of risk allows the organisation to make risk-based decisions, thus allowing itself to be guided in decision-making processes by the need to target an acceptable level of risk. In the case of the risk matrix shown in Figure 3.18, this means prioritising scenarios with a risk level close to the top right (very high impact). The proposed graphical representation guarantees a good communication of information to all stakeholders, while pursuing the principle of inclusiveness of risk management.

Figure 3.17 Prioritisation of risk given impact and likelihood.

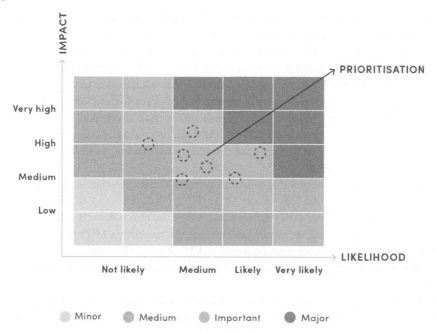

Figure 3.18 Risk prioritisation and the risk matrix.

3.8 The ALARP Study

ALARP stands for 'As Low as Reasonably Practicable', a concept that is strictly connected, in a risk management framework, with risk criteria in the risk evaluation phase and with the need or opportunity to further reduce risks in the decision phase. Given the fact that risks should be treated (accepted, modified, reduced or even increased in case risk is seen as an opportunity and obviously not when considering safety related aspects) and strategy may be defined on the basis of pools of control measures, it is fundamental to understand how much reduction should be obtained, which is the risk-reducing factor associated with each single control measure and the cost (effort not only in economic terms) associated with it. This approach should consider the existing controls, and it allows to judge the risks being ALARP or not and eventually define a further mitigation strategy (supporting the risk management over time). It is worth emphasising that the concept of ALARP, together with its effective practical application, is very often dependent on the effectiveness of the tools used in the risk analysis phase. Since risk matrices are a common tool in a variety of sectors, also considering the latest trends in employing performance-based approaches and risk-based thinking in several disciplines, their calibration and the risk criterion definition both play a fundamental role in subsequent ALARP studies.

Risk-based decisions are still an issue among the public, affected by a number of factors that involve several psychological dimensions: risk understanding, risk perception, risk communication and risk acceptability. We have recently observed this with the perception of risks associated with the 'Astra Zeneca' Covid-19 vaccine or even the risks associated with the new autonomous car vehicles since while regulatory authorities tend to introduce the ALARP approach in dealing with risk management for serious threats there are social and socio-economic implications, with the result of 'areas of concern about the validity of this approach'. Unfortunately, 'ALARP' is perceived as a synonym of a poor practice to maintain the risks as they are, without any risk treatment (or even discussion).

ALARP should be considered a fundamental tool in the risk assessment part of a risk management framework. Actually, it has proven effective for many years and also during design stages of engineering projects to define and optimise the global risk reduction strategy to be implemented, and is therefore an opportunity to support risk-based approaches in several risk management frameworks where the risk reduction associated with multiple measures can be achieved combining the risk-reducing factors associated with single controls having their economic and management over time impacts. ALARP assessment can then be applied to several situations even considering new emerging threats in industry that trends are going to face with risk-based approaches (e.g. functional safety, assets ageing, cybersecurity, etc.).

Risk reduction (considering the risks having negative outcomes as in the industrial safety and in the occupational safety fields) is mandatory. ALARP supports the risk assessment process and guarantees that it is possible to demonstrate with a structured approach that the residual risk cannot be mitigated more since there is a disproportion among the effort and the benefit. Even if ALARP application started in the United Kingdom, its use is becoming more and more popular in several other countries and in different sectors given the fact that residual risk from a risk assessment should be discussed. In my opinion, it exists since it is a method (approach) that complements risk reduction options' discussion and becomes a way to support with evidences and documents the continuous improvement demanded by several risk-based standards. It also exists to support policy and decision-making with a system thinking approach and the possibility to consider RAGAGEPs (Recognised and Generally Accepted Good Engineering Practices).

Each organisation should support ALARP studies, helping to develop a comprehensive risk mitigation strategy for scenarios deemed credible, identifying and prioritising any viable risk mitigation action that can reduce the risk associated with a scenario up to an ALARP level. In addition, for those scenarios without the opportunity to further reduce the level of risk, each organisation should provide a mechanism to document that the risk considered is already at a tolerable level (ALARP).

ALARP studies should be undertaken using a documented process and performed by personnel familiar with the details of the risk scenario being assessed. Typically, the personnel dedicated to ALARP studies are the same personnel who carry out the risk analyses.

Any potential risk mitigation opportunities identified within the ALARP process will be documented as follows:

- A recommendation.
- An option that is not recommended because of its impracticability in relation to the resources to be put in place (including those for maintaining the recommendation over time).
- An option that is not recommended because of other identified and deemed more favourable risk mitigation opportunities.

It should be noted that risk mitigation alternatives are often not mutually exclusive, so multiple recommendations could result from the ALARP study of a single risk scenario. Moreover, not all risk mitigation measures or recommendations in an ALARP study will generally be classified as control measures (barriers). Those recommended actions that cannot qualify as barriers, because they do not produce a reduction in the risk level by acting on the frequency of occurrence or severity of a scenario, will be collectively identified as 'other measures'. Examples of 'other measures' could be – depending on the specific context – warning signs or labelling. 'Other measures' are typically identified as actions to be applied to consequences whose risk is already acceptable.

An ALARP study aims to answer the following questions in a structured way for each risk scenario assessed:

- What alternatives are available to eliminate, reduce or manage the risk?
- What factors determine the feasibility of any risk mitigation alternatives?
    - How much risk mitigation is substantially achieved by the measure?
    - What resources are needed to implement the measure?
    - What synergies would be achieved if the measure is implemented?
    - Should the measure also be implemented in similar facilities/workstations?
    - Is the measure congruent with the current practices adopted in the reference work sector?
    - What is the technical and operational feasibility of the measure?
    - How long would the implementation of the measure take?
    - What other risks would be impacted by the implementation of the same measure?
    - What is the availability of the measure?

In general, the risk reduction strategy imposes a hierarchy of possible options in the following order:

- Eliminate the hazard (e.g. by changing the activity, process, system under consideration and source of the risk considered).
- Reduce the hazard (e.g. by reducing the amount of flammable substances stored in the warehouse).
- Control the hazard through additional measures (controls or barriers).

According to this hierarchy, for example, an action that eliminates the hazard will have a much greater benefit than the installation of an additional barrier to control it.

Before carrying out an ALARP study, an organisation might ask itself whether the action required to reduce a specific risk has already been taken with positive feedback from other assets in the same organisation or from other competitors. It is also necessary to ask first of all whether the suggested action is in fact a requirement arising from new mandatory regulations or standards or international best practice. In such cases, the organisation may decide not to carry out any cost–benefit analysis, deciding a priori to implement the new measure.

Although there are different methodologies to perform a cost–benefit analysis, for the purposes of this book it seems sufficient to mention qualitative analysis. It is based on the adoption of a matrix like the one in Figure 3.19. For each risk falling within the risk tolerability region (i.e. the ALARP region of the risk matrix in Figure 3.16), the analyst wonders what are the costs (e.g. in financial and/or time terms) and benefits (in terms of risk reduction) expected from the implementation of a particular option. The scale of benefits could for example be as follows:

- High: The risk leaves the ALARP region and becomes acceptable.
- Medium: The risk reduces in level but remains in the ALARP region of risk tolerability.
- Low: The measure does not reduce the level of risk.

		Expected benefits		
		High	Average	Low
Associated costs	High			
	Average			
	Low			

Figure 3.19   Matrix example for qualitative ALARP analysis.

MAXIMISING THE COST/RISK RATIO

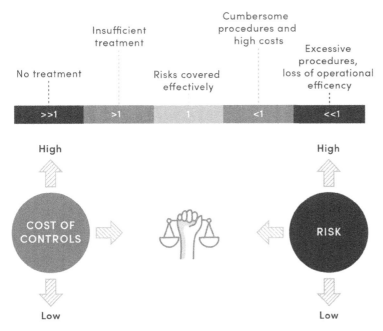

Figure 3.20 Achieving balance in risk reduction.

The definition of the cost level will typically depend on the sensitivity of the organisation and other strategic policy criteria, both in economic–financial and temporal terms. According to the example matrix in Figure 3.19, high-cost and low-benefit options will not be implemented, whereas high-benefit measures will always be implemented, regardless of the costs associated with their implementation.

When calculating associated costs, it is necessary to consider both the resources required for the design of the further control and those connected with its implementation and maintenance over time. It is also useful to consider not only direct costs for implementation and operation but also indirect costs (inspection, testing and periodic maintenance or information and training for use). In fact, like the risks that they intend to reduce, the controls have their own life cycle, within which the resources for maintaining efficiency over time could play a crucial role and be a discriminating factor in the selection of the most appropriate treatment measure.

In conclusion, the objective of the ALARP study is therefore to offer a cost/benefit assessment, which is summarised in Figure 3.20.

3.9 Risk Management over Time

Risk management does not end with a photograph taken at a certain moment but instead requires a dynamic approach. This is first performed by following the risk assessment phase with the treatment phase. The risks analysed will in any case be monitored and reviewed, as well as the entire structured approach to risk management (the framework) will be monitored and reviewed at predetermined intervals, thus reviewing the performance of the system.

3.10 Risk Treatment

The risk treatment often requires the choice of one or more options to modify the risks assessed during the previous evaluation phase and then implement these options. Once implemented, the risk treatment generally provides new control measures or changes to existing ones. This is a cyclical process consisting of the following steps:

- Assessment of risk treatment.
- Decision whether or not the residual risk, i.e. the level of risk obtained downstream of the implementation of the treatment, is tolerable.
- If not tolerable, generate a new risk treatment.
- Assess the effectiveness of this new treatment.

As discussed above, risk treatment options are not necessarily mutually exclusive or suitable for all circumstances. Generally, the possible options are as follows:

- Avoid risk by deciding not to start or continue the risky activity.
- Remove the source of the risk.
- Change the frequency of occurrence of the damaging event, with interventions aimed at reducing the frequency of its causes or reducing the probability of failure on demand of control measures.
- Change the level of expected consequences, for example by increasing the effectiveness of mitigating control measures.
- Share information on the risk level with other parties, including contractors and insurers.
- Accept the risk through an informed and aware decision.

Selecting the most appropriate risk treatment option often involves a cost–benefit analysis, already mentioned in the ALARP study previously introduced, always bearing in mind the legal requirements, the social responsibility and the environmental protection. Each organisation should also make decisions on risks that may require treatment that cannot be justified on economic basis, such as risks with very severe yet extremely rare consequences.

When choosing risk treatment options, the organisation should consider the values and perceptions of stakeholders by adopting the most appropriate means of communication with them, who should be involved in the decision-making process. Indeed, although equally effective, some actions may be more easily accepted by some stakeholders than others.

The corrective action plan should clearly identify the priorities in risk treatment. It should be noted that the risk treatment phase may itself introduce new risks, such as the failure or ineffectiveness of the treatment measures chosen. If new secondary risks are introduced during the risk treatment phase and need to be assessed, treated, monitored and reviewed, then these risks should be incorporated into the same treatment plan as the 'original' risks, avoiding that they are treated through a different process.

The objective of risk treatment plans is to document the treatment options chosen, clarifying how they are implemented. At least the following information should be contained in a risk treatment plan:

- The reasons for selecting a particular treatment option, including the expected benefits of such implementation.
- Who is responsible for approving the plan and who is responsible for its implementation.
- The proposed actions.

- The required resources (both instrumental and organisational–managerial).
- The performance measures and any constraints.
- The monitoring and documentation requirements (including the recording of evidence).
- The scheduling of actions according to a pre-established plan and based on the established priority.

Treatment plans (and not just the individual options contained therein) should also be integrated with the organisation's management processes and discussed with all stakeholders.

Decision-makers and other stakeholders should be aware of the nature of the residual risk after the treatment phase; therefore, the residual risk should be documented subject to monitoring and review and, where appropriate, further treatment. The risk treatment activities are summarised in Figure 3.21.

At the end of the treatment phase, the inherent risk is reduced to a lower level, named 'residual risk' (Figure 3.22). Residual risk can be defined as the risk that remains after the implementation of controls aiming to reduce the inherent risk.

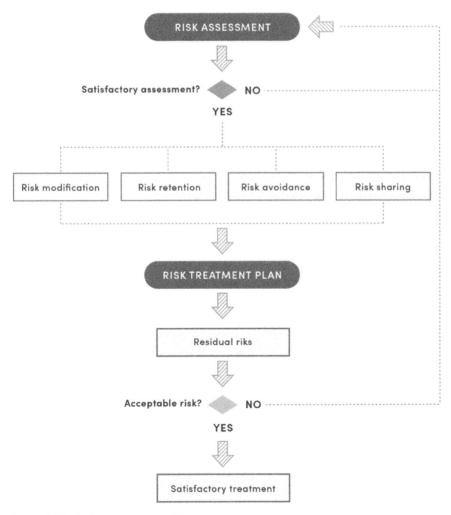

Figure 3.21 Risk treatment activities.

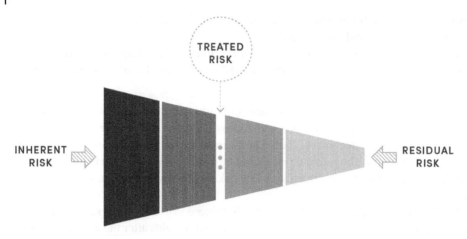

Figure 3.22   Residual risk.

3.11   Monitoring and Review

Both monitoring and review (also known as periodic review) are phases of the risk management process that involve, on a periodic or ad hoc predetermined basis, the verification and surveillance of the risks identified, analysed, evaluated and treated and the related control measures. The persons responsible for the monitoring and review phases must be clearly identified within the organisation. These phases should cover all aspects of the risk management process with the following aims:

- Ensuring that control measures are effective and efficient, both in design and operation.
- Obtaining further information to improve risk assessment.
- Analysing and learning lessons for improvement from events (including near misses and anomalies), changes, trends, successes and failures.
- Identifying changes in the internal and external environment, including changes to the risk acceptability and tolerability criteria and the risks themselves that may require a review of corrective actions and related priorities.
- Identifying emerging risks, including in relation to the results of an in-depth analysis of the root causes of adverse events.

The progress of the implementation of risk treatment plans provides a measure of performance to be monitored. The results can be incorporated into the management performance of the entire organisation, internal and external documentation and reporting measures and activities.

The results of the monitoring and review phases should be recorded and documented externally and internally when appropriate and should be used as input data for the review of the entire organisational risk management structure. This is in order to ensure that current risk management processes that are commensurate with the size and complexity of the company as well as suitable for the defined objectives are maintained over time.

In general, all risk management activities should be traceable. In the risk management process, data and evidence are the basis for improving methods and tools, as well as the entire process. The decision to retain this data must take into account the following factors:

- The organisation's needs for continuous learning.
- The benefits of reusing information for management purposes.
- The costs and benefits of creating and maintaining data records.
- Legal and operational requirements for data logging.
- How to access, retrieve and store multimedia data.
- The period of retention.
- The sensitivity of the information.

This includes periodic audit activities. The monitoring of systems through inspection (internal and second- or third-party inspection) guarantees the effectiveness and efficiency of the system.

3.12   Audit Activities

Regardless of whether or not third-party audits are applicable to the specific business context, the organisation intending to adopt a risk management system must conduct, at planned intervals, internal audits in order to receive information useful to verify whether the risk management system is

- compliant with the organisation's own requirements, including the risk management policy and the established objectives of the risk management system;
- compliant with the requirements of ISO 31000 in selected risk management paradigm;
- effectively implemented and maintained also in relation to changes that have occurred.

To verify this, the organisation must

- plan, establish, implement and maintain one or more internal audit programmes, where frequencies, methods, responsibilities, consultation, planning and reporting requirements are clearly identified, taking into account the importance of the processes involved and the results of previous audits;
- define the audit criteria and their scope;
- select auditors and conduct audits ensuring objectivity and impartiality of the inspection process;
- ensure that the results of audits are reported to relevant individuals such as managers, workers and, if any, workers' representatives;
- take action to address non-conformities arising from audits, thereby continuously improving the performance of the risk management system;
- maintain documented information containing the results of audits and the implementation of the audit programme.

The interested reader can refer to the ISO 19011 technical standard that defines guidelines for management systems audits, where the competency required for auditors is also specified.

3.13   The System Performance Review

Not only the specific risk management process must be subject to periodic monitoring and review, but also the entire 'corporate framework' that supports these processes must be subject to review. The structured approach to risk management, in order to be effective and provide constant support in achieving company performance, must be monitored and reviewed. In this sense, the organisation should perform the following actions:

- Measure the performance achieved in risk management through appropriate indicators, the appropriateness of which is periodically reviewed.
- Measure progress and deviations from expected targets on a regular basis.
- Periodically review whether the structure, policy and expected objectives are always appropriately defined, taking into account the external and internal context factors of the organisation and related changes.
- Document risk, progress against expected objectives, and how effectively the risk management policy is implemented.

On the basis of the results of the monitoring and any revisions, the organisation is asked to express its opinion on any decisions to improve the framework, the policy, and the expected objectives in terms of risk management. Such decisions should lead to tangible improvements in terms of risk management culture and, overall, to the effective and efficient implementation of the organisational model, with small but certain steps, to fulfil the organisation's policy and objectives (Figure 3.23).

The various technical regulations on management systems in any area identify this phase as a 'management review'. In this phase, senior management is required to periodically review the management system adopted. In doing so, it must also take into account the following factors:

- Status of actions resulting from previous reviews.
- Changes in internal or external context factors relevant to the management system, such as changed needs and expectations of stakeholders, change in legal requirements and change in business risks.
- Level of implementation of the policy and objectives of the risk management system.
- Information on the performance of the risk management system, including from the results of monitoring and measurement of performance standards.
- Adequacy of resources for the effective maintenance of the risk management system.
- Relevant communications with all stakeholders.
- Opportunities for the pursuit of the continuous improvement paradigm.

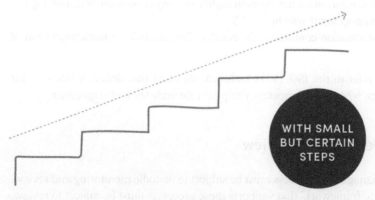

WITH SMALL
BUT CERTAIN
STEPS

Figure 3.23 Risk management process with continuous improvement.

Appropriate documentation must be kept as evidence of the results of management reviews.

In order to make the overall review phase of the system more objective and structured, it is useful to define numerical indicators and monitor each process under review, from which trends and the achievement of the set objectives can be inferred.

At the end of the process, the risks need to be documented in a report (Figure 3.24), whose content should contain the information recommended by ISO 31000, whose format is consistent (date of emission, version number, author clearly identified, date of approval and so on) and whose life cycle is properly managed.

The conclusion of the above discussion is dedicated to the person who is in charge of the risk management activities: the risk manager. Managing risk is a complex task and a risk manager should be appointed to ensure the required coordination. In this role, he/she should possess the set of skills and knowledge as identified in Figure 3.25.

In doing his or her job, regardless of the adopted work styles, the risk manager should not make the mistake of feeling that he or she is solely responsible for managing the process. He or she is certainly the leader, but the proper implementation of the risk management system requires a proper training of RM team members and a proper understanding of their responsibilities and the creation of a sound infrastructure, which can be achieved by only allocating the required resources (Figure 3.26).

CONTENT   FORMAT   LIFE CYCLE OF DOCUMENTED
                                        INFORMATION

Figure 3.24 Documenting the risk management process.

CRITICAL   PERSONAL   ADMINISTRATION   CREATIVITY
THINKING   INTEGRITY   AND MANAGEMENT

TECHNOLOGY   ECONOMICS   LAW AND   ACTIVE
                          AND ACCOUNTING   GOVERNMENT   LISTENING

Figure 3.25 Skills and knowledge for a risk manager.

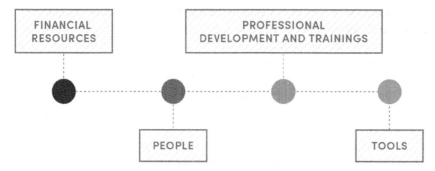

Figure 3.26 Resources to be allocated for an effective RM.

Figure 3.27 Understand the mission, objectives, values and strategies.

A good risk manager ensures the strategic alignment between the RM objectives and the organisation's goals (Figure 3.27). The consequent high level of consistency can be achieved only through a deep understanding of the mission, objectives, values and strategies of the organisation.

Finally, it is therefore worth highlighting how risk management is synonymous, in practical cases of risk treatment, with 'risk control'. Such risk control must follow the hierarchy and workflow that are summarised in Figure 3.28.

3.14   Proactive and Reactive Culture of Organisations Dealing with Risk Management

In order to understand the importance of the risk management culture, it is necessary to identify in advance its relationship with the processes to which it is (or should be) applied.

If, in fact, the purpose of an organisation is to pursue the maintenance and growth of its assets, whether material or moral, the quality level of risk management, and therefore the degree of commitment to its implementation, will be justified to the extent that it contributes value to the organisation, in terms of ensuring conservation or growth (Figure 3.29).

In summary, and however cynical the statement may seem, the cost of risk management – as the use of means and resources – and therefore its level of quality, must find 'remuneration' in the value it is able to guarantee to the organisation.

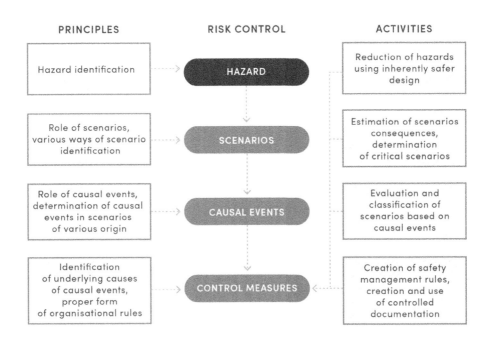

Figure 3.28 Risk control hierarchy in practice.

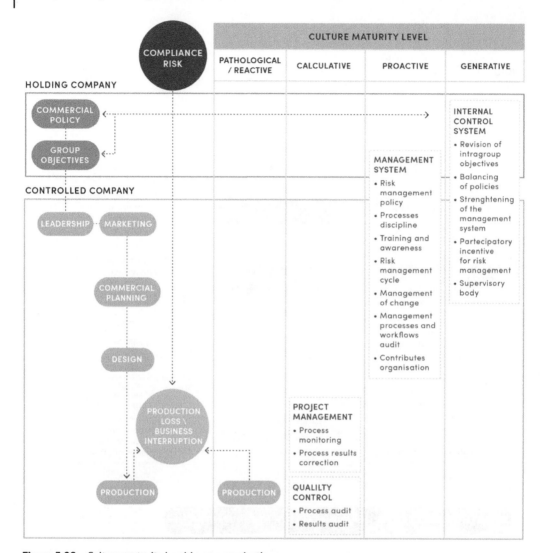

Figure 3.29 Culture maturity level in an organisation.

It is probably easier and more immediate to associate the concept in question with an entrepreneurial organisation where, through financial parameters, it can be easier to measure the cost of risk treatment and the expected and actual benefit on the managed processes.

The concept is indeed equally coherent for a moral organisation; consider, for example, the reputational value and the negative effects that the conduct of a representative could generate if it is not in line with the institutional policies of the organisation itself, even more so if the organisation is not able to remedy it in a preventive or corrective manner.

From the above considerations, it is possible to derive, as a first and immediate consequence, that risk management can, or even better should, be framed as a normal component of all the technical and managerial management processes of the organisation, rejecting any 'ethical' approach to risk and excluding that its treatment is limited to dedicated sectors of the organisation but often 'separate' from its ordinary operating cycle.

Talking about 'risk management' as a separate process with respect to the functioning of the organisation is in fact a risk in itself:

- On the one hand, it ends up depriving risk analysis of the technical contribution and vision in which a given objective is framed, focusing on the result and relegating the risk profile to an external factor to 'have to' consider, perhaps on the basis of regulatory constraints.
- On the other hand, in the dialectic of the organisation, it polarises the role subjectivities creating useless, if not potentially harmful, functional antinomies with respect to the achievement of the result.

Managing the processes of the organisation, in an integral and inclusive way of risk profiles, both in the organisational and operational phases, allows to enrich the perspectives considered in the management process and, above all, to balance the treatment of the risk with all the profiles of the organisation, triggering a virtuous mechanism of the development of the quality of management.

Let us try for example to consider the process of recovery and commercial development of a building or the design and management of an industrial production plant.

From the first point of view, that inherent to the implementation of the process, both activities, although technically different, appear as extremely complex processes involving the treatment and convergence of multiple competences and organisational roles in the sub-processes respectively entrusted.

The progress of the activity requires the identification of profiles that give order to the complexity and promote choices and priorities in the comparison of alternatives.

It is also evident how the treatment of the risk profiles analysed in relation to the alternatives considered, in the current implementation and management perspective, allows the enrichment of the process and the optimisation of the choices.

Consider, for example, the management repercussions of the design choices in the fire prevention field with respect to the development of the projects or the operating costs associated with these choices; it is quite clear that during the design phase, the interest of the project team could favour the respect of delivery times or ease of implementation, leaving the management with heavy inheritance in terms of operational or financial risk.

On closer inspection, in addition to the risk of process fragmentation, there is also the risk that the number of factors considered (including risk factors) may slow down the process; however, in application of the considerations in question, the treatment of this risk at the organisational level can appropriately assess and mitigate the profile, foreseeing intermediate phases, i.e. the organisational figures, among those involved, responsible for promoting the overcoming of possible decisional deadlocks.

In any case, the process is significantly improved as any solution adopted will have been assessed at least with an awareness of the degree of risk taken on and the commitments to be made to mitigate its impact.

From the second point of view, the inclusion in the process management, already in the organisational phase, of the treatment of risk profiles allows, on the one hand, to anchor the analysis to the specificity of the initiative and, on the other hand, to stimulate the contribution and responsibility of all the actors involved, making the treatment of risk an added value of the organisation.

In fact, the specificity of risk analysis, integrated in the management process and in the awareness of the actors involved, mitigates, at least at an organisational level, the risk of missing or inadequate assessment of operational risks or those of overestimating them, negatively constraining the planning and management of the initiative undertaken.

Therefore, in the above terms, the treatment of risk profiles is integrated within the reference process, both organisational and operational, as a transversal qualitative requirement in all phases of the process itself, losing the connotation of separateness and speciality that often distinguishes it and is, at least in power, a factor of maintenance or creation of value in the organisation and results.

With the considerations represented, we wanted to affirm that the treatment of risk contributes to the conservation and increase of the value of the organisation, and in coherence with this contribution risk management can be a qualitative component of the organisational and operational processes of the organisation (Figure 3.30).

In this perspective, risk management on the one hand implements and enriches the contents of the organisation's processes and, on the other hand, creates the conditions for a participatory approach to risk management that mitigates the risk of subjective polarisation in the implementation of operational processes and, above all, creates the conditions to produce further value through the recovery of existing production.

Let us now try to make this last profile explicit and let us do so by applying the concept with respect to the path of growth of a human being, being aware of all the approximation of the comparison but confident of the support that an empirical reference to a case of common knowledge can offer.

When a few years old, a child starts walking; he stumbles or collides with objects, either stationary or moving, on his path because they are not properly considered.

Progressively, with a series of falls and some crying, the child increases his skills, reacts by learning to recognise and evaluate objects on his path to avoid them.

Therefore, he achieves a degree of awareness in which he observes the rules dictated on the basis of experience and walks with reasonable confidence in the home environment.

Furthermore, the child learns to apply the rules of experience autonomously and to interpret heterogeneous signals for walking in environments other than domestic and habitual ones.

In the end, the child pursues his or her will to walk by applying independently and automatically the rules of experience acquired and increases his or her skills by walking, running and playing.

The example in question, whose summary and approximate character is reiterated, is however certainly useful to focus on some concepts mentioned in the previous pages.

You learn not to fall, to walk indoors, then outdoors and then again to run and play, processing together the objective pursued and the associated risks.

In other words, an entrepreneurial organisation pursues economic objectives set out in the articles of association and does not a priori manage risk; however, it is clear that the treatment of the risks associated with the objective is a component of the management process aimed at achieving the result.

Treating the risk is impossible only in association with the objective identified and the process necessary to pursue it. We can say that risk does not exist without the objective of which it remains as a sort of negative predicate.

In this perspective, the treatment of risk brings value to the organisation.

In the child's developmental process, as simplistically described, objective and risk are treated in direct association, initially unconscious and mutually interfering, in the result of enabling the child to achieve the objective and setting the conditions for further development of his/her abilities.

In this path, objectives and risks continue to be treated jointly and to feed each other in a process of capacity and awareness growth that continuously recovers the bases of experience to rework them with respect to the new objectives and to identify new risk profiles to be evaluated with respect to the new objectives.

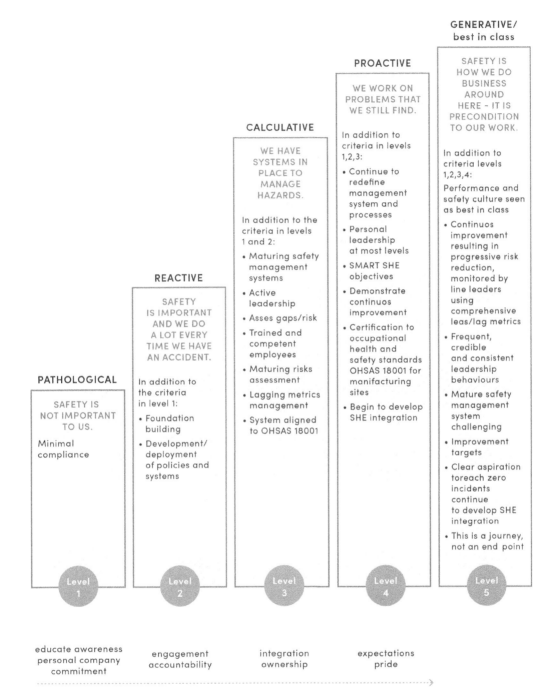

Figure 3.30 Safety culture levels.

In this perspective, at a methodological level and even more so in an organisation, the treatment of risk integrated in the management process brings the added value of fluidification and process efficiency through preventive and corrective remedies already integrated in it.

Finally, we cannot but observe how, in the organic unity of our child, the rules of experience are acquired and made available to all the sensory and cognitive abilities of the person, evidently affected by the same developmental path.

It certainly does not belong here to dwell on the interactions of this unity, but we cannot note how the same perspective applied within an organisation, or even more so within complex and mutually interfering organisations, can further contribute to the growth of the organisation through the recovery and diffusion of the rules of experience, the application of the same rules on different management processes of the same organisation and the stimulus to the identification of new signals or new scenarios.

In other words, the treatment of risk, as a speculative predicate of the objectives pursued, as a process culture can offer further added value to the organisation.

We therefore try to apply risk treatment in the developmental stages related to the child to an organisation, classifying a series of 'standard' and very well recognisable conditions, as shown in Figure 3.31.

1) The 'pathological' condition

The 'pathological' condition corresponds to the absence of preventive risk treatment, a condition in which the organisation promotes its objectives through processes aimed exclusively at obtaining results.

The organisation identifies the objective and develops in a serial manner the activities deemed necessary to obtain it without consideration of the interfering factors – if not limited to those that can offer opportunities for direct maximisation of the result – and therefore without the identification of preventive or subsequent corrective factors.

The management process is necessarily poor because the lack of risk treatment prevents the implementation of the analysis of the operational process, jeopardises the possibility of developing synergies between company processes and is characterised by the lack of information disseminated throughout the organisation.

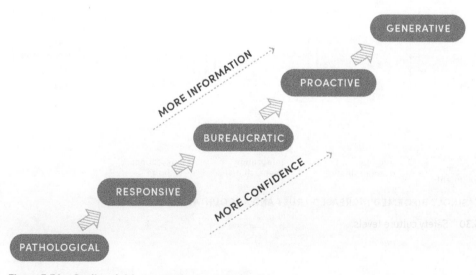

Figure 3.31 Quality of risk management approach.

The identification of the objective itself is the result of an organisational process that vertically declines the concrete objective and the expected result through a functional and polarised dialectic on the figures of responsibility without contributions of critical analysis; after all, the pathological condition pervades the same organisational structure, preventing a priori the preliminary analysis necessary to define the process and the result.

In other words, the organisation wants to 'walk' and starts to do so without considering the ability to sustain the necessary commitment over time, without identifying the obstacles on the path or alternative trajectories or without verifying the opportunity to achieve the same objective in different and alternative ways.

The organisation does not treat the risk until it manifests itself in a dangerous condition that results in damage that jeopardises the achievement of the result.

The organisation reacts within the limits of what is necessary with respect to the incidence of the damage with respect to the organisation and the restoration of the operation of the process without, however, re-evaluating its suitability and prosecutability and, above all, without 'capitalising' the event by transforming it into a rule of experience to be evaluated with respect to the organisation and the complexity of its processes.

One might conceptually wonder whether this is acceptance of the risk or whether the organisation ignores it a priori.

The pathological condition is typical of organisations operating on processes and scenarios characterised by limited technical production constraints, market scenarios with a low competitive level and a low level of outsourcing of operational processes and regulatory contexts that tend to be prescriptive, which would make the value contribution of risk management appear negligible.

On closer inspection, in reality, the risks of business continuity or the risk of inadequate calibration of production objectives, for example, are quite significant even in expected monopoly markets that could generate significant revenue losses or penalties with respect to the failure to meet demand.

In short, in order to answer the question about the organisation's awareness, we can consider that the pathological condition – which, as indicated above, pervades the organisational structure and processes – is at the same time the result of a risk acceptance that is not fully aware by an organisation that, even knowing it, does not accept the stimuli and does not use them to evolve, evidently endorsed by a technical, regulatory and market context that is not particularly demanding.

It is certainly possible to affirm in this sense the correlation between the degree of quality of risk management in the organisation and the reference context (Figure 3.32).

2) The reactive condition

If an obstacle is hit when trying to walk, the reaction will be all the more important the more significant the damage and pain suffered.

This in brief can be a description of the 'reactive' condition in risk management.

This is a condition immediately superior – if possible identifying a line of qualitative development of the risk management culture – to the 'pathological' condition.

The organisation not only suffers damage for which it may even be sanctioned, as in the pathological condition, but in the face of the importance of the damage suffered also reacts immediately by making significant efforts to eliminate the damage suffered and align the process with the management of the risk that has occurred, integrating the process itself with corrective measures that had not been previously assessed.

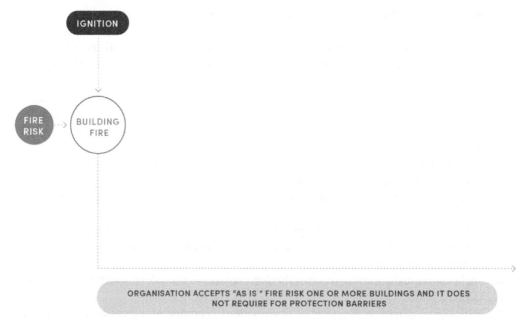

Figure 3.32 The pathological condition.

This is a strong reaction, the result of the awareness acquired by the organisation as a result of the damage suffered but limited to the event that occurred.

The reaction does not feed the evolution of risk management neither with respect to organisational nor operational processes.

As with the pathological condition, it can be assumed that, in the reactive condition, the organisation has accepted the risk although with the greater awareness reflected in the intensity of the reaction.

After all – if in the organic unity of the person, evolution is dictated by the natural ability to elaborate stimuli, the satisfaction of needs and the action of external educational and training factors – the ability of the organisation to understand, acquire awareness, react to stimuli and gradually evolve to 'superior' conditions is the result of an artificial construction strictly dependent on organisational policies, the subjectivities responsible for management roles and the technical, regulatory and scenario constraints in which the organisation operates.

In this perspective, it is possible to affirm that the reactive condition can be endorsed by prescriptive regulatory contexts in which the cases of possible non-compliance are typically outlined and identifiable and to which the strong but precise reactive approach of the organisation in case of accident is adapted.

This context probably does not favour the development of risk management as a generative process of added value for the organisation (Figure 3.33).

3) The bureaucratic condition

In the 'bureaucratic' condition of risk management, we can certainly identify a first concrete step in the line of qualitative development of risk management in the organisation.

In the bureaucratic condition, the risks are consciously treated by the organisation which, according to their identification, dictates the risks as preventive corrective factors.

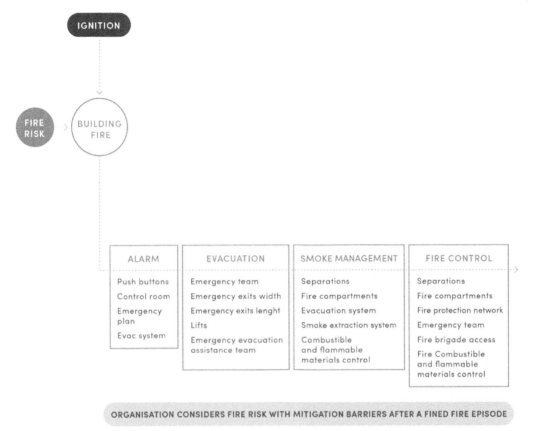

Figure 3.33 The reactive condition.

Not only that, the assumption of the rules by the organisation implies that they are disseminated to all members of the organisation who must comply with them in order to prevent non-compliance and accidents.

The qualifying step of the bureaucratic condition lies in the affirmation of a risk management policy by the organisation which in turn informs and shapes the operational processes.

It is only through the affirmation of a policy that risk management takes on a systemic character in organisational and operational processes, attesting to the awareness of the organisation and removing the treatment of risk from the individuals who make it up.

The bureaucratic condition definitely overcomes the grey area of pathological and reactive conditions in which risk acceptance is based on the limited awareness of the organisation.

Of course, it is not that the individual members of organisations defined as pathological and reactive cannot be able to identify and assess the risks of the activities from a technical and managerial point of view; simply, that type of organisation does not require it.

In the bureaucratic condition, the organisation manages the risk, identifying the profiles and dictating the rules that everyone must implement to avoid the dangers.

In other words, an organisation in bureaucratic condition is like a child who has acquired the necessary rules to walk inside the house; it is an organisation that moves with relative safety within a given perimeter and as long as all its actors respect the rules.

On closer inspection, however, it is an organisation whose risk management is not flexible and does not adapt to changes in the scenario, does not produce added value, at most it guarantees the preservation of existing value as long as the context does not change.

Furthermore, it is an organisation that does not seize the opportunities associated with the commitment required of its actors to comply with the rules and misses the opportunity to take and treat critically external stimuli to enrich processes or simply exploit the synergies between homogeneous processes in dealing with the corresponding risks.

Finally, precisely because of this rigidity, such an organisation would not be safe in those regulatory contexts that base responsibility on risk management and that introduce the adequacy of the evaluation with respect to the accentuated complexity of the current technical and management processes, ending with the admission of an a posteriori judgement on the goodness of the management process and therefore on the conduct of the actors involved (Figure 3.34).

4) The proactive condition

In the proactive condition, there is a widespread attention of the actors in the organisation to the danger signals.

The organisation has taken an important step forward compared to the conditions described above. In the bureaucratic condition in fact

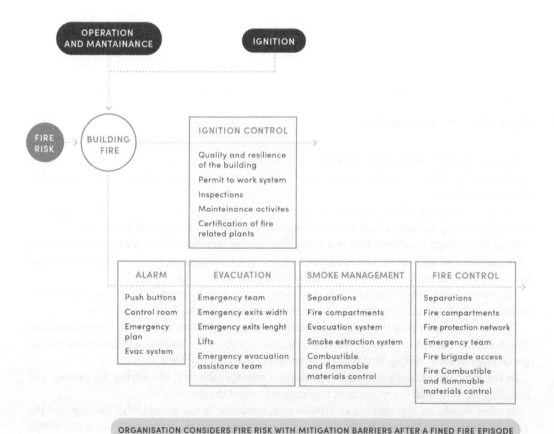

Figure 3.34 The bureaucratic condition.

- the policy has enabled awareness of the organisation of risk and its treatment to be taken into account;
- awareness generated the analysis and assessment of the risk which in turn led to the definition of the rules;
- the rules are deliberately spread throughout the organisation and everyone must follow them to avoid dangers.

In the proactive condition, on the other hand, risk awareness extends beyond the boundaries of the organisational process and informs operational processes: the actors not only follow the rules to avoid dangers and accidents, but are also attentive in advance to danger signals.

They identify the signals and react by bringing them back into the management process for their treatment and for implementing the risk management and knowledge assets.

If the precondition of the proactive state lies in the existence of a risk policy on the part of the organisation (as already in the bureaucratic condition), the maturation towards the proactive condition is dictated by the organisation's need to grow.

As the child, after learning to walk in the house, ventures outside into an unfamiliar environment, the proactive organisation needs to identify danger signals to learn to move in new scenarios, markets, constraints and contexts, developing the process and handling of the associated risk.

One might wonder whether the detection of the danger signal is the implementation of a rule or the result of the organisation's stimuli.

Probably, the delimitation between the sphere of having to do it and wanting to do it is not identifiable in a clear demarcation line.

More likely, observing an organisation in concrete terms, one would end up observing a range of behaviours in which they both alternate in profiles but whose orientation over time, in size towards one or the other of the spheres in question, could certainly be an indication of the organisation's maturity.

Of course, to speak of 'will' is to be understood in any case within the framework of a system of rules dictated by the organisation; it would be in other words a will progressively induced or easily obtainable by the organisation as a result of the development and implementation of a policy oriented towards sharing and widespread participation in the organisation.

In any case, it is clear that, as shown at the beginning, risk management in the proactive condition is able to generate value for the organisation (Figure 3.35).

5) The generative condition

The generative condition represents in some way the complete maturation of the organisation, the one in which risk management is an integral part of company processes and a factor competing with the production of value in the perspective indicated at the beginning of this discussion.

In the condition in question – taking up again the comparison with regard to the duty of care and voluntary behaviour – the actors of the organisation, whether they are involved in organisational or operational processes, naturally consider risk management to be an integral part of their activities, working, on the basis of the sensitivity induced by the organisation and the training factors provided, to identify risk profiles, to implement containment actions and to contribute to updating the analysis and evaluation and therefore the processes.

The organisation naturally develops a path of continuous regeneration based on the widespread contribution of the actors, which ends with self-feeding.

The generative condition allows the organisation to adapt to changing scenarios, to deal with regulatory development with greater flexibility and to overcome the technical constraints of operational processes.

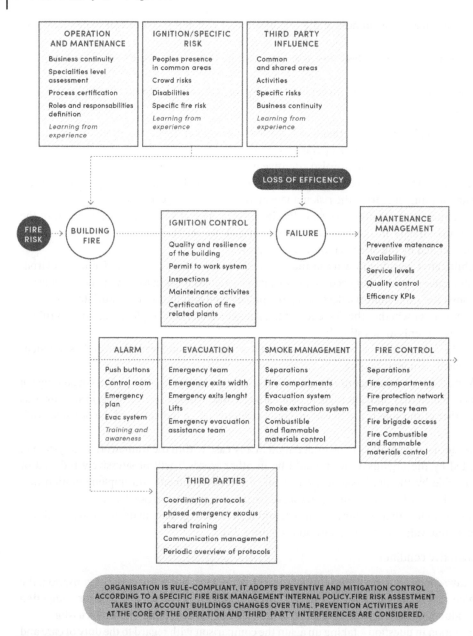

Figure 3.35 The proactive condition.

But above all, the generative condition elevates the organisation by depersonalising it from the actors that make it up and characterising it in terms of subjective synthesis superior to the policy pursued.

In other words, through policy and generative maturation, the organisation protects itself from the risk of application deviations that can ultimately characterise the individual conduct of its actors.

In conclusion, the descriptive articulation of the degree of culture that characterises risk management in organisations through the proposed classification is obviously a mere methodological expedient aimed at affirming the diversity and orientations that can be revealed in practice.

However, it is immediately evident that these are not immanent conditions but rather orientations of the organisation in a more or less fluid condition from one to the other, depending on the degree of consolidation of the policy in its adequacy of application within the organisation.

Nevertheless, the conditions described can be read as a term of reference or comparison that can offer support to the analysis of concrete situations from which to move to orient the development of a given organisation towards the opportunities offered by the correct application of risk management in the perspective of value creation (Figure 3.36).

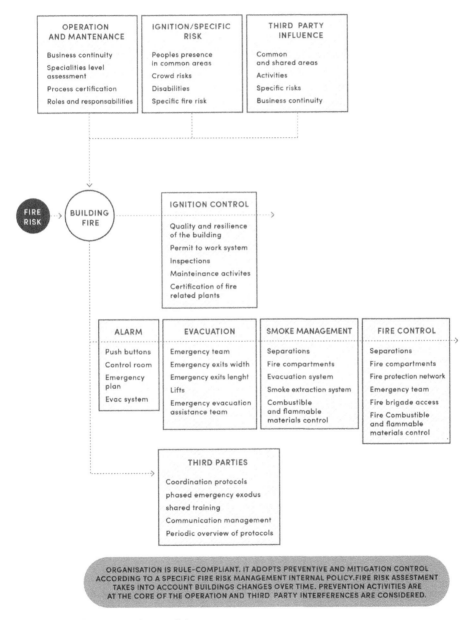

Figure 3.36 The generative condition.

3.15 Systemic Approach to Fire Risk Management

In conclusion of this chapter, it emerged that dealing with fire risk management is not dissimilar from other types of risk management. The deterministic, linear, prescriptive approach is now replaced with a continuously improved, circular, performance-based approach, whose centre is the Plan–Do–Check–Act (PDCA) (or Deming) cycle, as shown in Figure 3.37.

Fire risk assessment is also crucial to perform an exhaustive enterprise risk management: indeed, fire scenarios and related risks also have a significant impact on other types of risks, including those shown in Figure 3.38: financial, strategic, compliance and operational risks.

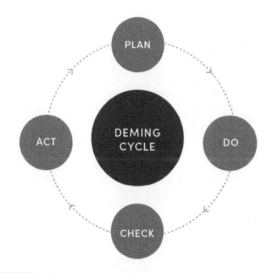

The same approach applies to all managment systems:
ISO 31000 - Risk Managment;
ISO 9001 - Quality;
ISO 45000 - Occupational Health and Safety at work;
ISO 14000 - Environment;
Major Hazard Managment System (Seveso EU Directive);
Major Risk Managment System (Offshore EU Directive);
...

Figure 3.37 The Deming cycle PDCA.

Figure 3.38 Enterprise risks affected by fire risk.

4

Fire as an Accident

Fire safety design and fire safety management over time are complex activities that require the knowledge and awareness of the stakeholders about the fire scenarios that may arise and develop according to the fire threats associated with their assets.

The degree of knowledge and competency must be appropriately proportionate to their roles and responsibilities.

It is clearly beyond the scope of this chapter to present the principles of combustion and fire dynamics, topics that require more detailed study. However, it is useful to point out that each application context is characterised by clearly different fire risks. These fire risks are associated with a series of peculiarities of the domain of interest and find their expression in the development of more or less complex scenarios as well as being determined, in their evolution, by the strategy adopted and its partial/complete success. In any case, it is important to emphasise that fire and explosion scenarios in the civil construction sector are particularly different from those observed in the industrial sector, especially in the presence of hazardous substances. In the industrial sphere, the same severity can lead to major accidents in many countries regulated by specific regulations. In the following sections, an attempt is made to briefly illustrate the types of accident scenarios in the industrial and civil sectors. Their description starts from identifying the physical phenomena associated with them. This is often used to characterise the risk component known as magnitude or severity.

4.1 Industrial Accidents

Compartment fires are only a single example of the fire accidents that may take place. They are typical of the civilian occupancies, while in the industrial sector a number of different events may be referenced as fires or explosions. This is utmost valid for the chemical process industry, in which fires and explosions are one of the incident scenarios.

An incident may be defined as an unusual or unexpected event that either resulted in or had the reasonable potential to cause an injury, release, fire, explosion, environmental impact, damage to property, interruption of operations, adverse quality affecting or security breach or irregularity.

Commonly, the incident scenarios that affect the process industry are classified into three main types:

Fire Risk Management: Principles and Strategies for Buildings and Industrial Assets, First Edition.
Luca Fiorentini and Fabio Dattilo.
© 2023 John Wiley & Sons, Inc. Published 2023 by John Wiley & Sons, Inc.
Companion Website: www.wiley.com/go/Fiorentini/FireRiskManagement

- Fires (any combustion regardless of the presence of flame; this includes smouldering, charring, smoking, singeing, scorching, carbonising or the evidence that any of these have occurred).
- Explosions (including thermal deflagrations, physical bursts and detonations).
- Toxic releases.

The consequences of an accident span from fatal to minor injury and damage only (economic loss), in a scale of magnitude that is not uniquely predefined. Similarly, the likelihood that an adverse event will happen again spans from certain to rare. Talking about fires, explosions and toxic releases, their likelihood and magnitude are summed up in Table 4.1.

It may be of interest to cite the results of some statistics (Figure 4.1) about the most common causes at the origin of industrial accidents (Figure 4.2) since they highlight the critical factors on which the attention of the consultant should have to be primarily addressed. It is essential to keep in mind that these results are purely indicative.

What is commonly referred to as fires and explosions actually belong to a wide range of accidental phenomena that are in many cases only apparently similar. The common factor is the release of large amounts of energy into the environment in a fairly short time. A fire in a confined space (a warehouse or a building) or a partially confined space (an industrial installation) may last for hours or days, and an explosion (in the broadest sense, their classification is discussed below) or a blaze (in the sense of a fire in a cloud of flammable vapours) occurs in the space of seconds or tenths of a second.

Table 4.1 Incident typologies and correlated potentiality and magnitude.

Type of incident	Likelihood of occurrence	Fatality potentiality	Economic loss potentiality
Fire	High	Low	Medium
Explosion	Medium	Medium	High
Toxic vapours release	Low	High	Low

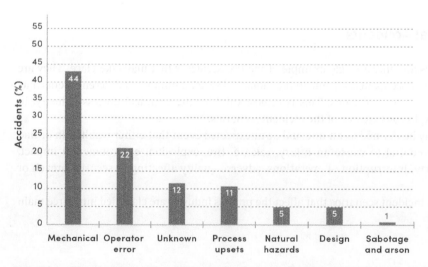

Figure 4.1 Causes of industrial accidents in chemical and petrochemical plants in the United States in 1998.

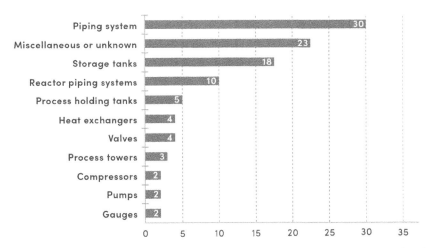

Figure 4.2 Components related to the industrial accidents in chemical and petrochemical plants in the United States in 1998.

Industrial accidents involving fires or explosions can be different in nature; consequentially, their evolution follows different dynamics. Generally, process plants handle a significant amount of liquid or gaseous fuel that, after uncontrolled releases (so-called 'loss of containment' events) may result in an accident. Depending on the type of the release, it is possible to identify some typical scenarios, considering the loss of containment of the hazardous substance that may lead to consequences (Figures 4.3 and 4.4). Physical effects include toxic effects, thermal effects and blast/fragment effects, or all of them depending on the involved chemicals. In any case, it is worth remembering that the effects recorded can be traced back to both the effects generated by the primary event and those generated by domino events, which, in some cases, can lead to a particularly severe escalation that is greater than that associated with the effects of the primary loss of containment.

4.2 Fires

Combustion in uncontrolled ways of combustible material goes by the current name of fire. Combustion is a strongly exothermic and relatively fast chemical reaction in which one of the reactants (fuel), reacting with an oxidising agent (oxidiser), oxidises, producing thermal energy. For the process to give rise to what is commonly called *combustion*, the heat output per unit volume must be sufficiently high.

Depending on the typology of a fire, specific threats may arise given the prominent effects (Figure 4.5).

4.2.1 Flash Fire

It is a sudden blaze with a limited duration of few seconds. It is caused by the ignition of solids, vapours or gases. A rapid and subsonic flame front is its main feature. A graphic description of the effects is given in Figures 4.6–4.8.

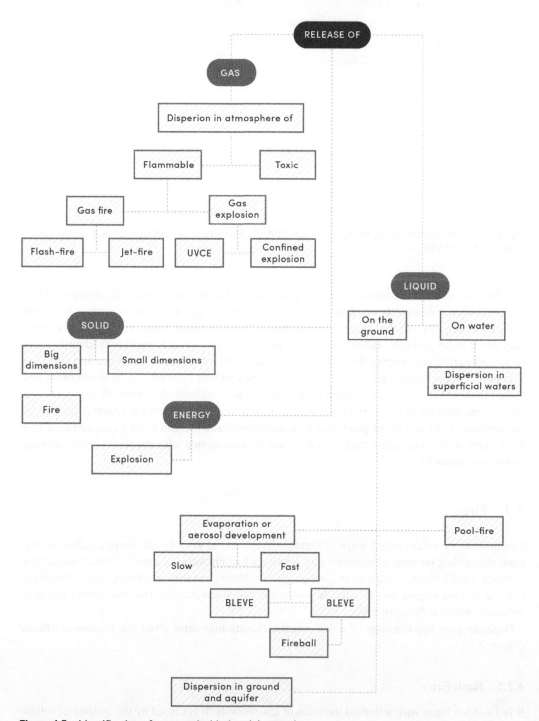

Figure 4.3 Identification of some typical industrial scenarios.

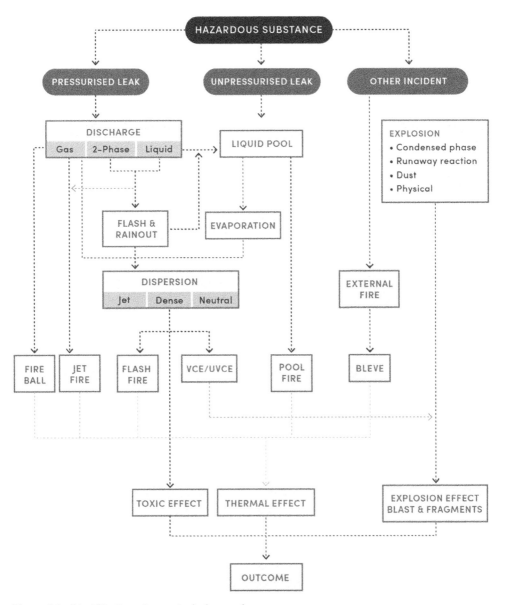

Figure 4.4   Identification of some typical scenarios.

FIRE TYPE	ENGULFMENT	RADIATION	INSIDE BUILDING
FLASH-FIRE	YES	NO	POSSIBLY
JET-FIRE	YES	YES	YES
POOL-FIRE	YES	YES	YES
FIREBALL/BLEVE	YES	YES	POSSIBLY

Figure 4.5   Major threats from different fires in industrial environments.

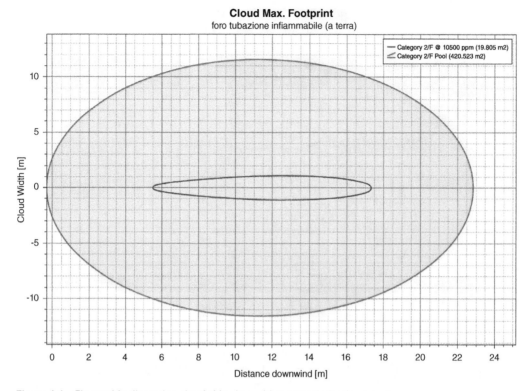

Figure 4.6 Flammable dispersion cloud side view without terrain impingement.

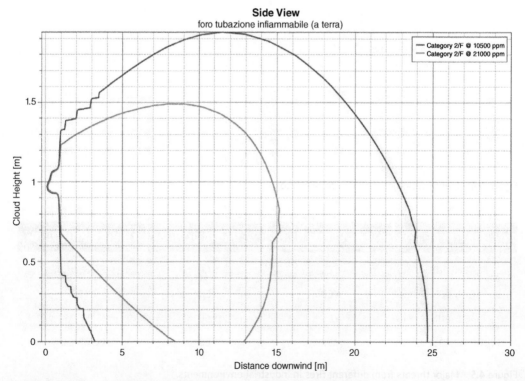

Figure 4.7 Flammable dispersion cloud side view with terrain impingement.

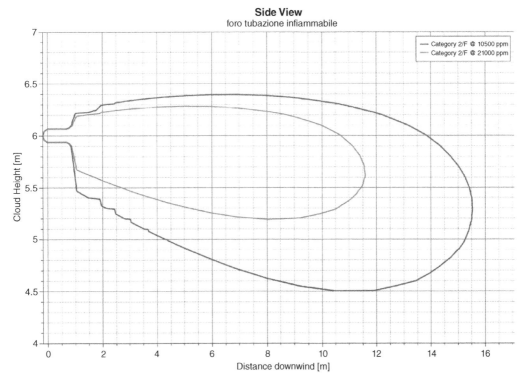

Figure 4.8 Flammable dispersion cloud footprint.

4.2.2 Pool Fire

A pool fire occurs when there is a fire from a flammable or combustible liquid spill or a tank fire (in which case it is often more specifically referred to as a *tank fire*). A *pool fire* consists of a flame that has a horizontal dimension equal to that of the spill and a higher height, generally around twice its diameter. A *pool fire* may be confined or unconfined depending on whether the liquid spillage is contained (e.g. from a tank, a tank or a containment basin) or occurs on open ground or on a body of water.

The *pool fire* can give rise to the *boilover* phenomenon if the fuel is above a layer of water. In this case, sudden boiling of the water underneath may occur due to radiation, drawing a large amount of fuel into the flame and increasing the power of the phenomenon.

The horizontal dimension of the flames is comparable with the spill dimension while the height is almost the double. It may be confined or not, depending on whether the spill occurs in a tank or on the unconfined ground. When the fuel is spilt on water, the pool fire may produce, under specific conditions, the so-called 'boilover'. It is the boiling of the underlying water with the indirect involvement of a significant quantity of fuel. This sudden phenomenon produces an increase of one order of magnitude in the combustion rate. To evaluate the radiant power emitted by the pool fire, it is necessary to know

- geometry of the flame;
- features of the flame;
- combustion rate;
- radiation (related with distance as shown in Figure 4.9) and the geometrical related factors.

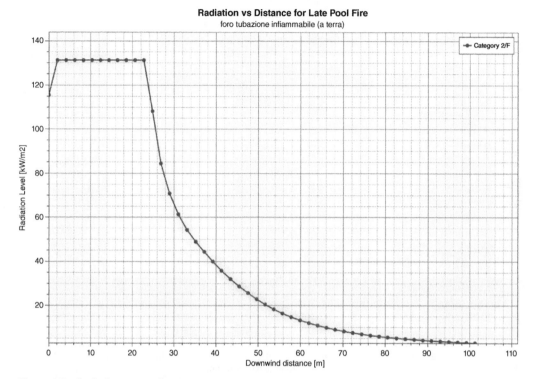

Figure 4.9 Radiation versus distance.

4.2.3 Fireball

The fireball or fireball generally occurs as a result of the collapse of a tank containing a liquefied fuel gas (liquified petroleum gas [LPG] is typical), with immediate vaporisation and subsequent ignition of the contents (Figure 4.10). The rapid evaporation creates a cloud of fuel with a concentration above *the upper flammability limit*, which therefore burns with a diffusive flame. The evolution of the phenomenon, apart from the initial phases governed by the momentum of the released vapour, is governed by the buoyancy phenomena caused by density differences between the air and the hot combustion gases.

Diffusive flame

Figure 4.10 Schematic representation of a fireball in the stationary stage.

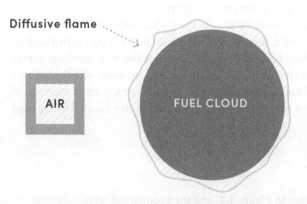

Usually, they can be the consequence of two events:

- The collapse of an LPG tank, with the consequent vaporisation and ignition of the containment.
- The ignition of a cloud of gaseous fuel (rarer, generally it generates a flash fire).

In both the cases, the cloud of fuel burns with a diffusive flame. Modelling a fireball allows estimating the produced thermal radiation. To do so, it is necessary to evaluate the following factors:

- Involved mass of fuel.
- Diameter and duration (Figure 4.11).
- Radiation (Figure 4.12) and geometrical related factors.

Three of the models widely adopted to calculate the emitted power per unitary surface of the fireball (thus an estimation of the radiation on the surroundings) are the following:

- *Point source model*: It considers that all the energy is produced from one single point, i.e. the centre of the fireball.
- *Solid flame model*: It estimates the radiating power by assuming that the flame is equivalent to a grey body.
- *Flame emissivity*: It evaluates the emitted power (Figure 4.13) starting from some peculiar parameters like temperature, dimension and composition of the flame.

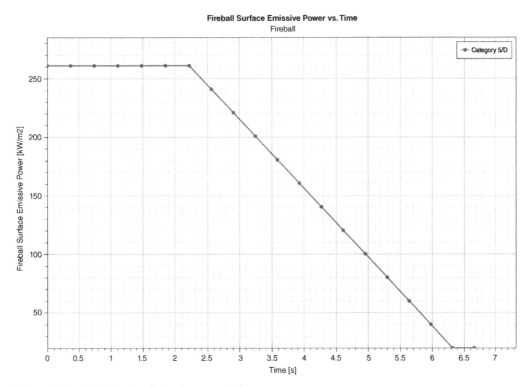

Figure 4.11 LPG fireball – dimensions versus time.

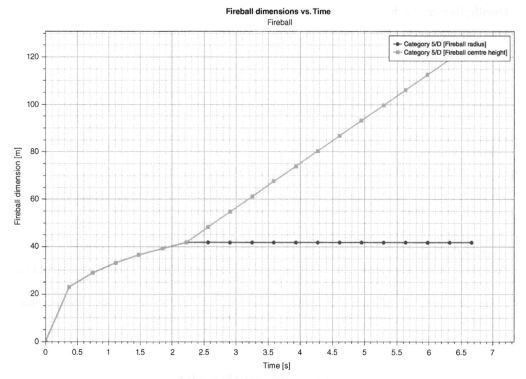

Figure 4.12 LPG fireball – radiation versus distance.

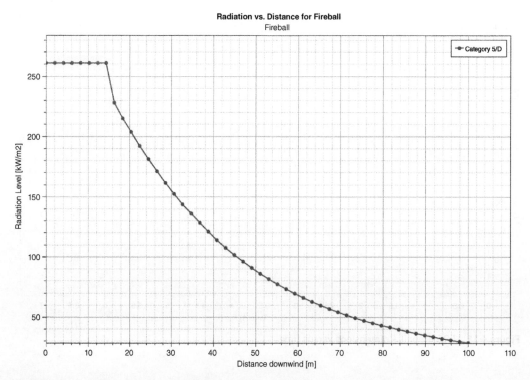

Figure 4.13 LPG fireball – surface emissive power versus time.

4.2.4 Jet Fire

The term *jet fire* means *ignited jet* and generally occurs as a result of the non-catastrophic rupture of a pipe containing a liquid or gaseous fuel under pressure, from which a jet escapes, the dynamics of which are primarily governed by the momentum of the fluid itself. The triggering of the jet produces a turbulent flame. It is typically a triggering phenomenon for more complex accident dynamics: when the *jet fire* directly hits (or affects with its thermal radiation) other equipment, causing what are known in the jargon as *domino effects* on structures, equipment, piping, etc. In this case, the thermal effects cause direct damage and trigger a chain of even more serious incidents. The direct and indirect consequences of a jet fire on possibly exposed people are very serious (see, for example, the accident that occurred at the Thyssenkrupp factory in Turin in December 2007).

A jet fire has a significant diffusion and power. The release of a gaseous substance, from a tank in pressure or piping, is its main cause. It may have different shapes (horizontal, vertical or inclined jet) depending on the local conditions, such as the presence of wind and the geometrical configuration.

There are several parameters to evaluate the radiated power, eventually in relation with the distance of the observer (Figure 4.14). Some of them are as follows:

- Quantity of fuel taking part in the reaction.
- Distance covered by the jet.
- The distance of the defeat (i.e. the crack) from the source of ignition.

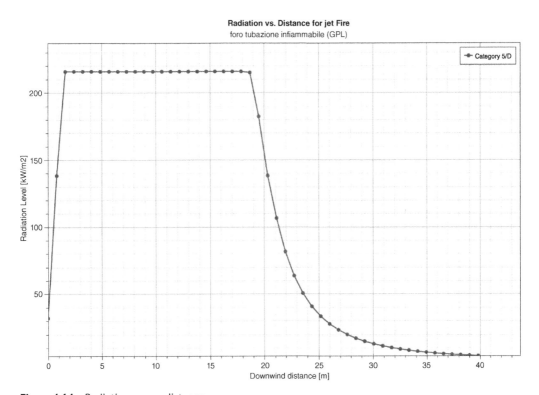

Figure 4.14 Radiation versus distance.

Those parameters are then correlated to other conditions such as the dimension of the crack and the inner pressure of the tank. Several parameters contribute to model the jet fire in a very complicated way.

4.3 Boiling Liquid Expanding Vapour Explosion (BLEVE)

It occurs when a tank, containing a liquid under pressure, suddenly collapses causing the rapid depressurisation and the subsequent evaporation of the fluid, resulting in an extremely dangerous explosion (Figure 4.15).

The most frequent cause is the engulfment in flames of a tank. Flames warm the upper part of a tank (the one above the liquid level), while the lower part remains at a lower temperature because heat is transferred to the liquid inside that changes its phase. This may result in a decrease of the mechanical resistance of the metal above the liquid level and a parallel increase of the inner pressure because of the boiling liquid. Therefore, from a structural point of view, actions increase and strength decreases. When the crack appears, the subsequent prompt evaporation, due to the sudden depressurisation, causes the catastrophic rupture. A famous example is the incident of the Mexico City occurred on 19 November 1984, which caused about 500 victims and more than 7000 injured people. In that occasion, the rupture of a piping containing LPG was the reason for a release of a flammable cloud that was ignited by the torch of the plant. The vapour cloud explosion (VCE) and the jet fire that followed were the two contributors of the first BLEVE of an LPG spherical tank. Other 15 BLEVEs then followed, causing the complete destruction of the plant. To describe the effects of a BLEVE it is possible to consider the overpressure radius (Figure 4.16), the overpressure versus distance effects (Figure 4.17) and the impulse (Figure 4.18).

4.4 Explosion

An explosion involves the rapid release of a large amount of energy, usually in a limited space and time, which is followed by the rapid expansion of the gases involved, causing a pressure wave. The energy developed and also dissipated by secondary mechanisms, e.g. thermal radiation and the launching of fragments or parts of equipment, is known as *missiles*.

In most cases, the energy is chemical in nature, i.e. it results from the rapid unfolding of a highly exothermic reaction. The most frequent explosions are attributable to combustible gases/vapours or combustible dusts; sometimes there are phenomena of significant violence following the loss of control of process plants and chemical reactors. Such phenomena are referred to in Anglo-Saxon literature as *runaway reactions*.

Figure 4.15 Sequence of events leading to a BLEVE.

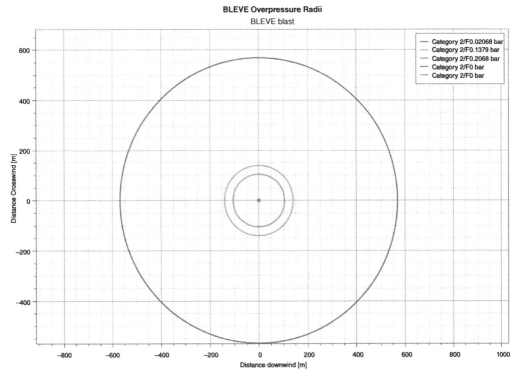

Figure 4.16 Overpressure.

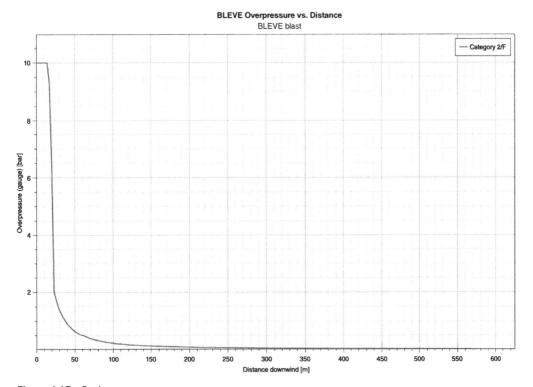

Figure 4.17 Peak overpressure.

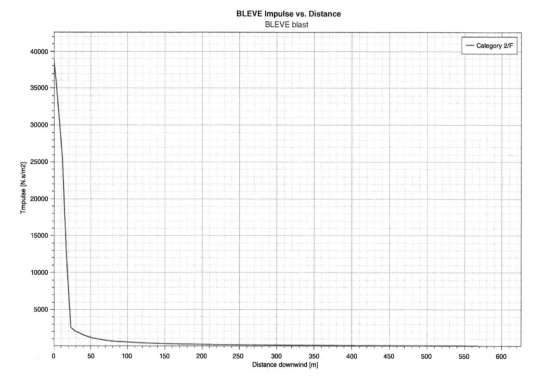

Figure 4.18 Impulse.

More rarely, explosions of explosive solids occur in the industrial sector. On the other hand, we speak of mechanical explosions when the energy is not derived from a chemical reaction (e.g. the explosion of a compressed air tank).

Explosions can be classified given their resulting effects.

4.5 Deflagrations and Detonations

A flame front is defined as the portion (layer) of the mixture in which the combustion reactions take place. The speed at which the flame front propagates towards the unburned portion is called *the flame front speed*; if this is lower than the speed of sound (e.g. of the order of m/s), *deflagration occurs*.

Generally, it is observed that the flame front speed does not coincide with the laminar speed of combustion, as it advances not only due to the kinetics of oxidation reactions but also, above all, as a result of the expansion of the combustion gases; furthermore, the fluid dynamics of the flame front may be turbulent in nature, which may result in very high-speed propagation, especially if it occurs under highly confined conditions, e.g. inside a duct.

In such cases, the speed can exceed that of sound and reach values of thousands of m/s, resulting in a detonation. This is of course also the case with explosives.

4.5.1 Vapour Cloud Explosion

A *vapour cloud explosion* is caused by the release of a large quantity of flammable gases and/or vapours into the open air, which, not being immediately ignited, have time to mix with the oxygen in the air. Under these conditions, following an ignition, a non-confined explosion occurs, known as VCE, which leads to the very rapid development of a pre-mixed flame which, in some cases, has a detonating character.

Accidents resulting from large magnitude VCE are numerous in history. If the development field of the event is reasonably unconfined, such an event is referred to as an unconfined VCE (*unconfined vapour cloud explosion*, UVCE).

Unconfined vapour cloud explosion (UVCE): It is an unconfined explosion of vapour in the atmosphere, even if partial confinement is usually the real context because of natural obstacles, the presence of buildings or simple openings that deeply influence its dynamics. This kind of explosion is worldly considered among the most dangerous in the chemical industry sector. It is generated after the release of a significant amount of vapours in the atmosphere. The ignition and the subsequent combustion create a flame front and an expansion of the combustion products, causing a shock wave. One of the most sadly famous examples is the one that occurred in Flixborough in 1974, where 40 tonnes of cyclohexane were released from a reactor, generating a cloud of half a million of cubic metres whose ignition caused 28 victims, the destruction of the plant and an economic loss of $150 billion.

Confined or partially confined explosion: In this kind of explosion, energy is released inside a containment structure, such as a tank, a reactor, a room or a building. If the explosion, involving piping or a vessel, is generated by a flammable gaseous mixture, it is possible to have deflagration or detonation. In the former case, a pressure safety valve (PSV) may be an effective safeguard; the latter may result more severely, and PSV are considered ineffective for this scenario. Instead, it is not clear enough if detonation may be generated from a dust explosion, in the context of an industrial plant. Confined explosions produce pressure waves that may cause, in an interconnected system, the so-called 'pressure piling'. The phenomenon is caused by an increase in both temperature and pressure inside a tank, determining a similar growth in the connected system, thus generating further increases. To avoid this complex behaviour and isolate the various systems, it may be useful to install rapid depressurisation valves. Explosions are described considering the overpressure radius at several distances (Figure 4.19) and overpressure peaks (Figure 4.20).

4.5.2 Threshold Values

The vulnerability of exposed elements to the types of fires and explosions described above is a function of the hazard characteristics of each. In some cases, in compliance with specific requirements from the regulation, damage to persons or structures is correlated to the exceeding of a threshold value of the physical effect, below which it is conventionally considered that damage does not occur, above which, conversely, it is considered that damage may occur. In general, the physical effects deriving from the accident scenarios that can be hypothesised may lead to damage to persons or structures, depending on the specific type, intensity and duration. In particular, for assessments connected with the prevention of major accidents, typical of industrial installations, the possibility of damage to persons or structures is defined through the exceeding of the threshold values expressed in Table 4.2, given as an example. Same thresholds are used to describe the effects as in the examples given in the previous paragraphs.

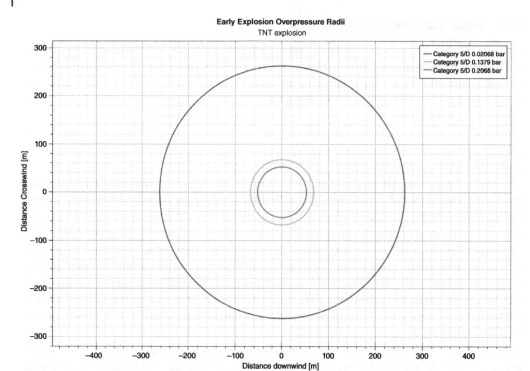

Figure 4.19 Overpressure radius.

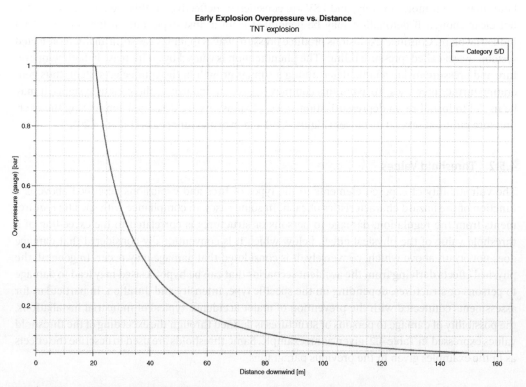

Figure 4.20 Explosion peak overpressure.

Table 4.2 Threshold values for impacts of incident scenarios.

Incidental scenario	High lethality	Start of lethality	Irreversible injuries	Reversible injuries	Damage to structures/domino effects
Fire (stationary thermal radiation)	$12.5 \, kW/m^2$	$7 \, kW/m^2$	$5 \, kW/m^2$	$3 \, kW/m^2$	$12.5 \, kW/m^2$
BLEVE/FIREWALL (variable thermal radiation)	Fireball radius	$350 \, kJ/m^2$	$200 \, kJ/m^2$	$125 \, kJ/m^2$	200–800 m*
VCE(peak overpressure)	0.3 bar (0.6 open spaces)	0.14 bar	0.07 bar	0.03 bar	0.3 bar
Toxic release (absorbed dose)	LC_{50}	–	IDLH	–	–

*According to combustion type and tank.

Figure 4.21 PEM activities at the core of scenario description.

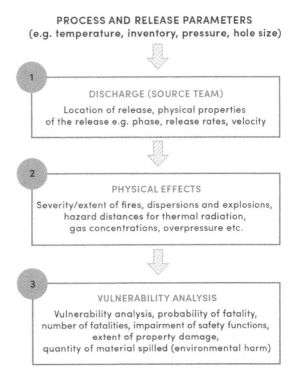

PROCESS AND RELEASE PARAMETERS
(e.g. temperature, inventory, pressure, hole size)

1 DISCHARGE (SOURCE TEAM)
Location of release, physical properties
of the release e.g. phase, release rates, velocity

2 PHYSICAL EFFECTS
Severity/extent of fires, dispersions and explosions,
hazard distances for thermal radiation,
gas concentrations, overpressure etc.

3 VULNERABILITY ANALYSIS
Vulnerability analysis, probability of fatality,
number of fatalities, impairment of safety functions,
extent of property damage,
quantity of material spilled (environmental harm)

4.5.3 Physical Effect Modelling

In order to appreciate the different events and to conduct fire and explosion risk assessment, it is fundamental to understand and estimate the physical effects coming from the identified scenarios. This activity is known as physical effect modelling (PEM), and it is an integral part of scenario description (Figure 4.21).

Given process and release parameters (e.g. temperature, hold-up, pressure and hole diameter) and considering the source term definition, it is possible to estimate physical effects of interest in order to conduct vulnerability analysis on the basis of defined threshold values.

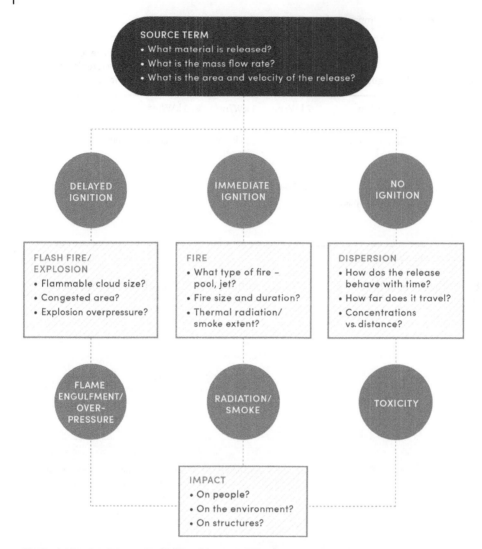

Figure 4.22 Conditions affecting the scenario effects and general dynamics and escalation.

PEM activities should consider both the source terms and all the events that may determine the resulting scenario (Figure 4.22), including the ignition source, the ignition timeline, duration of LOC, duration of the scenario and congestion in the area. All these parameters may deeply affect the resulting effects and, in certain cases, the resulting event (including any potential escalation).

4.6   Fire in Compartments

The objective of this section is to provide the basic tools to understand the dynamics of a fire and especially the peculiarities of the environment in which it breaks out and develops with potential involvement of occupants and property.

A fire in a confined space (e.g. a room or compartment) can develop in a multitude of different ways, mainly depending on the geometry of the compartment, the type of ventilation, combustible material present, and the size and type of surface. The succession of events that generally characterise the development of a confined fire from ignition is presented below.

After ignition, the fire produces a variety of toxic, non-toxic, irritant gases and solids, and increasing amounts of energy, most of which can be attributed to the spread of the flame; in the early stages, the fact that the environment is open or confined is essentially irrelevant to combustion because the fire is fuel-controlled. The hot gases in the flame are surrounded by cold gases, and the mass of the hotter, less dense ones rises upwards due to the difference in density or rather due to its 'buoyancy'; this mass contributes, including any flames, to form the plume of the fire.

As hot gases rise, cold ones are also drawn into the plume, and this mixture of air and combustion products reaches the ceiling, generating the layer of hot gases.

Only a small portion of the mass of gas stratified on the ceiling originates from the fuel; in fact, the largest portion of this comes from the fresh air dragged laterally into the plume, which continues to move the gases towards the ceiling. As a result of this entrainment, the total flow in the plume increases, while the average temperature and concentration of combustion products decrease with height.

When the plume hits the ceiling, the gases spread out in a circular jet, the velocity and temperature of which is important to know in order to make estimates of the response of any smoke and heat detectors or sprinkler connections near the ceiling, for compartments where such protection systems are present.

Subsequently, the jet reaches the walls of the compartment and is forced to move downwards along the wall; however, the gases in the jet are still hotter and consequently possess greater buoyancy than the ambient air, so the flow still turns upwards.

A layer of hot gases is formed near the ceiling and a less hot layer, which we call cold, below, in contact with the floor. We see that this characteristic of the confined fires typical of rooms and compartments with modest extensions and regularity in plan (such as the vast majority of rooms in civil dwellings) is the main assumption of the so-called 'zone' calculation codes.

The upper layer is composed of the mixture described above, while the lower layer is composed only of air, and the simplification whereby their composition and temperature are assumed to be homogeneous over time is used; this does not apply to the dimensions. The 'field' models, which are discussed below, make it possible to further discretise the volumes coinciding with these two zones of interest by subdividing them into smaller control volumes that allow the general behaviour to be revealed from the behaviour of the fire quantities measured in these elements, with a noticeable improvement in the degree of detail that can be obtained in proportion to the density of the elementary elements under study but with a significantly greater burden of calculation. The degree of detail and, more generally, the calculation model, must be congruent with the purposes of the analysis.

As the fire continues, the plume continues to draw in air and get bigger, causing the upper smoke layer to grow with it and thus bringing the interface between the two areas ever closer to the floor. Generally, it is in the lower part of a room affected by fire that the most breathable air conditions for occupants persist.

As the hot smoke layer descends and the overall temperature rises, heat transfer increases by both convection and radiation. The transfer of thermal energy to the fuel does not only occur through flame contact, but a very important, if not decisive, role is played by the radiation emitted by the hot layer itself and by the hot walls of the compartment, which generate an increase in the combustion rate of the fuel present and the overheating of other possible igniters.

The presence or absence of ventilation plays a key role in this phase. Ventilation means the escape of hot fumes from the compartment through openings along the walls, typically doors or windows, as soon as the hot gases touch the ceiling. Above all, it must be taken into account that the heat developed by a fire can shatter windowpanes or lead to cracks, creating ventilation that was not initially considered. In any case, it is worth remembering that in most cases the presence of door and window frames, even if these are closed, is not fully sealed, so that, in relation also to the pressure regimes that are established, ventilation regimes, albeit limited, are often guaranteed by these leaks.

The fire may continue to grow either due to an increase in the speed of combustion, propagation to a neighbouring object or even the ignition of fuel at a distance, in any case the temperature will continue to rise and the radiation from the hot layer downwards will be so strong that it will reach a stage where all the material in the compartment is burning. The sudden increase from a growing stage to a fully developed fire is called 'flashover'.

Eventually, in the fully developed phase, the flames spread through the openings and all combustible material is involved in the fire, which can burn as long as the fuel and sufficient oxygen (the oxidant) are present to sustain combustion.

The very interesting case for confined fires is that of a lack of oxygen: if there are no openings or only a small heat dispersal area, the hot layer soon descends towards the flame region, but the air drawn into the combustion zone now contains little oxygen, leading the fire to extinction.

Generally, in a burnt room, the minimum volumetric concentration of oxygen that is still capable of sustaining a combustion process with a flame is approximately 12%; it follows that a fire with a flame can proceed until the amount of oxygen in the room has been roughly reduced to almost half of that initially available. The absence of this concentration, at certain points, may lead to partial combustion, even at singular points in the compartment, which are apparently not congruent with a more extensive degree of damage in even neighbouring areas of the same compartment or in other compartments.

The dotted line in Figure 4.23 shows that a fire can reach point 'A' and from there begins the transition period to flashover, but due to oxygen deficiency, which has occurred in the meantime as combustion continues, the rate of energy release decreases, as does the temperature of the gases.

Even if the rate of energy release decreases, a pyrolysis effect may continue[1], which in turn results in the significant accumulation of unburned or partially combusted gases in the environment. If at this point a window suddenly breaks (or there is another collapse or there is any alteration of the environment that brings the system to a state favourable to flame combustion), hot gases will escape through the top of the opening and cold, fresh air will enter through the bottom of the opening. This may decrease the overall heat load of the compartment, but the fresh air could cause an extremely sudden increase in the rate of energy release, which would result in a resumption of fire growth back towards the room flashover but at a faster rate than the first observed growth phase, as shown by the dotted line in Figure 4.23.

In the worst case, the inflowing air may mix with unburned pyrolysis products generated by the oxygen-starved fire. Any source of ignition can trigger the resulting flammable mixture leading to an explosion or very rapid combustion of the gases; the expansion caused by the heat of combustion will cause the gases to violently escape from the openings, generating the phenomenon called 'backdraft'.

In Figure 4.23, this phenomenon would be illustrated by a straight, vertical line passing directly from point C and reaching very high temperatures. Usually, a backdraft will only last for a very short time, of the order of seconds (backdrafts lasting minutes have, however, been observed); however, its danger is also due to the fact that it is very often followed by a flashover since the thermal

---

1 Pyrolysis is a process of thermal decomposition of solid fuel in the complete or almost complete absence of combustion agents, characterised by the absence of flame and the production of highly non-combustible gases.

peak tends to ignite all the fuel in the compartment, leading to a fully developed fire once again. Having said this, it is clear that if in the first phase of a fire, the fire growth dynamic (in terms of the speed at which thermal energy is released into the environment concerned, often connected to the type of ignition and the possible presence of combustion accelerants) is a determining factor in the subsequent development in the environment and the possible propagation to other communicating or neighbouring environments, it is certainly true that the presence of a high fire load (combustible) in the flashover and backdraft phases determines a significant severity of the effects.

When unburned gases from an under-ventilated fire manage to penetrate an enclosed space bordering the fire site, they can mix very well with air to form a combustible gas mixture. Even a small ignition is then sufficient to cause up to a flash fire or an explosion depending on the mass quantity of vapours in the flammable range of the mixture, in the latter case with the generation of over-pressure waves.

Given the characteristics of a confined fire, as discussed in the previous sections, it is generally possible to describe such events using a standard representation that moves its steps from the recognition of four clearly identifiable phases. The evolution over time of a fire in a confined environment can therefore be defined as the transition between these phases and the observation of the modification and speed of modification of the quantities of interest, including, first and foremost, temperature.

The schematic subdivision of the fire dynamics in a confined space into five phases leads to the representation of the dynamics on the same curve as Figure 4.23 in Figure 4.24.

Given the above representation, around which there is, in fact, a universal international consensus, further considerations can be made regarding the particularly representative moments of a confined fire:

- *Ignition*: It is a process that generates an exothermic reaction characterised by an increase in temperature to values above those of the environment involved. It can be triggered, with a match, spark, etc., or spontaneous if the fuel accumulates sufficient energy.
- *Growth*: This can be more or less rapid depending on the type of fuel, access to oxygen (comburent) and the interaction between flame and environment. The fire can be described in terms of the rate of energy released (heat release rate [RHR or HRR]) and the flue gases produced, and these two variables are not closely related. The growth phase can also last a long time and be interrupted without necessarily leading to subsequent phases.

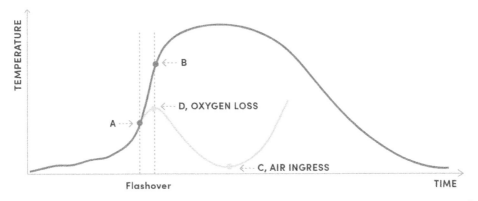

Figure 4.23 Possible evolution of a fire in a confined space in terms of gas temperature versus elapsed time.

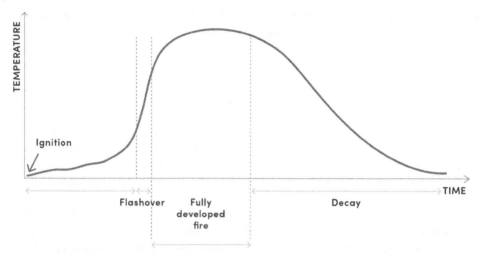

Figure 4.24 Phases of a fire in a confined space.

- *Flashover*: This has already been introduced as the abrupt transition between the growth and fully developed state of the fire, which cannot be defined as a mechanism but rather as a phenomenon associated with thermal instability. From it the 'pre-flashover' and 'post-flashover' phases are derived, which in turn are characteristic in the application of fire safety methodologies.
- Wanting to quantify flashover in some way, one can refer to data introduced by the literature and judging its attainment at compartment temperatures of 500–600 °C or an equivalent of radiant energy to the floor in the order of 15–20 kW/m². It is worth stating that these conditions are obviously not compatible with human survival, so while much lower thresholds are considered for occupant safety in the event of a fire, these thresholds for flashover and their modification in subsequent phases until complete extinction may make practical sense if the purpose of the analysis is to investigate aspects such as property safety and fire safety of load-bearing structures.
- *Fully developed fire*: In this phase, the energy released into the room is maximum and is in many cases limited by the availability of oxygen. This phase is called ventilation-controlled combustion because it is assumed that the oxygen needed for combustion enters through the openings. In the ventilation-controlled phase of the fire, it is the case that unburned gases collect at ceiling level and in contact with the air coming in from outside, they burn, generating flames that protrude outside the compartment through the openings present. The average temperature of the gas in the compartment during this phase is often very high, varying in a range between 700 and 1200 °C.
- *Decay*: This is the last phase of the fire, where the fuel is fully consumed and therefore the rate of energy release decreases and so does the average temperature of the gas in the compartment. The fire in this phase again goes from being controlled by ventilation to being controlled by the amount of fuel remaining.

The description of a fire is in fact the recognition and description of the phases that characterise it by means of the quantities that determine them. Therefore, within the scope of investigation and forensic engineering activities relating to real fires that have actually occurred, through the recorded evidence it is possible to highlight the phases and from there be able to formulate hypotheses down to the first moments of ignition by means of an in-depth study of the dynamics associated with the evidence.

The transition between phases as well as, more generally, the actual presence of phases (up to the coincidence of entire phases at given moments in time) can be influenced by a number of factors.

The heat transfer phenomena occurring in a compartment involved in a fire are due to convection and radiation. The thermal power transmitted by the fire to the structural, closure and partition sub-systems that describe the overall system coinciding with the compartment being studied (or investigated) is a function of the surfaces, the emissivity of the burning materials, the flames and gases produced, the angles of incidence of thermal radiation and the surface heat transfer coefficient. A part of this power is absorbed by the (structural) components of the compartment and raises its temperature, so much so that at a certain point even the surfaces bordering the room become radiators of a non-negligible entity; a part passes through the components and is transmitted outside the room; finally, a part is reflected and contributes to increasing the thermal level of the environment subjected to the fire. A considerable contribution to the energy evolution of the fire is made by the products of combustion, which, finding venting through the openings, remove considerable energy. It has been ascertained that in a civil fire, on average, 60% of the thermal energy is removed by the gases escaping to the outside, 10% is irradiated towards the surrounding environment and 30% is absorbed by the surfaces delimiting the compartment involved. The propagation mechanism is significantly influenced by the arrangement of the furniture, the presence of combustible coverings and their position. The latter behave differently if they are applied to the ceiling rather than the walls, if they are directly in contact with a non-combustible substrate rather than spaced by a gap.

Having said this, it is evident that, in the case of using calculation codes based on fire engineering methods, which will be discussed in the following section, the greater the possibility of taking these factors into account, the better the accuracy of the estimate (even a posteriori), of the fire dynamics. It is equally important to state that, even in the absence of detailed information (a situation that is very common in investigation and forensic engineering activities as well as in fire prevention activities in the initial phases of the design of a new building), it is possible to make direct estimates of the general trends (the dynamics) of a fire in relation to the hypotheses for a real fire or representative fire scenarios for a fire in a compartment under study.

Heat release rate (HRR or RHR), along with temperature, is one of the parameters that best describe a fire. A fire can be briefly described as a volumetric energy source, i.e. a kind of 'burner' that releases heat (heat release rate – HRR), particulates (combusted material in a solid state) and gases. The HRR represents the 'identity card' of the fire and is one of the main input information for fire simulation calculation codes. While simplified calculation codes operate by considering a defined thermal release curve, more accurate calculation codes possess the predictive ability of the HRR curve trend throughout the duration of the fire considering all the control parameters anticipated in the previous point, including ventilation, fuel present and its location, heat transfer surfaces and so forth. In evaluating the temperature values that may be reached in a compartment during a fire in the pre-flashover phase, the HRR value as input data is more reliable than that of the fire load (in this phase, generally, only part of the fuel participates in combustion).

If the fire load (which is in fact the physical representation of the maximum energy that can be released into the environment as a result of total combustion) were to be used in this phase, the estimation of the temperature values in the environment following the fire would be excessively conservative, as it would be assumed that in an extremely short time all the fuel in the entire environment could participate in the combustion process. This is only theoretically permissible for assessments in a post-flashover phase.

The rate of fire growth depends on the ignition process, the spread of flames and the rate of combustion. For fires involving furniture or various goods in storage (such as in a civil building), the phenomenon is describable with usually very complex formulae. In any case, each object involved in combustion defines a fire growth rate. The reference growth time is defined as the time required to obtain a peak of the fire's thermal energy release rate of 1 MW.

The propagation or growth phase lasts approximately 10–30% of the entire phenomenon.

The course of the curve in this phase can be approximated with the relation

$$Q = \alpha \cdot t^n$$

where Q is the thermal release rate HRR, α is the fire growth coefficient and n is an exponent that experimentation over many years allows us to set, with absolutely unanimous consensus, equal to 2.

In Table 4.3, some values of α are shown as defined in the technical literature.

It is also possible to assess fire growth through the maximum rate of heat release produced in $1 \ m^2$ under controlled fuel conditions (kW/m^2) called RHRf. In this case, reference is made to table E5, here Figure 4.25, of Eurocode EN 1991-1-2. In industry, different values may be referred to based on experimental testing.

The RHR or HRR in this case is calculated with the relation

$$Q = 10^6 \left(\frac{t}{t_\alpha} \right)^2$$

Max Rate of heat release RHR_f			
Occupancy	Fire growth rate	t_α [s]	RHR_f [kW/m^2]
Dwelling	Medium	300	250
Hospital (room)	Medium	300	250
Hotel (room)	Medium	300	250
Library	Fast	150	500
Office	Medium	300	250
Classroom of a school	Medium	300	250
Shopping centre	Fast	150	250
Theatre (cinema)	Fast	150	500
Transport (public space)	Slow	600	250

Figure 4.25 Fire growth rate and RHRf.

Table 4.3 Values of α for different growth rates according to NFPA 204M.

Speed of growth	α (kW/s^2)	Time (s) to reach 1055 kW
Ultra-fast	0.19	75
Fast	0.047	150
Medium	0.012	300
Slow	0.003	600

where Q is the thermal release rate HRR and t_α is the time required to achieve a heat release rate of 1 MW.

In Figure 4.26, similarly to what was done previously with the quantity 'temperature', a schematic representation of the HRR growth curve of a fire and its phases as a function of time is shown.

The RHR (or HRR) curve provides a simplified mathematical schematisation of the so-called natural fire, i.e. the actual fire that has developed. The three phases, initial (quadratic), intermediate (constant) and final (linear), describe in a simplified (de facto normalised) way the course of a fire in a confined space, usually governed by ventilation. The curve is associated not only with the fire load but also with the dynamics of combustion (slow fire, fast fire, ...).

The area subtended by the RHR curve with time (s) in the abscissa and thermal power (kW) = (kJ s^{-1}) in the ordinate represents the fire load (kJ), the energy available to be released (essentially the fire load).

In Figure 4.27, the representation of the growth curves in Table 4.3 is shown.

The terminology introduced is used in the description of the phases:

- *First phase* (*pre-flashover*): A direct function of combustion speed and the amount of fuel, and therefore energy, available. There is a quadratic growth with a greater or lesser slope depending on the characteristics of the material and its physical conditions; the slope indicates the speed.
- *Second phase*: In an indoor environment, after a certain amount of time, a temperature rise is reached that causes all combustible materials to catch fire; at this point, the determining factor becomes the available ventilation, and the amount of material that can burn will only depend on the greater or lesser amount of ventilation available. We have a horizontal diagram, with the maximum power that can be developed remaining constant over time, limited by the ventilation present.
- *Third phase*: Linear, representing the progressive switch-off.

Characterisation in such phases is also employed in the investigative and forensic fields, as even particularly severe and/or complex events can be described by means of such discretisations of the dynamics.

For the purposes of this section, reference will be made almost exclusively to the 'pre-flashover' phase since the objective of the present study is to analyse the safety condition for which the

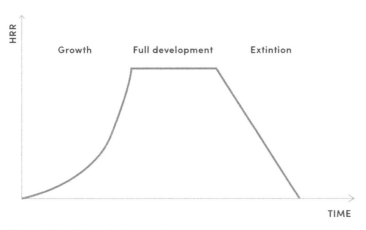

Figure 4.26 Thermal release rate.

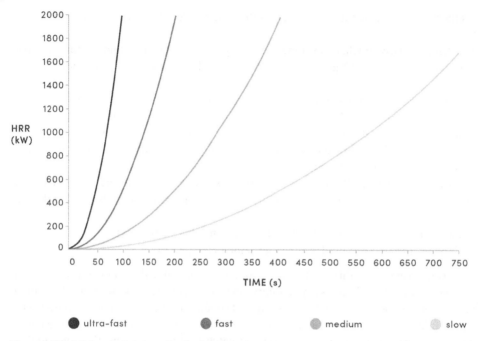

HRR (kW)

TIME (s)

● ultra-fast   ● fast   ● medium   ○ slow

Figure 4.27   HRR growth curves for the first fire phase.

protection of the lives of the occupants present in the compartments is guaranteed; therefore, reference will be made to the first phases of the fire, flashover (and the conditions for its establishment in a room) being a condition that is absolutely not compatible with the presence of occupants.

The study using internationally recognised standard fire curves makes it possible to eliminate any uncertainty associated with outbreaks at the earliest stage of ignition since the generalised event is a function of the type of curve selected. As discussed later, 'field' calculation codes allow for the precise evolution of the fire curve during the event, calculating the power and release curve over time from actual conditions.

As anticipated and explained in the following section, current calculation codes allow for an excellent representation of a complex phenomenon such as fire and enable a series of in-depth investigations, within the scope of investigation and forensic engineering activities, aimed at identifying fire dynamics.

It is worth stating that both fire engineering methods and advanced calculation codes are today widely recognised not only by the extensive technical literature of reference but also by the standards pertaining to the body of regulations on fire prevention in multiple countries (e.g. in Italy with the Ministerial Decree of 3 August 2015 as supplemented and amended by Ministerial Decree of 18 October 2019, also known as the 'Fire Prevention Code' or 'Horizontal Technical Regulation', recently referred to also by the new regulatory acts, linked to Legislative Decree 81/2008, on fire safety and fire risk analysis in workplaces for activities not subject to the controls of the fire brigade) and in the field of prevention of major accidents (D.Legislative Decree 105/2015 also known as the Italian transposition of the 'Seveso III' Directive, a community act that from the very first edition emphasised the importance, within the scope of risk analysis activities, of conducting numerical calculations for the quantitative description of complex phenomena including fires and explosions and for the identification of reference quantities whose exceeding of thresholds set by the standard itself determines conditions of damage to persons and property in relation to the extent of the effects). In particular, the aforementioned 'Fire Prevention Code' dedicates, in its 'Section M-Methods', three chapters to the methods of so-called fire engineering respectively dedicated to general methods, fire scenarios for performance design and safeguarding life with performance design. These include references to the relevant technical literature from which the considerations in the previous section have been removed. Since the physical phenomenon is the same, the methods can be applied effectively both in the design phase of the fire safety strategy (the purpose of the Fire Prevention Code) and in the study phase of events that actually occurred.

5

Integrate Fire Safety into Asset Design

The process of designing and managing a building, regardless of its size or intended use, occurs in interrelated stages according to a well-defined logical sequence. The Royal Institute of British Architects, through a graphic scheme better known as the 'RIBA Plan of Work',[1] has broken down the design process into individual sequential stages, as depicted in Figure 5.1 where design main phases are listed in the traditional execution order.

The first two (Phase 0: strategic definition; Phase 1: preparation and briefing) are those actions that do not require the intervention of architects and designers at the technical level and are rather conducted by the client with their consultants. In these phases, market surveys are conducted with the aim of understanding whether or not to initiate a project and end with the definition of a programme called project brief. Although this phase never explicitly mentions the definition of a building or object, it has already introduced a measure on the site and also a preliminary 'collection of health and safety information'. Some information can be generated very roughly, such as the main proportions and impact of the building type. This phase, also called 'pre-design', is officially outside the design activity, but it contains many definitions that define additional constraints. Imagine, as an example, a brief defining of a possible healthcare facility in a narrow site: safety challenges arise due to changes in an established mostly horizontal layout of this typology: this raises issues even before its detailed definition and further modelling.

The next steps in the design process (Phase 2, Phase 3 and Phase 4) are reflected in almost all countries with slight differences, especially in definitions and details, but based on substantial similarity.

All that precedes the bidding process leading to production and construction (Phase 5).

Each design phase involves a specific interaction with fire safety.

In the case of conceptual or schematic design, the design brief roughly defines the layout and main areas of the buildings, including certain parameters such as occupancy, type of space, access and potential secure areas. This is the phase where layouts are subject to change, even radical changes, in an effort to make the design brief find its best fit based on all constraints and regulatory inputs. This phase involves the involvement of all professional figures (architectural designer, structural designer, plant engineer and fire protection professional) with the goal of developing a project that is financially sustainable during both construction and future operation, environmentally efficient, and safe for end users.

---

1 The Royal Institute of British Architects (RIBA) Plan of Work is an industry standard document outlining the process of briefing, designing, constructing and operating building projects.

Fire Risk Management: Principles and Strategies for Buildings and Industrial Assets, First Edition.
Luca Fiorentini and Fabio Dattilo.
© 2023 John Wiley & Sons, Inc. Published 2023 by John Wiley & Sons, Inc.
Companion Website: www.wiley.com/go/Fiorentini/FireRiskManagement

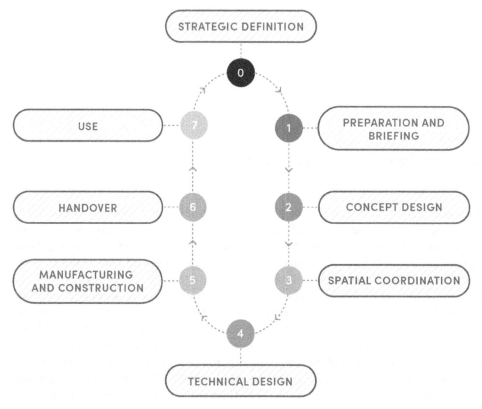

Figure 5.1 Phase design process.

Figure 5.2 depicts all aspects that are part of the design process.

Detailed design (Phase 4) is when the already defined layout becomes essentially final and materials come into play in a more precisely defined distribution. The design at this stage is submitted for approval and issuance of building permits. At this stage, normally specialist consultants play an active role in sizing structural construction details, plant engineering or fire requirements. During detailed design, the continuous integrated collaboration between professionals becomes crucial as at this stage the construction choices are made that can simultaneously meet both regulatory and performance requirements. Relative to fire safety aspects in this phase, specific insights are made, combined with more details, on the contribution of active measures such as smoke extraction and control, different sprinkler systems or active water diffusion.

The next phase (Phase 5), on construction documentation, mainly focuses on material specifications and construction details. Again, the verification of the suitability of each choice is subject to comparison with usually widespread codes and requirements for material use in relation to its class of components (e.g. curtains and ceilings) combined with information on their placement in the environment. These performance parameters are then added to the special specifications for subcontractor bids.

The above design process, representative of the circle of life of a building, must be fielded by all the professionals involved, who, with their knowledge, must make their continuous contribution at all stages from the initial design to the future use of the building.

In Figure 5.3, a diagram comparing the visions of planners is shown, mainly focused on buildings (aesthetics and efficiency) and the vision of the fire professional, which aims at occupant safety.

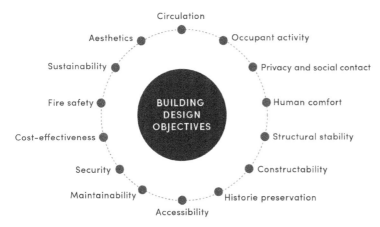

Figure 5.2 Building life cycle.

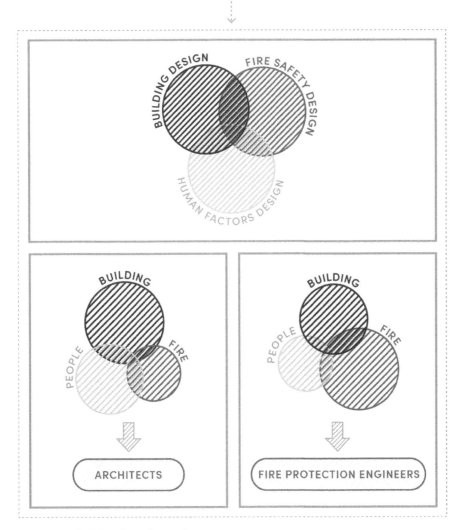

Figure 5.3 Building fire safety performance.

Optimising the relationship between aesthetics, the 'efficiency of the systems and the safety of the occupants' are the continuous challenges that professionals must constantly pursue, with the cultural integral approach described in Figure 5.4.

The interaction between fire safety and architectural design must therefore be considered the 'heart' of a project, that is, the definition of its typology and consequent layout. We have just shown how this relationship passes through the traditional stages of the process and finds form in each of them. At the schematic design stage, however, when a project is just identified in its basic components, i.e. the definition of rooms, floors, stairs and spaces, with their relative occupancy, location, typology and relationships in a three-dimensional form, a fire risk assessment analysis could already show the critical paths and also the possibility of beginning to customise them in the safest way. The early involvement of a fire expert at this stage may enable the designer to model and locate safe areas around the building and its immediate surroundings, including terraces and open spaces, as well as ventilation strategies, which could be combined with passive comfort management measures.

Actually, fire safety is a much more complex issue that has a great possibility of interaction with the development of architectural design, thanks to the change offered by the method of so-called

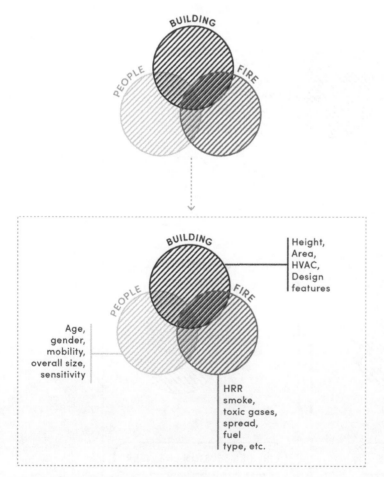

Figure 5.4 Building fire safety performance.

'fire engineering'. The method, extensively illustrated in this book, is based on the actual simulation of fire effects and smoke distribution in three-dimensional space. It causes the discussion of safety to become spatial and dynamic, allowing greater freedom in the definition of layouts and the relationship between spaces. A freedom that is obviously balanced by accountability based on actual simulation of different scenarios. The 'fragmentation of design' is indeed widespread socially to an even greater extent than observed at the time, and even small architectural design firms have to deal with the constant negotiation of professional relationships and exchange of information with a network of colleagues. Fire safety is no exception in this regard. But specialisation cannot be separated from the possibility of rooting collaboration on the fire engineering possibilities opened up by performance-based codes and no longer fixed prescriptions. This opens up specialised performance consulting and the development of a specific solution that goes beyond traditional prescription-based measures. Figure 5.5 schematises the fire safety analysis process.

The use of fire safety engineering provides a perfect example of how early involvement and interaction between a fire safety specialist and a designer can absolutely be fruitful.

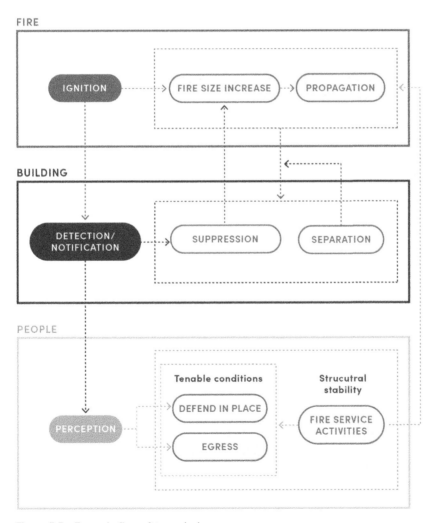

Figure 5.5 Example fire safety analysis process.

The idea that arises from this collaboration might seem narrow just because of its peculiarity, but, in reality, it only demonstrates how safety can be considered an integral part of space design, and this can only benefit the clarity of space layout, the serenity of the paths indicated, working on the mobilisation phase and not just the escape phase.

Between the 1990s and the early 2000s, the world of architecture witnessed the pioneering eras of the so-called 'digital revolution', which through new technologies and software, embraced by a new generation of architects, allowed buildings to be imagined and experienced through cinematic animation.

One of the founding characteristics of these new processes can be identified in the 'anticipatory effect' that digital simulation can have on the designed building, which sees drawings approaching reality before any physical construction. The digital model prompts 'building before building'.

Virtual prototyping has become a rule in many disciplines that has caused perceptions to change, towards objects and environments that have always been perceived as static and in their virtual version become subject to perhaps endless updates. In architecture, this ability to anticipate is often seen as a major risk of taking some of the 'magic' out of the creative process, while for others it finally brings to the table the opportunity to work fully on the form that gives true meaning to architecture.

In any case, this anticipation, this virtual prototyping, no matter what form it is taking, is changing the process intrinsically. Collaboration and new technologies can not only challenge design habits but also overcome many limitations or rather perceive limitations and discover unexpected and unusual solutions. The main challenge is to use the digital environment to support innovation in professional practice.

A similar approach can be adopted for industrial assets. While the fire and explosion risks are different and usually assessed by using specific (often quantitative) methodologies that include estimation of associated frequencies and severities (including the physical effects), risk assessment should be an integral part of the design process. Design process, due to the complexity and peculiarities of these assets, is a staged approach with a different level of detail given the design phase of the asset.

Main phases are illustrated in Figure 5.6.

Especially for some plants, such as those in the process industry, risk assessment analysis is now recognised as a fundamental pillar of both plant design and operation. In general, it is the responsibility of the plant owner to correctly define the risk analysis to be conducted at each stage and to define the requirements in terms of content, leaving discretion in the selection, in relation to the specific activity and its complexity, of the most appropriate method or methods.

- It is possible to define specific design stages that correspond to as many stages in the design process: conceptual design.
- Front-end engineering design – process design.
- Detailed engineering design.
- Construction design verification.
- Pre-commissioning safety review.
- Post-start-up.

To these stages nowadays other two stages are usually added: inherent safety design and decommissioning design phase that consider risk reduction in both the conceptual phase and in the dismantling phase at the plant end of life.

Given this complexity and duration of time of the entire design process, required safety assessment takes its steps already during the design process and allows the design itself to be

PROJECT PHASE   **STAGES**

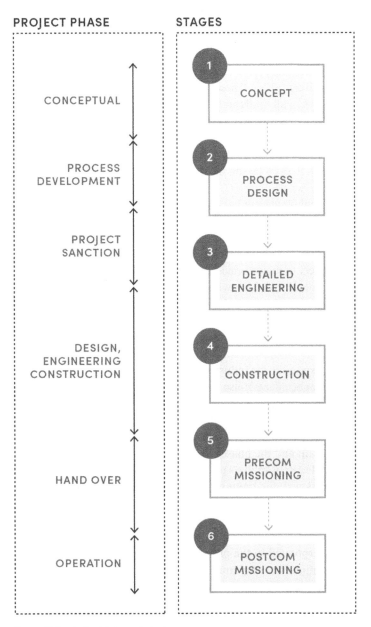

Figure 5.6 Industrial asset design process.

adjusted according to the findings of the risk assessment in order to prevent and mitigate the occurrence of industrial accidents and the accidental hypotheses that promote them.

For each step of the safety analysis associated with each step of the design life cycle, one or more, possibly alternative, methodologies are to be used. The selection of the most suitable methodology is a function of the type and complexity of the plant to be analysed, as well as the substances present, their methods and conditions of use, and any accidental events that may occur, including the occurrence of potential major accidents.

In the areas of activities where flammable substances are present in the form of combustible gases, vapours, mists or dusts, the fire risk assessment must also include the risk assessment for explosive atmospheres. As discussed in this book, several methods are available to conduct fire and explosion risk assessment activities in industrial premises, ranging from qualitative to quantitative methods, based on physical effect modelling (PEM). These methodologies have been used worldwide for many years in process industries of all kinds and are well codified in international standards and technical literature. The choice of methodology is undoubtedly a function of a number of aspects, including the design phase (conceptual, basic engineering and detailed engineering) or the life cycle of the plant (even on existing plants, a review of the safety level can be carried out in order to assess possible improvements to it, even taking into account a period of operation of the plant itself).

It is evident that the design of industrial plants can be considered a complex activity and that it has a particularly important influence on the future safety level of the plant.

The degree of safety is a function of both the design activity and the user's operation of the plant.

Fire and explosion risk assessment is a fundamental element in the design of an industrial plant, regardless of whether it is formally subject to major-accident prevention legislation requirements (such as COMAH, Seveso and OSHA), but clearly plants where a deviation from the intended operation can cause serious harm must be subjected to a safety assessment as early as the design or basic engineering phase in order to highlight any shortcomings and allow the adoption of useful measures to improve the level of safety for operators, third-parties and contractors, people, environment and industrial assets.

Fire and explosion risk assessment, especially for industrial plants, given their complexity and potential to produce accidents (and major accidents), also in relation to the presence of hazardous substances, is not an activity that is generally developed 'once and for all', but for new plants it is in fact a structured process that is developed considering all the phases of the design since it is a process made up of specific phases, interrelated, in sequence.

In relation to the complexity of the plant and the phases that characterise it, it is the designer's responsibility to identify the most suitable safety analysis process phases, each supported by a clear identification of the objectives to be achieved, the input elements, the expected results, methods and assessment tool.

Safety assessment (including of course fire and explosion risk assessment) represents a systematic analysis of a project by experienced and competent personnel, at defined stages of its development, to ensure that the safety standards incorporated in the project meet legislative obligations, good engineering and rule of art standards, national and supranational technical reference standards, as well as the project's criteria for guaranteeing safety (and major accidents prevention).

Safety assessment, a requirement of certain regulatory areas, ensures that the designer has useful information to understand any critical issues associated with the project in good time, prior to its construction and certainly before its operation by the end user who commissioned it.

In any case, the designer, in relation to the characteristics of the installation and the safety criteria identified, selects the most appropriate type of study or pool of studies that will allow ensuring safety in the subsequent commissioning, start-up and operation phases.

As part of the design development, it is appropriate to conduct a series of safety checks, the purpose of which is to verify, from the point of view of plant safety:

- the congruity of the engineering choices and design documents with the general safety criteria adopted;
- the modifications required as a result of the safety studies carried out;
- the assumptions and hypotheses underlying the safety studies themselves.

These checks also involve reviewing the design documentation in order to identify and anticipate areas of inadequacy and to initiate corrective actions to ensure that the final design responds appropriately to all the defined safety criteria. In particular, it is desirable that the design is reviewed prior to construction. This review includes

- verification that the design meets the requirements identified through the preliminary safety studies;
- verification of compliance with all safety requirements.

A key role also, in terms of verification, is the last (in chronological order) verification that is generally conducted as part of the construction of a plant, known as the Pre-Start-Up Safety Review (PSSR).

As part of this verification, the existence of all safety analyses is verified, with the appropriate degree of detail and the existence of all safety documents (including the classification of areas with explosion hazards) as well as their congruence with the design and with what is actually constructed in the field ('as-built').

In particular, a complete and proper PSSR includes:

- ascertaining that the requirements identified in the detailed safety studies have been implemented and that any modifications are justified;
- ascertaining the existence of documentation relating to the requirements;
- ascertaining the availability of adequate operating procedures for normal operation, start-up, planned and emergency shutdowns and for maintenance and reclamation;
- ascertaining the availability of adequate emergency procedures;
- ascertaining the existence of documentation collected and produced according to as indicated in the risk control plan;
- ascertaining that the initial operational phase is consistent with the safety objectives set in the project;
- the identification and recording of any difficulties in operation and maintenance.

6

Fire Safety Principles

While for a fire engineer and for the stakeholders involved in fire safety design activities the importance of the knowledge of the principles of the fire event and associated fire scenarios (usually described by their physical effects over time from the ignition or the threat rise, such as the loss of containment) has been shown, it is equally fundamental to develop a knowledge on the fire safety strategy elements. Fire safety design and maintenance over time is the tool to guarantee an acceptable fire risk given a selected pool of fire scenarios. This is a performance that can be achieved by adopting a combination of fire controls (preventive and mitigative barriers) composed of a strategy that considers the vulnerabilities. A structured thinking may be supported considering two fundamental National Fire Protection Association standards (550 and 551) developed and maintained by the Fire Risk Assessment Methods Technical Committee. These standards are nowadays an important example of the holistic approach that can be used by fire engineers and stakeholders to address the complexity of the socio-technical systems we should protect from the fire threats.

6.1   Fire Safety Concepts Tree

As anticipated, National Fire Protection Association (NFPA) has played an important role in describing the elements composing the fire safety in simple and structural terms.

The NFPA is an international non-profit organisation established in 1896 in the United States, with a membership of more than 75 000 people from a hundred nations. The output of the NFPA is considerable (more than 300 codes and standards) and covers many of the issues that affect human safety (the codes contain criteria for construction, workmanship, design, service provision and installation).

The organisation is primarily concerned with drafting and publishing codes, guidelines and safety standards whose ultimate aim is to define minimum requirements, norms and techniques for fire prevention and extinction. They are not always transposed and incorporated into legislation but are nevertheless always considered a reference standard of the highest quality and authority by fire professionals around the world. They indeed are considered 'state of the art' in several countries, and in many cases they influenced local perspective regulations.

The NFPA's numbering and standardisation methodology, which has been approved by the American National Standards Institute (ANSI), is peculiar: all documents issued by the organisation go through a consensus development process in which only volunteer technicians and experts (around 6000) participate. The NFPA in no way audits the work of the volunteers, nor does it pass

Fire Risk Management: Principles and Strategies for Buildings and Industrial Assets, First Edition.
Luca Fiorentini and Fabio Dattilo.
© 2023 John Wiley & Sons, Inc. Published 2023 by John Wiley & Sons, Inc.
Companion Website: www.wiley.com/go/Fiorentini/FireRiskManagement

judgement on the results they achieve, but merely administers the consensus development process (approved by ANSI) and sets the rules.

Two of the standards produced by the NFPA specifically deal with fire risk assessment and these can be considered fundamentals:

NFPA 550 ('Guide to the Fire Safety Concepts Tree'), which defines a fire safety analysis tool;
NFPA 551 ('Guide for the Evaluation of Fire Risk Assessment'), which deals with the evaluation process of conducted risk analyses.

Both, in their simplicity, are a fundamental answer from which to benefit in one's professional activity. The two standards are now maintained by one of the organisation's most important technical committees, the 'Fire Risk Assessment Methods' Technical Committee. These standards underline the pillars of the fire safety and the workflow of performance-based fire engineering.

6.2 NFPA Standard 550

The NFPA 550 Standard has its origins in a 1985 work, the 'Guide to the Fire Safety Concepts Tree', developed by a specific NFPA Committee called the 'Committee on System Concepts'. The Committee was set up with the aim of developing systems, concepts and criteria for fire protection in civil and industrial structures. The Committee's work firstly led to the development of a visual model of fire safety engineering concepts and then to the drafting of the corresponding application manual.

The Committee was disbanded in 1990, but its work underwent several evolutions, reconfirmations and revisions, the latest of which is the version released in 2007 and issued by a technical committee of international experts (where Society of Fire Protection Engineers [SFPE] is also represented).

The fire engineering concepts, originally modelled by the 'Committee on System Concepts', are organised according to a hierarchical tree model, structured in such a way that each element of the tree corresponds to a concept specific to fire safety and can be successfully and easily employed in the vast majority of cases, as well as extended to ensure applicability in more specific and non-routine cases. The applicability is very extensive since the expressed concepts can be applied to a number of different facilities across multiple sectors.

The aim, which was fully successful, was to identify different possible strategies for fire prevention and fire damage control (the fire risk control barriers), and then integrate them into a single visual model, characterised by an extremely intuitive graphic notation, which clearly shows the relationships between the elements identified. The result of the work is represented by a functional structure, called the 'Fire Safety Concepts Tree', which is easy to interpret and consult and can be used in various areas related to fire safety, including brainstorming and stakeholder engagement.

The definition of this structure is provided in Chapter 4 of the publication, dedicated to the description of the 'Structure of the Fire Safety Concepts Tree', where it is stated (see Article 4.1) that *the Fire Safety Concepts Tree shows the relationships between fire prevention strategies and fire damage control strategies.*[1]

The tree is constructed according to some very strict logical and graphic rules, similar to Boolean algebra. From the point of view of visual impact, the relationships between the various elements of fire safety are evident as they are represented by two types of graphic elements: lines and nodes.

---

1 The Fire Safety Concepts Tree shows relationships of fire prevention and fire damage control strategies.

'Lines' only relate different firefighting concepts to each other and relate them to a single concept hierarchically superior to them, recomposing them into it. 'Nodes', on the other hand, express logical operations and serve to establish how these concepts relate to each other.

This 'constructive' approach proves to be very useful, compared to other analysis tools, because it correlates different aspects of fire safety which are often still analysed separately today (cf. Art. 4.1.1), with the consequent loss of an overall view: *Traditionally, fire safety characteristics, such as the type of construction, combustibility of the materials contained, protective devices and occupant characteristics, have been considered independently of each other. This can lead to unnecessary duplication of protection. On the other hand, when these characteristics are not co-ordinated, gaps in protection or a lack of the desired redundancy may occur.*[2]

In Article 4.1.2, it is emphasised that 'the obvious advantage of the Fire Safety Concepts Tree lies in its approach to fire safety. Rather than considering each fire safety characteristic separately, the Fire Safety Concepts Tree examines them all and demonstrates how they influence the achievement of fire safety goals and objectives'.[3]

Therefore, by using the Fire Safety Concepts Tree, it is possible to analyse the potential impact of different strategies, to identify the problems inherent in adopting certain strategies over others and to highlight how some of them contribute to improving the reliability of the entire safety system.

The elements of the Fire Safety Concepts Tree are represented by rectangles, containing within them the title of the concept represented, with simplicity and effectiveness.

The higher-level concepts are normally abstract (logical elements), e.g. 'Fire management' or 'Combustion prevention'; the further down the tree structure one goes, the more concrete and specific the elements become, e.g. 'Eliminate combustible substances', 'Report the presence of fire', 'Provide protected escape routes', etc., each concept essentially expresses an objective. Going from the top to the bottom of the tree means defining objectives of increasing specificity, starting from the general objectives, typically constituting the founding elements of the organisation's fire safety policy, and ending with the prevention and protection measures that guarantee the practical implementation of the action to be taken in compliance with the definitive objective.

The top root of the tree is the element called 'Fire Safety Goals' and represents the ultimate goal to be pursued: therefore, fire safety is ensured if and only if its 'goals' are fully achieved.

Given this approach, the structure is aimed at supporting performance-based approaches starting with the identification of basic fire threats.

The elements of the tree are connected to each other by logical operators and in particular by

an 'OR' type operator (which performs the logical union of concepts);
an 'AND' type operator (which performs the logical intersection of concepts).

The 'OR' operator represents a situation in which any of the concepts placed below the symbol results in the concept placed above that symbol. In Figure 6.1 for example, the concept 'A' is verified if at least one of 'B1', 'B2' or 'B3' is verified.

---

2 Fire safety features, such as construction type, combustibility of contents and protection devices, and characteristics of occupants traditionally have been considered independently of one another. This can lead to unnecessary duplication of protection. On the other hand, gaps in protection or lack of desired redundancy can exist when these features are not coordinated.

3 The distinct advantage of the Fire Safety Concepts Tree is its systems approach to fire safety. Rather than considering each feature of fire safety separately, the Fire Safety Concepts Tree examines all of them and demonstrates how they influence the achievement of fire safety goals and objectives.

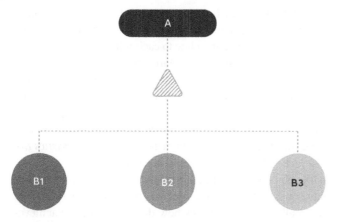

Figure 6.1 Typical scheme of an OR operator.

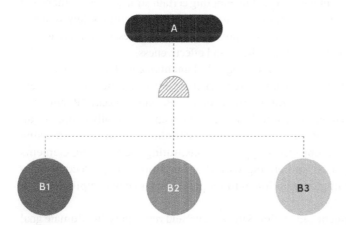

Figure 6.2 Typical scheme of an AND operator.

If, for example, rectangle 'A' identified a protection system and rectangles 'B1', 'B2' and 'B3' the components of that system, it would be effective if at least one of the components was present and active. Differently, the protection system 'A' would only be effective if all three of its components were absent or inoperative.

The 'AND' operator, on the other hand, represents the situation in which all concepts placed below the symbol are equally indispensable to achieve the concept placed above that symbol. In Figure 6.2, 'A' is only verified if all 'B1', 'B2' and 'B3' are verified.

As an example, consider that the elements in the figure represent the conditions that must occur for an automatic protection system (the rectangle 'A') to activate and be effective: this would mean that the system is only effective when all the underlying conditions are verified. If even one of them is not verified, system 'A' could not operate or operate effectively.

The tree proposed by NFPA is very extensive and detailed, the result of a thorough analysis aimed at defining a tool with broad applicability. In Figure 6.3, only the highest levels of the structure are presented; the complete scheme can be found in the NFPA Standard.

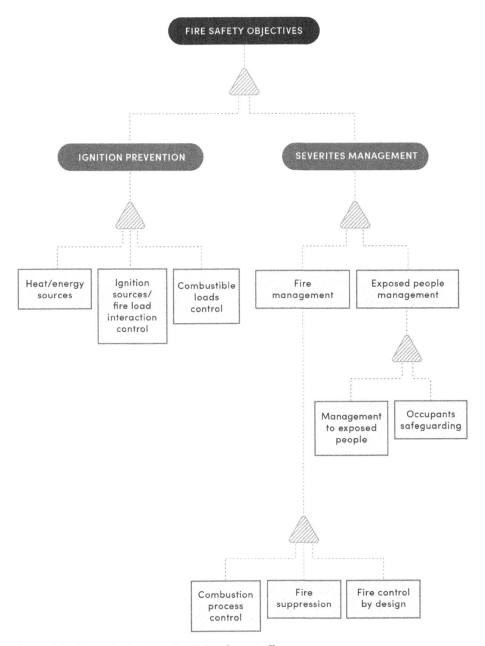

Figure 6.3 Higher levels of the Fire Safety Concepts Tree.

As can be deduced from Figure 6.3, the objectives of fire safety, which represent the highest level of the tree, can be pursued essentially in two ways: by preventing a fire from developing rather than by coordinating appropriate measures to limit its damage.

The first line of action is all strategies that fall under the category 'Fire Prevention'; the second is everything that can be categorised as 'Fire Impact Management'.

It is also evident from the diagram that these two categories are alternative and competing with each other: theoretically, it is sufficient to ensure the complete reliability of one of the two to completely satisfy the safety objectives. Usually, however, in real practice both fire prevention and impact management principles are correctly applied at the same time in order to increase the probability of achieving fire safety objectives and thus exploiting the redundancy of different applicable strategies pertaining to each of the two aspects.

The 'fire prevention' category consists of measures to control ignition sources, combustible substances and the processes during which these two elements may come into contact (hazard control activities).

Therefore, the simplest fire prevention strategy to pursue is obviously the elimination or control of one of the two elements indispensable to starting a fire: the ignition sources and the combustible substances present in the structure. The limitation of this strategy, however, lies in the impossibility of preventing new, unforeseen dangers from arising once all known dangers have been eliminated.

For example, it would serve no purpose to eliminate all ignition sources in a building where combustible substances are normally found if other sources could somehow penetrate from outside the controlled area and subsequently ignite the substances present there. Therefore, the obvious solution to this type of problem would be to eliminate combustible substances and ignition sources simultaneously, which would greatly reduce the likelihood of a fire occurring.

Other strategies aim instead to control the interaction that could occur between combustible substances and ignition sources. In this case, it is of paramount importance both to secure the transport phases of materials and sources (the two elements must never come into close contact) and to control all processes during which energy transfer could occur.

Of course, in order to be truly effective, all fire prevention measures require constant periodic verification by the owners of the facilities, resulting in the importance of fire risk management over time.

Differently, the category of 'Fire Impact Management' can be achieved through two distinct strategies identified by the subcategories 'Fire Management' and 'Management of Exposed Elements'.

The term 'fire management' is used to identify a series of strategies that aim to reduce the dangers associated with the growth and spread of fire in order to contain its effects. This can be achieved by, for example:

acting on the fuel or the environment to reduce the heat and smoke produced;
controlling the combustion process, trying to extinguish it manually or automatically;
controlling the spread of fire through containment or smothering.

The 'management of exposed elements', on the other hand, is implemented by coordinating all those measures that may affect the protection of property, activities, people's health or in any case all those vulnerable elements that may be included in the objectives of fire safety. The strategy can be carried out by limiting the number of elements exposed to the danger of fire rather than by ensuring that they are in adequately protected places or, where this is not feasible, that they can be quickly taken to safe places.

The Fire Safety Concepts Tree is a qualitative guide that can be used successfully in many areas. NFPA 550 itself takes care of, by way of example, some of the most common and specific uses such as verification of fire safety and installations in existing buildings;

fire safety design of new buildings and installations (as well as assets temporary or defined modifications);

the assessment of changes having an impact on the level of safety that arise when the occupancy, use or fire protection of a building or installation is changed;

the qualitative assessment of the reliability of the fire safety provided (also by assessing the reliability of individual protection or prevention systems);

communication with architects and other professionals involved in the design and management of buildings;

the identification of equivalence between firefighting strategies;

research and classification of firefighting strategies;

performance-based assessment and the breakdown of higher-level objectives into sub-objectives that are easier to observe and measure but with full respect for the organisation's overall objectives defined in the fire policy.

The determination of fire safety strategy is one of the most immediate applications of the Tree both because its elements well represent the requirements to be met in order to achieve fire safety objectives and because the structure helps to verify the actual coverage of the requirements selected as participants in the fire strategy.

The Concepts Tree can consequently be used as a checklist in order to verify the degree of fire safety achieved, becoming a means to identify existing alternatives and ways in which redundancy (and thus reliability) can be implemented, to keep track of changes that have occurred (compared to the initial design) during the life of a building, and to assess their impact on safety. With more detailed assessments it is possible to verify different and alternative strategies with a similar risk reducing factor.

It should be remembered that codes and standards usually require those who use them to have a good knowledge of the principles of fire safety, which are themselves articulated over several disciplines. In contrast, the Fire Safety Concepts Tree provides a very simple visual representation of everything contained in the codes and standards irrespective of the technological specifications of individual systems. Precisely for this reason, the Tree can become both a tool for basic fire protection engineering teaching and a means of communication between fire safety specialists and others during the identification phase of specific requirements (e.g. in the context of design group meetings).

Furthermore, when addressing the problem of identifying equivalent fire safety strategies, the Concepts Tree can become a very effective tool, albeit in a qualitative way, and by understanding the Boolean logical operators between elements it is possible to verify the absence of common failure causes. The standard itself states that (see Art. 5.3) *the Fire Safety Concepts Tree has a more specific application in supporting code definition. An important feature of code definitions is the 'Equivalence' clauses. Equivalence clauses state that alternatives to the requirements specified in the codes are acceptable if they provide an equivalent degree of fire safety to that of the code. The Fire Safety Concepts Tree provides guidance in identifying design strategies that can provide an equivalent degree of safety. Operators of type 'OR' indicate points where, within the Tree, there is more than one way to implement a strategy. A decrease in the quality or quantity of an input of the 'OR' operator may be balanced by an increase in one of the operator's other inputs. In order to determine whether a particular design strategy provides an equivalent level of security, an engineering analysis would be required; nevertheless, the Tree provides a guide to determining the concepts to be evaluated.*

The *SFPE Engineering Guide to Performance-Based Fire Protection Analysis and Building Design* by referring directly to the NFPA tree identifies a process for developing and evaluating test design strategies in order to decide whether they provide a satisfactory level of safety.[4]

In the case of using the Tree to decide which strategies to set up, it should be borne in mind that it has a purely exhaustive nature, i.e. it attempts to show all the strategies useful for achieving the final objective. This objective, however, thanks to the 'OR' operators, can be reached by following various paths without requiring the total coverage of all the strategies contained within the structure; that is, although all the strategies that can actually be pursued are listed in the Tree, this does not imply that all of them must be implemented. Furthermore, the essentially logic-based mode of construction allows insiders the extension of the Tree and non-insiders the understanding of the Tree regardless of the extension.

Speaking of 'performance-based assessment', it is finally recalled that (cf. Art. 5.7) 'the purpose of performance-based fire safety assessment is to ensure the achievement of a set of specified objectives. Fire safety is the overall result to be achieved with regard to fires'.[5]

The use of the Concepts Tree is characterised in any case by certain limitations. Given its hierarchical nature, in fact, the Tree does not allow the representation of multiple interactions that occur when a concept at a given level constitutes the input element of different concepts at a higher level. Nor does the Tree allow for the representation of interactions that may occur between elements at the same level.

A further limitation consists in the impossibility of representing the temporal factor. In fact, the Tree of Concepts is not able to specify whether events must occur at a certain speed, nor whether the elements input to the 'AND' operators must occur in a precise chronological sequence (think of the manual firefighting strategy, which requires various operations to be carried out in a certain sequence to guarantee the efficiency and effectiveness of the intervention). Furthermore, it is not able to represent the simultaneity of the various elements (e.g. if people and combustible products are not present in the same place at the same time, there is no risk of a fire causing casualties).

A further and perhaps less obvious limitation is the fact that each tree can only visually represent one objective at a time. It often happens, however, that there are multiple fire objectives to be achieved within a building, resulting in a situation that cannot be managed or rather not easily managed. The only solution is to design different Tree structures, one for each identified objective, with the consequent increase in analysis time and activities aimed at maintaining the representations congruent with changes in the building and fire design over time.

---

4 *Code equivalency*: A more specific Fire Safety Concepts Tree application is as an adjunct to building codes. An important feature in building codes is the provision for 'Equivalencies'. Equivalency clauses state that alternatives to specified code requirements are acceptable if they provide a degree of fire safety equivalent to that of the code. The Fire Safety Concepts Tree provides a guide to identifying design strategies that may provide an equivalent of safety. 'OR' gates indicate where more than one means of accomplishing a strategy in the tree is possible. A decrease in the quality or quantity of one input to an 'OR' gate can be balanced by an increase in another input to the same gate. Determination as to whether a particular design strategy provides an equivalent level of safety would require an engineering analysis; however, the tree provides guidance on which concepts to assess. The *SFPE Engineering guide to Performance-Based Fire Protection Analysis and Design of Buildings* identifies a process for developing and evaluating trial design strategies to determine whether they provide a satisfactory level of equivalency.

5 *Performance-based evaluation*: The purpose of performance-based fire safety evaluation is to ensure attainment of a set of stated goals. Fire safety is an overall outcome to be achieved with regard to fire.

Finally, it is important to emphasise that, due to the abstractness of the concepts it contains, the Tree does not lend itself to providing quantitative data (e.g. probabilities of occurrence), which are instead typical of different graphical analysis tools (e.g. fault trees and event trees).

In any case, given these limitations, the tree is an invaluable starting point for any further considerations having a greater level of detail as well as the preferred tool to summarise the results.

NFPA Standard 550 provides an example of a bottom-up procedure to be used in order to examine a tree structure and assess its fire safety.

The first step in the procedure is to define the objectives; during this phase, the security strategies to be applied are established.

Then, starting with the lowest elements of the tree, those without any input elements, the degree of coverage that each element is able to achieve is assessed, also taking into consideration the reliability that they must possess within the firefighting system and overall protection strategy.

The degree of coverage identified is then defined as either non-existent or below/congruent or above requirements.

Going back up the Concepts Tree, an evaluation is then carried out for the elements of the levels above, respecting the two simple rules set out below:

where the evaluated items constitute an input for an 'OR' type operator, the resulting value will be at least equal to the best category assigned to the input;

in the case where the evaluated items constitute an input for an 'AND' type operator, the resulting value will be equal to the worst category assigned to the input.

Proceeding in an exhaustive and disciplined manner, the root of the Tree is reached and all elements are rated. When all elements have been evaluated, the entire Tree can be examined to determine where improvements can be made. The evaluation should also include the overall reliability check associated with the graphic structure developed, coinciding with the fire strategy developed.

Given the importance of this tool, in this book, the tree will be demonstrated having a great flexibility, including its use for highlighting the failures observed in fire strategies from real fire events and for deriving lessons learnt.

6.3 NFPA Standard 551

The NFPA Standard 551, also known as the 'Guide to Fire Risk Assessment', contains criteria for assessing the completeness and correctness of the method used for fire risk analysis.

The Standard is addressed both to the authorities having jurisdiction (to whom it is primarily addressed) and to all those individuals and organisations, such as building owners or insurance companies, who, in various capacities, deal with these assessments and have to evaluate them.

The standard provides the specific elements on which to focus the review, so it follows that these elements can also be used by the risk analyst to support the preparation of his own analysis in a proactive manner.

The NFPA:551 Standard neither lists or describes in detail the methods that can be used for 'fire risk assessment' nor does it provide guidelines for how this assessment should be conducted. It merely describes the technical review process that a fire risk assessment undergoes and defines the documentation that must accompany it in order for it to be reviewed.

Fire risk assessments find application in many areas, such as demonstrating the adherence of the design of a new building to the required performance, verifying fire safety improvements in a workplace, estimating the fire risk in a structure, determining the protective measures required by a standard or code, etc.

The primary aim of conducting a fire risk assessment is to identify the level of risk and/or the methods useful to reduce it within the limits considered acceptable. Normally, the achievement of the first objective, the identification of the risk, is only the natural premise to the achievement of the second and more important objective represented by the elimination or, in any case, the containment of the identified risk through the control resulting from the alteration of a firefighting strategy.

Referring to risk, the Standard specifies that, although by definition the fire risk to be assessed corresponds to the sum of the risks[6] associated with each possible loss scenario, in practice it is inconceivable to exhaustively examine all possible hazards and scenarios. This aspect is of fundamental importance in fire engineering. Numerous standards and regulations specify this; for example, a standard devoted entirely to the identification and classification of fire risk for the purpose of selecting design fires, such as ISO /TS 16733, emphasises the fact that the number of all possible fire scenarios in any given reality can easily tend to infinity.

In addition, the quantification of risk may vary significantly due to different viewpoints of those involved in the assessment process, i.e. all those who have economic, public safety, regulatory or occupant safety interests related to fire risk, who may give different importance to property, personal safety and economic damage caused by business interruption, depending on whether they are tenants or shareholders of a business rather than operators and employees, external regulatory or insurance authorities, or members of communities or neighbours who may be affected by a fire.

It is important that all stakeholders potentially involved in the collaborative fire risk assessment process are identified at an early stage of the process, especially those stakeholders whose interests may be conflicting. This will ensure that, as these stakeholders will be called upon to establish the objectives of the assessment, the resulting results are representative of each point of view and provide a good starting point for the decision-making processes to be put in place downstream of the assessment.

It is also essential that all the variables involved in estimating the risk are measurable (in terms of monetary values, number of people injured, etc.) and comparable with reference values (in terms of frequency of occurrence or number of events over a given period of time) in order to establish whether one falls, wholly or partially, within the risk acceptability criteria identified above.

The concept of the quantification of variables is often reiterated in fire engineering, especially with regard to the measurability of reference values (e.g. in ISO 16732, cf. Art. 8.2), which, if they were not such (i.e. if the assessment were conducted exclusively in a qualitative manner), could not allow an easy final decision on the acceptability of the estimated risk.

Finally, as in the tradition of all NFPA documents, this one also contains a series of definitions, many of which have become commonplace in the language of fire safety engineering, relating to the concepts covered by the Standard. The most significant ones are listed below:

event (cf. Art. 3.3.4): 'the occurrence of a particular set of circumstances, whether certain or uncertain and whether singular or multiple';[7]

---

6 Calculated as the product ($l \times p$) of the potential loss (l) times the probability (p) that the loss is actually incurred.

7 *Event*: The occurrence of a particular set of circumstances, whether certain or uncertain and whether singular or multiple.

fire scenario (see Art. 3.3.7): 'a set of conditions that characterise the development of a fire, the spread of the products of combustion, the reaction of people and the effects of the products of combustion';[8]

risk (see Art. 3.3.13): 'the probability and possible consequences of undesirable events associated with a given structure or process'[9];

scenario clusters (cf. Art. 3.3.14): 'a group of scenarios having some (but not all) characteristics in common',[10] clusters are usually introduced with the aim of limiting the actual number of analyses to be performed, as only one analysis per cluster will be performed instead of one for each scenario;

acceptance criteria (see Art. 3.3.1): 'the units and threshold values against which a Fire Risk Assessment is judged'[11];

Fire risk assessment (see Art. 3.3.6): 'a process used to outline the risk associated with a fire scenario or fire scenarios of interest, their likelihood and their potential consequences. Other documents may use other terms to designate 'fire risk assessment', such as fire risk analysis or fire hazard analysis assessment'.[12]

Both the process of drafting the fire risk assessment and the subsequent review process can be broken down into distinct phases, as illustrated in the diagram in Figure 6.4.

The initial stages of the process can be traced back to the definition of the problem, the objectives and the acceptance criteria; having a correct identification of all the issues at stake, the actors involved and the risk elements lead to the appropriate choice of the analysis method that will be used.

The definition of the problem consists of identifying and documenting the reasons why the assessment is being carried out (e.g. proposing methods to reduce the level of risk in an existing building); identifying the physical and chemical factors, such as fire, explosion, smoke and toxicity of products, that may cause damage; and identifying, within the locations subject to assessment, the elements that are potentially exposed to risks such as (cf. Art. 4.4.2.1) 'people (occupants, employees, the public and emergency workers); property (structures, systems and components of the built environment); the environment (national parks, monuments and hazardous materials); the mission, in the Anglo-Saxon meaning of the term (legacy, continuity of work, information and communication)'.[13]

---

8 *Fire scenarios*: As used in this document, a fire scenario is a set of conditions and events that characterises the development of fire, the spread of combustion products, the reactions of people and the effect of combustion products.

9 *Risk*: The paired probabilities and consequences for possible undesired events associated with a given facility or process.

10 *Scenario cluster*: A group of scenarios having some, but not all, defining characteristics in common.

11 *Acceptance criteria*: Acceptance Criteria are the units and thresholds values against which a fire risk assessment is judged.

12 *Fire risk assessment (FRA)*: A process to characterize the risk associated with fire that addresses the fire scenario or fire scenarios of concern, their probability and their potential consequences. Other documents may use other terms, such as fire risk analysis and fire hazard analysis assessment, to characterize fire risk assessment as used in this guide.

13 The exposed target at risk should be identified. This may include any or all of the following:
   1) people (occupants, employees, general public and emergency responders);
   2) property (structures, systems and components of the built environment);
   3) environment (national parks, monuments and hazardous materials);
   4) mission (heritage, business continuity and information/communication).

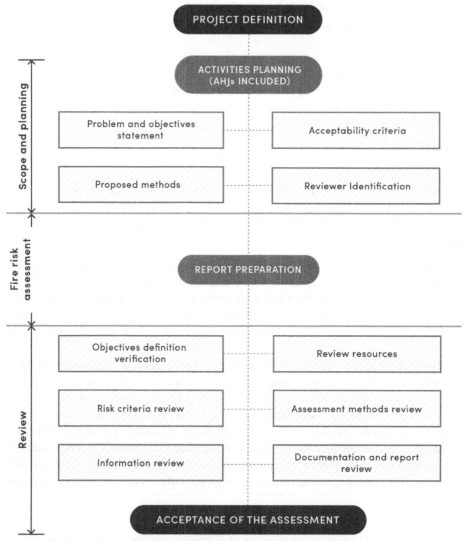

Figure 6.4 Overview of the revision process.

In addition to this, it is necessary to define the metrics of the variables in play for risk calculation purposes and the reference values to be considered as risk acceptability limits. It must be borne in mind that acceptability criteria must always be defined (relying on performance standards, regulations issued by the authorities and guidelines dictated by standards and guidelines, etc.) at the planning stage of the assessment work, taking care that the set of criteria selected can cover all the areas covered by the risk analysis.

The criteria depend on the type of problem being analysed but in turn influence the choice of method one decides to use to conduct the fire risk analysis.

Once the problem has been defined, the risk analysis proceeds with the selection and application of the evaluation method, an aspect that is definitely not secondary, given that the evaluation of the correct application of the methods to the type of problem to be ascertained represents a crucial point within the subsequent review process. Precisely in virtue of this aspect, the Standard reserves the entire Chapter 5, 'Selection and Evaluation: FRA Methods', to the description of the main characteristics of the existing methodologies and useful recommendations for those who must apply or evaluate them.

The authorities with jurisdiction must be involved in all these process steps (problem definition, definition of acceptance criteria and choice of methods) as well as in the subsequent process review and final approval. The authorities with jurisdiction are directly and unequivocally responsible for the final approval phase, which can only be granted after verifying that both the premises and the data used in the analysis phase (e.g. the characteristics of the buildings, the occupants and the types of fire considered) reflect the actual situation at the locations being assessed.

The methods for conducting a fire risk assessment differ mainly in the way they estimate the probability of occurrence and the magnitude of the expected consequences should the event take place.

For the estimation of both aspects, it is possible to resort to experience, using qualitative methods, rather than to calculations and models of various kinds, using quantitative methods; much depends, of course, on the degree of knowledge one has of the events to be estimated, the level of applicability of the individual methodologies to the scenarios to be analysed and the ultimate goal to be achieved through the application of the methodologies (typically, do you want to estimate the impact on safety of certain changes by varying their probability of occurrence or do you work on the side of limiting the expected consequences?).

Examples of qualitative techniques are those based on risk matrices or risk, 'What–if?' type analyses, or the Fire Safety Concepts Tree itself illustrated above and contained within the NFPA 550 publication.

Quantitative techniques include those based on statistical analysis (e.g. to select fire scenarios), event trees with their respective probabilities of occurrence (e.g. to calculate the probabilities associated with the occurrence of a given sequence of events) and probabilistic or deterministic mathematical fire models (e.g. to estimate fire losses) often contained within software packages.

In this book, examples of both are provided along with some case studies.

With reference to fire models, it is emphasised that although probabilistic models are considered by their nature to carry a certain degree of uncertainty (since they are state models in which the change from one state to another can occur with a certain degree of probability), deterministic models are nonetheless so, despite their name. This is due to the fact that deterministic models depend both on the goodness of the input data available, in the same way as probabilistic models, and on the approximations and assumptions made (both by the models themselves and by the user) with regard to the variables in play such as the geometry of rooms and buildings, the influence exerted by interior materials, and the mathematical models used to describe the activation and operation of firefighting equipment.

Depending on the way in which the probability of occurrence and the magnitude of the expected consequences are estimated (qualitatively or quantitatively), four categories of increasing complexity can generally be identified, to which a fifth must be added that emphasises the costs and benefits of the various solutions implemented:

qualitative methods, i.e. all those that employ exclusively qualitative techniques for risk assessment;
Semi-quantitative methods for assessing the probability of occurrence, which address only the latter aspect with numerical estimation techniques, e.g. through the use of probabilistic event trees;
Semi-quantitative methods for the assessment of consequences, which only address the latter aspect with techniques capable of providing a measure of damage, e.g. through the use of models to represent the evolution of a fire, and the consequent estimation of consequences, through the

calculation of parameters such as the extent of the fire, smoke generation and ambient temperature;

quantitative methods, i.e. all those that employ exclusively quantitative techniques for risk assessment in both aspects;

cost/benefit assessment methods, which also provide an assessment of the expected costs associated with a particular fire safety project; these costs include both the costs of installing and maintaining firefighting equipment and the expected losses in terms of damage to the structures of buildings subject to fire risk and damage to the property contained therein; these costs complement the fundamental assessment of existing risks to people, an assessment that is always carried out for the other types of assessment methods as well.

The choice of method to be used in risk assessment must be guided by considerations regarding: the operational scope and the objectives to be achieved (the type of objectives may either allow the use of qualitative methods or, on the contrary, force the adoption of more complex methods); the regulations and laws in force; precedents (if certain methodologies have proven to be particularly suitable in certain areas, they will be reused subsequently); the degree of uncertainty that exists; the capabilities of those who must carry out the assessment process; and the target audience. The methods used must in any case possess certain important basic requirements, which can be summarised as easy availability (non-public or poorly documented methods are to be avoided), quality, judged on the basis of conformity with fire engineering methods and applicability to the specific purposes of the analysis.

The documents accompanying the risk assessment must explicitly state the method used and the reasons supporting the correctness of the choice in order to achieve the objectives of the assessment process. A brief description of the solution method and the numerical calculations used should also be provided.

With regard to the data used with the selected method, it becomes necessary to also indicate how they were chosen, from which sources they were taken, what, if any, are their limitations and what assumptions were made about them or other parametric default values used.

It is evident that all this information is not requested solely in order to achieve a documental completeness that may be useful to those who will later have to deal with fire safety from a practical or design point of view: it is explicitly required by NFPA 551 precisely to enable the authorities with jurisdiction to complete the verification of the quality of the fire risk assessment process adopted.

The goodness of the evaluation performed is a function

of the input values;

of the methodology employed;

of the analyst's competence.

It is therefore crucial to make sure that a given method is appropriate for the fire risk analysis of a given situation.

It must also be remembered that, notwithstanding the skill of those who carry out such an evaluation and notwithstanding the quality of the evaluation processes put in place, all the methods, models, hypotheses formulated and data used cannot fail to be characterised by a certain level of uncertainty that renders the results obtainable less certain; to take this aspect into account, the NFPA Standard 551 requires that the evaluation document quantify this degree of uncertainty and be capable of providing a (cf. Art. 4.5) *reasonable assurance that the acceptance criteria are nevertheless met*, even considering the worst case permitted by such inherent variability.

The degree of uncertainty can, of course, be more or less easily quantifiable, depending on the number of both the variables involved and the fire protection concepts or systems that need to be evaluated. It should be remembered in this regard that the most complex fire assessments are those that must consider the impact on the overall level of risk of a set of changes made to any one of the protection systems (meaning both active and passive systems) or factors such as alarm systems present, type of building occupants, training provided to personnel, etc.

It may be useful to remember that uncertainty is divided by the Standard into two types, the epistemic uncertainty, attributable to the very structure of the models used to represent a fire event, and the aleatory uncertainty, attributable, for example, to the behaviour of individuals in the face of the fire event.

The natural scope of the fire assessment is the fire scenario, which must, however, be fully characterised by a series of distinctive elements: the conditions of initiation and development of the event, the way in which the products of combustion are spread, the possible reaction of people and authorities affected by the event and the possible final effects of the combustion process. The characterisation of a scenario must be conducted in a precise manner: any approximations introduced in the characterisation phase or in the phase of assessing the contribution made to the overall fire risk, while permissible, must be fully justified in relation to the context to which they apply. This is justified by the fact that, although it has been illustrated above that it is not possible in practice to exhaustively examine all possible hazards and scenarios, it is important to prevent the risk of certain fire scenarios from being erroneously minimised because, by definition, the fire risk to be assessed is the sum of the risks associated with each possible scenario involving losses.

The standard dedicates an entire article to illustrate how (see Art. 5.1.4) 'depending on the problem under consideration and the objectives of the Fire Risk Assessment, the method employed may need to explicitly evaluate the effects of an alternative design on each of the events in the fire scenario in order to assess the overall risk associated with the alternative'. The following are examples for a typical fire scenario related to human safety:

1) *Start of fire: This is often based on the most likely event defined by a set of circumstances, such as a fire started in a living room by a cigarette butt. Prevention education would reduce the likelihood of this event and the resulting risks.*

2) *Fire growth: This is based on all probable developments in a fire, from combustion without flame development to the occurrence of flashover. Fire protection systems such as sprinklers, compartmentalisation and door closing mechanisms can help to contain these fires and reduce the resulting risks. Reducing the level of risk depends on the reliability and effectiveness of fire control systems.*

3) *Smoke diffusion: This is based on the diffusion of smoke to critical escape routes and to other parts of a building. Fire protection systems, such as smoke control and pressurisation of stairwells, can help contain smoke and reduce the resulting risks. Reducing the level of risk depends on the reliability and effectiveness of smoke control systems.*

4) *Occupant exposure: This is based on the blocking of escape routes by fire and smoke. Fire protection systems such as fire alarms, voice communications, well-marked escape routes and refuge areas can help provide occupants with early warnings and guide them to evacuate the building or seek refuge in certain areas. Reducing the level of risk depends on the reliability and effectiveness of the warning and evacuation systems.*

5) *Non-response of the fire department: It is based on lack of response or delayed response. A proper notification procedure and adequate fire department resources would help to free trapped*

occupants or control the fire. Risk reduction depends on the reliability of the notification procedure and the adequacy of resources of the fire department.[14]

The NFPA Standard 551 deals with the technical review process that a fire risk assessment undergoes and the definition of the documentation that must accompany it in order for it to be reviewed, implemented by retracing each analysis step and assessing the congruence between input data and the findings of each step.

Since, in some cases, it is the authorities with jurisdiction that have to review the assessment, they have the task of verifying the availability and quality (e.g. the validity of the reference criteria used, the statistical reliability of the data and the elements of uncertainty that characterise them) of all the information that was used to carry out the assessment, whether this consists of data from technical literature, technical drawings and documents or output data obtained by computerised calculation methods used in the process, etc.

Specifically, the authorities must judge whether the data used allow for subsequent verification, also taking into account that changes easily lead to the need to revise the measures put in place and, consequently, to re-examine the data already used in the past. In order to facilitate this requirement, all data used should be easily available (i.e. public or private but easily accessible) and well catalogued. In addition to this, it is obvious that the data must be appropriate to the situations under consideration and the methods to be applied since inevitably each context presents specific problems and risk profiles and may require a different level of investigation and validation (e.g. in the case of critical facilities, such as nuclear power plants or oil installations).

Less obvious, but equally important, is the attention to be paid to the cultural and geographic context in which one operates, as the Standard itself emphasises that data on equipment reliability or maintenance frequency may heavily depend on the climatic conditions in which one operates (operating temperatures, if particularly cold, influence failure assumptions, during the development of the risk analysis), the economic value attributed to losses resulting from fires (where less importance is attributed to damage, it is very likely that the relative data available are of a lower number or quality) and the amount of maintenance carried out in a company (lower economic availability leads to less maintenance activity and consequently to an underestimation of the data relating to failures).

In all those cases where data are derived from statistical observations (e.g. by sampling) or where data from scenarios or studies that are not strictly related to the case being assessed are used, it is

---

14 *Fire scenarios*: Depending on the defined problem and FRA objectives, the FRA method may need to explicitly assess the effect of a design alternative on each event in the fire scenario in order to assess the risk associated with the alternative. The following are examples for a typical life safety fire scenario:*Fire ignition*: It is often based on the most probable event in a particular setting, for example, cigarette and ignition of a cough in a living room. Prevention education would reduce the probability of occurrence of this event and the consequential risks.*Fire growth*: It is based on all probable developments of a fire, from smoldering to flashover fires. Fire protection systems such as sprinklers, compartmentation and door closers may help to contain these fires and to reduce their consequential risks. The reduction in risk depends on the reliability and effectiveness of the fire control systems.*Smoke spread*: It is based on smoke spread to critical egress routes and other parts in a building. Fire protection systems such as smoke control and stairwell pressurisation may help to contain the smoke and to reduce its consequential risks. The reduction in risk depends on the reliability and effectiveness of the smoke control systems.*Exposure of occupants*: It is based on smoke and fire blocking egress routes. Fire protection systems such as fire alarms, voice communication, clear egress routes and refuge areas may help to provide early warning to occupants and to direct them either to evacuate the building or to seek refuge in certain areas. The reduction in risk depends on the reliability and effectiveness of the warning and evacuation systems.*Failure of fire department to respond*: It is based on no response or late response. Proper notification procedure and adequate fire department resources would help to rescue trapped occupants or to control the fire. The reduction in risk depends on the reliability of the notification procedure and the adequacy of fire department resources.

important to specify how and when such data were derived: in this way it will always be possible to assess the correctness and relevance of any initial assumptions, extrapolations made and approximations or manipulations introduced. Such alterations in the data can easily result from not observing events or samples that are considered insignificant, such as minor incidents, failures that do not lead to significant consequences and events that remain below the observers' threshold of interest: all these phenomena lead to alterations in the frequencies of occurrence of the observed data, which therefore require adjustments, before they can be used to make a correct fire risk assessment. The same computational tools (typically modelling software and related input data) chosen for fire scenario simulations may rely on assumptions and approximations that must be verifiable by the authorities having jurisdiction.

If an approach based also on cost–benefit analysis is used, a complete and detailed assessment of costs obviously becomes essential, capable of analysing not only the costs of designing and implementing firefighting measures but also those resulting from proper verification and maintenance over time, as well as those that can be estimated in terms of human lives and occupant safety or economic losses, associated with any losses resulting from the occurrence of a fire.

In summary, with reference to the methods of analysis used, it should be reiterated that these must prove to be correct and correctly employed: the consequences and objectives of the methods must be measurable, the elements of uncertainty must be contained and in any case identified and documented and the basic assumptions for fire risk assessment (when present) must always be made explicit.

NFPA 551, for the reasons already mentioned in this same section, attaches great importance to the documentation that accompanies a fire risk assessment and takes care to detail all the types of information that must be included in the final documentation: in order to be considered complete, this must include operating and maintenance manuals, analysis documents and fire design summary diagrams.

The documentation should also cover the scope, purpose, entities and persons involved, the scenarios (including the reasons that may lead to the exclusion of some of them or the grouping of others into clusters of scenarios) and the objectives and performance criteria of the evaluation.

Other crucial elements of the documentation concern the methods and results of the analysis, the limits of its validity, the risks identified, information on how it was conducted, the verification and validation of the methods used and the calculation models, the prerequisites assumed, the points left unresolved, the manuals provided to accompany the analysis and the corrective actions taken.

The type (qualitative, semi-quantitative, etc.) of analysis method chosen to conduct the risk analysis determines both the results that can be obtained and the limits to which they are subjected, and this indirectly influences the type of documentation that can be attached to the assessment. In the latest versions of the Standard, this aspect has been emphasised and expanded compared to the previous version.

An important part of the documentation is undoubtedly that relating to the acceptance criteria that will be used in order to establish the risk tolerance level. Since acceptance criteria are a key element in verifying the adequacy of existing fire safety structures and measures, their determination must be made as early as possible, in the initial stages of planning, must be adequately documented, and must serve to facilitate the decision-making processes that will be necessary following the conclusion of the fire risk assessment and possible future changes.

It should be noted that (cf. Art. A.7.2.4) 'documentation of the assumptions made during the definition of the required performance ensures that future changes can be captured. These changes (such as changes to specific maintenance procedures), which could inadvertently alter key elements or features critical to the performance of the building and its systems, must be taken into account in order to maintain the existing level of safety in the face of the deteriorating changes'.[15]

Every fire risk analysis has its limits of validity since changes that may subsequently be made to any of the factors (in the broadest sense of the term, thus also including standards, legislation, the behaviour of the people involved, technological progress and organisational changes) that contributed to the outcome of the analysis may be such that they render it out of date. This problem is addressed by precisely documenting the initial validity limits and subsequent changes, which can be achieved through periodic inspections and a continuous process of change detection, unless, as one would logically assume, it is preferable to inhibit some of these changes in advance, e.g. through appropriate training of those involved, so as to prevent the overall fire risk from increasing.

Authorities or third parties (e.g. independent auditors, insurance company technicians, etc.) involved in reviewing the fire risk assessment must judge whether all the assumptions made during the risk analysis process reflect the conditions actually present at the locations subject to

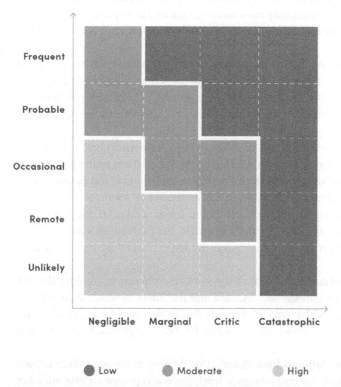

Figure 6.5 The Risk Matrix Method applied to fire risk.

---

15 Documentation of the assumptions made in deriving the required performance ensures that future modifications can be captured. These modifications, which could inadvertently change the key elements or features critical to the intended performance of the building and its systems, such as changes in specified maintenance procedures, must be accounted for in order to maintain the level of safety before the implementation of the detrimental modifications.

assessment; at the same time, they must review the method chosen to conduct the analysis by validation (e.g. by comparing the results with other similar models, with test data, etc.) or verification (which requires a proper demonstration that the model used always produces consistent and predictable results).

The auditors' fundamental tool always remains the documentation supporting the risk assessment, on the basis of which they are required to decide whether the scope, purpose, objectives, methods used, performance criteria chosen, outcomes and validity limits of the analysis, and corrective actions to be taken after its conclusion have been fully addressed, identified and documented.

6.3.1 The Risk Matrix Method Applied to Fire Risk

In Appendix A of NFPA 551 (more precisely, in paragraph A.5.2.5), the Hazard Matrix Method applied to specific fire risk assessment is extensively explained.

This in the light of the fact that this methodology is in fact the one applied in the vast majority of valuations (also due to the immediacy of understanding the underlying graphical notation). Moreover, since it is often associated with qualitative estimates, a certain degree of skill and

Table 6.1 Probability levels.

Probability	Description
Frequent	Likely to happen frequently, with a probability $p > 0.1$
Probable	It will occur many times during the life of the system ($p > 0.001$)
Occasional	Unlikely to happen during system operation ($p > 10{-6}$)
Remote	So unlikely, that it can perhaps be assumed that this danger will not materialise ($p < 10{-6}$)
Unlikely	The probability of this happening is indistinguishable from 0 ($p \sim 0.0$)

Table 6.2 Severity categories.

Severity	Impact
Negligible	The impact of the losses would be so insignificant as to produce no significant consequences.
Marginal	The loss will cause an impact on the plant, which may have to suspend operations briefly. Some cash investment may be required to restore the plant to full operation. Minor personal injuries may occur. The fire may cause localised environmental damage.
Critic	The loss will cause a significant impact on the plant, which may have to suspend operations. Significant investment may be required to restore the plant to full operation. Personal injury with possible loss of life could occur. The fire could cause significant but reversible environmental damage.
Catastrophic	The fire will either cause the loss of one or more lives, or have a disastrous impact on operations and lead to a long or permanent shutdown. The plant would immediately cease operations when the fire breaks out. The fire could cause irreversible and significant environmental damage.

caution is also required on the part of the verifiers in ascertaining the conditions that led to certain considerations.

The full description of the method is given below, quoting the text of NFPA 551 in full, while an example of fire risk matrix is given in Figure 6.5.

'The Risk Matrix Method, currently documented as MIL-STD-882D, was developed in the 1960s as a system safety technique, specifically for military systems. Following this approach, each hazard is assigned a probability level and a severity category'. Tables 6.1 and 6.2 are adapted from the respective tables in MIL-STD-882D.

A Risk Matrix uses probability levels and severity categories to represent the axes of a two-dimensional risk matrix, as shown in Table 6.1.

7

Fire-Safety Design Resources

This chapter intends to present some technical standards widely known and used worldwide by fire consultants and experts to design fire safety. These standards integrate the requirements from the local fire safety regulation (in the majority of cases having a prescriptive essence). It is structured into four parts, depending on the organisations that authored the standards.

Presented documents are an invaluable reference since they can be applied, in their main principles, to a different range of sectors, being also effective for fire engineers at the beginning of their careers.

7.1 International Organisation for Standardisation (ISO)

The International Organisation for Standardisation (ISO) is an important and globally recognised organisation that deals with the definition of technical standards at international level.

The name ISO is not an acronym but draws its inspiration from the Greek word 'isos', 'equal': in fact, its founders, aware that acronyms are changed from country to country ('IOS' in English, 'OIN' in French, etc.), decided to uniquely define the organisation's name.

ISO was founded in London in 1946 by delegates from 25 nations with the clear intention of *facilitating international coordination and unification of industry standards*. The official seat was established in Switzerland, in Geneva, and became operational in 1947.

It currently has more than 18 000 regulatory documents to its credit, dealing with both traditional business, industrial and technological activities, and issues of 'good management' and service provision.

ISO members are represented by national standardisation bodies; as members, they have the right to participate in any ISO technical committee set up within which topics of interest to them are debated.

Commissions are mostly responsible for drafting, following the rules set out in the ISO/IEC Directive, Part 2, drafts of international standards, which they must then circulate to the commission members in charge of voting on these standards in order to approve them (the standard must achieve at least 75% of the votes to be approved).

Normally, other international organisations, both governmental and non-governmental, also take part in the work of the commission.

Commissions may also decide to publish, in cases of urgency, documents that cannot be considered as standards but still have a strong indicative value for markets and manufacturers; these

Fire Risk Management: Principles and Strategies for Buildings and Industrial Assets, First Edition.
Luca Fiorentini and Fabio Dattilo.
© 2023 John Wiley & Sons, Inc. Published 2023 by John Wiley & Sons, Inc.
Companion Website: www.wiley.com/go/Fiorentini/FireRiskManagement

documents are the 'ISO Publicly Available Specifications (ISO/PAS)', technical specifications resulting from agreements reached between technical experts (with at least 50% of the votes), and the 'ISO Technical Specifications (ISO/TS)', specifications resulting from agreements between the members of a technical committee (obtained with at least 66% of the votes). Each ISO/PAS or ISO/TS must be reviewed after three years to decide whether it can be renewed for a further three years, whether it should become a standard or whether it should be abolished.

ISO closely works with the International Electrotechnical Commission (IEC) for all matters concerning electrotechnical standardisation.

The IEC is a worldwide operating organisation for international standardisation in which all national electrotechnical commissions participate (for our country, the CEI). The IEC's main objective is the promotion, cooperation and dissemination at international level of standards, technical reports, technical specifications, etc. relating to the electrical and electronic fields. This activity is carried out through technical commissions and benefits from the agreement at international level with ISO, an agreement that leads to close cooperation between the two institutes on specific topics of common interest.

It is significant to point out that the initiatives conducted and underway within ISO on the subject of standardisation and standardisation in risk analysis have been judged by CEN at European level as a reference for our continent. This assessment was carried out on the side-lines of the 'CEN Conference on Risk Assessment' held on 16 June 2006 in Brussels and attended by representatives of numerous European countries, including Italy. At the conference, the need for a major harmonisation action in the area of risk analysis emerged.[1] This need was identified by conducting a specific survey prior to the conference, which was attended by 33 stakeholders divided into government agencies (37%), large companies (18%), standardisation and unification bodies (6%), consultants (30%) and other stakeholders (including research organisations, 9%), from 15 countries. The survey and the discussions held at the European conference emphasised the need for harmonisation and standardisation with regard to the following topics: risk qualification, combination of risks (natural and technological above all), risk identification activities, methods for estimating frequencies and consequences, presentation and communication of the estimated risk level. In fact, it appears that

the terms *risk analysis* and *risk management* are not well understood (apart from experience in specific fields, such as industrial and technological risk and food risk);
the fundamentals of the risk management process (identification, analysis, evaluation and treatment) are known in general terms but organisations are not able to apply the flow of risk management activities in a comprehensive manner to their projects, processes, products, etc.;
risk analysis activities are not carried out with due regard to operational experience and historical analysis even though the importance of proper management of accidents and near misses is well known.

CEN identifies the contents of the publications already prepared and being prepared by ISO as the reference level for the required harmonisation with regard to both definitions and the identification of a correct risk identification and management workflow. This harmonisation is of considerable importance since the use of congruent definitions and methodologies promotes comparison between areas of different applications of the concepts of risk, risk assessment, risk management, etc. CEN advocates the adoption of the documents already developed by ISO and in particular ISO/IEC

---

1 Vollmer G., Repussard J., Pirlet A., Brun-Maguet H., 'Standards and best practices in risk assessment. Needs and expectations of harmonization in the risk assessment field', AFNOR Normalisation, Département Développement, 17 January 2007, France.

Guide No. 73 and Standard No. 31000, both under development at the date of the European Conference, and now among the most prominent standards on the fundamentals of risk management.

The following sections describe the most important ISO documents on fire risk analysis and management, hazard identification, selection of reference scenarios, and linking and coordinating fire engineering activities with preliminary risk assessments.

7.1.1 ISO 16732

The technical specification ISO 16732-1, 'Fire safety engineering – Guidance on fire risk assessment', provides the basic conceptual elements of conducting fire risk analysis.

Since the elements set out in the specification have the fundamental characteristic of being applicable to all types of fire scenario, it is extremely useful to have a single framework that provides the cues for both fire risk classification (understood as quantification of the level of risk) and for correctly interpreting the risk associated with fire incidents.

The risk level interpretation phase is fundamental not only to immediately verify the acceptability of the calculated risk with respect to a pre-established value, but also to be certain that the risk class identified (or, in the case of quantitative approaches, the numerical value that represents it) is congruent and comparable with the reality that one intends to examine. The importance of the risk level interpretation phase is also evident when trying to understand the benefits expected, in terms of risk level reduction, from the introduction of technical and/or organisational-managerial measures aimed at improvement, whether they are preventive or protective.

Guidelines or technical specifications with general applicability, such as ISO 16732, become a useful tool for both analysts and authorities with jurisdiction: for analysts, obviously with due customisation, they allow them to consistently use the same approaches to several cases and this allows them to obtain common references between cases derived from different realities, for example; for authorities, on the other hand, they provide the possibility to view and consider analyses with the same evaluation method.

The technical specification presents an extremely extensive list of definitions that, given the international nature of ISO publications, can be considered official for all fire risk analysts. Some of the most important definitions are those associated with the terms 'engineering judgement', 'fire risk' and 'fire risk assessment'.

Fire risk, in particular, whether it relates to one or more fire scenarios, is a measure that is in most cases estimated by considering the probability of occurrence of an event (or scenario) and the expected consequences related to it; the risk may possibly be associated with a fire design in order to assess, with the same methods, alternative fire prevention and/or protection solutions that may be a set (pool) of technical measures and specific technical-managerial measures for the mitigation of one or more fire scenarios.

Given that a 'fire scenario' is defined by ISO itself as a 'qualitative description of the course of a fire, possessing key time steps that characterise the fire and differentiate it from all other possible fires', the technical specification places particular emphasis on the grouping of scenarios that possess homogeneous characteristics. Such groupings, known as 'scenario clusters', as also defined by other guidelines such as the SFPE, are useful for 'isolating' scenarios that can be considered representative and reasonably conservative for the purposes of conducting fire risk analysis. The groupings of scenarios can then be graphically represented by special matrices (the 'fire risk matrix', the use of which is suggested by the ISO technical specification) that correlate the interval of frequency of occurrence and the interval of the magnitude of the expected damage underlying them.

The technical specification emphasises that conducting a fire risk analysis is important (obviously independently of any regulatory constraints specific to each country) in all those cases in which the characteristic spatial dimensions of a fire are not sufficient to quantify the severity of the consequent damage, as the 'negligibility' of the values assumed by the parameters typical of the deterministic approach is poorly suited to characterise the 'significance' of the effects. An example is the case in which the damage caused by a small fire that affected high-value assets rather than characterised by the need to be available at all times must be quantified.

The analysis is also important in cases where the frequency of occurrence of certain scenarios is low, but the consequences that may result are extremely significant. This includes all situations where there is a large number of vulnerable people, where there is the possibility of a fire spreading very quickly, or where there is a high fire load in critical areas such as the escape routes of a building or in areas with load-bearing structures.

It is important to emphasise that, according to ISO 16732, fire risk analysis is an essential tool in all cases in which fire-safety engineering with a strong deterministic character (hence the extensive use of predictive codes of fire evolution) cannot fully represent certain fire scenarios that are characteristic of specific realities, not even qualitatively. This situation typically occurs when the deterministic development of a limited number of scenarios is not sufficient to fully represent the totality of scenarios that characterise the overall fire risk in reality.

This is completely in line with what is stated in local regulations in different countries.

Analysis is also essential in cases where:

- the availability of certain technical preventive and/or protective measures is a key parameter due to the variability characteristics inherent in system reliability;
- the variability of the input data to the deterministic approach is such that the resulting data is equally variable so that it is not possible to fully represent all fire hazard situations potentially representative of a specific reality.

The latter case is easy to understand if one considers, for example, the propagation of a fire in a multi-storey building with a complex layout: unless there are specific compartments that can subdivide the environment into logical units with a complete classification of the fire risk that characterises them, a deterministic analysis of the evolution of a scenario is difficult to pursue. In a situation of this type, a probabilistic approach is more desirable, which can consider the high variability of layout conditions (state of doors and ventilation openings, availability of fire detection and extinguishing systems) as well as the possible high variability of vulnerability (presence of various types of occupants, differences in the number of occupants in consideration of time slots, etc.).

Indeed, in these cases, it will hardly be possible to conduct a safety assessment based solely on a deterministic analysis of all conceivable fire scenarios as the above parameters change.

The technical specification emphasises the concept of 'fire risk management', which must follow the risk classification obtained and which, subsequently, may also lead to a second, more in-depth assessment phase in the event of any specific peculiarities being identified to be subject to targeted control.

Managing risk means interpreting it, assessing its acceptability against a criterion that may be defined by law rather than by a decision shared by stakeholders and then communicated to all interested parties.

This risk management process is to be considered fundamental for the constant verification of the level of fire safety detected in the initial analysis, a level that may increase or decrease over time in consideration of any changes that may have occurred.

It is important to note that the ISO technical specification formally introduces the concept of 'treatment of the identified risk' with a different meaning from the previously mentioned management of risk over time. With 'treatment' of the level of identified risk, in fact, the technical specification designates that process of selection and subsequent implementation of measures that make changes to the level of risk or to the class of risk possibly associated with one or more compartments. The measures may consist of management-type changes, or technological implementations, which must necessarily be recorded and made the subject of specific design and subsequent review of the fire strategy.

In an innovative way, therefore, fire risk analysis, through risk treatment activities, is linked to all aspects of 'operational control', which is a fundamental element of a modern and effective safety management system.

The fire risk analysis proper (referred to as 'fire risk assessment'), on the other hand, consists of two main elements that are separable from each other, but both are necessary for a classification activity: the estimation of the risk and its subsequent evaluation.

The estimation phase involves determining the two characteristic elements of a risk, namely

the expected consequence, assessed with respect to the identified vulnerabilities; these vulnerabilities may be independent of each other since, as also provided for by the recent normative acts on performance-based fire engineering, it is possible to conduct analyses aimed at assessing the risk associated with different objectives; for example, one may wish to assess both the degree of safety of the occupants' escape in an emergency and the degree of fire resistance of the structures;

the probability associated with the occurrence of the event, or more appropriately its occurrence interval, with reference to all the groupable scenarios and the re-composition of the frequencies associated with them, possibly making use, as also suggested by other guidelines, of graphical visualisation tools such as EP and F-N curves.

The evaluation phase ('evaluation'), on the basis of the ISO/TS 16733 technical specification, must necessarily follow the risk estimation activity for a scenario, an activity that may possibly be completed by the deterministic analysis of its evolution.

The assessment, taking into account operational experience and the feasibility of technical improvements according to the 'ALARP' concept, must be based in the first place on a comparison with a suitably established risk acceptability criterion, possibly referable to a risk value already obtained for similar cases that can be taken as a reference.

The standard reports a further refinement of the concept of risk acceptability, namely the definition of three different 'zones' of risk acceptability in a graphical representation (of risk matrix) that describes the frequency of the scenario as a function of its magnitude, where:

- the first region (at the far left) defines the region of acceptability;
- the central part is the region where the risk is as low as reasonably possible (ALARP);
- the third region (right-hand end) defines the unacceptability of the assessed situation.

It is equally true that if the resulting value fell in the ALARP region, it would not be clearly defined a priori whether the risk was actually acceptable or not, unless further criteria of acceptability were adopted, such as comparing the 'technically' feasible with the 'economically' sustainable in relation to a case under consideration. If a firefighting strategy is technically unfeasible, then it could be discarded; similarly, a strategy that disproportionately increases costs in relation to the benefits of its implementation should be discarded.

Necessarily, a strategy involving the opposite situation, i.e. a disproportionate increase in the magnitude of the risk against a reduction in the costs incurred for its implementation, should be discarded.

The ISO technical specification helps the risk analyst by providing several tools (such as the adoption of safety factors or margins of safety) to overcome uncertainty in the choice of discriminating criteria for the inclusion of different fire scenarios in the assessment, the identification and characterisation of which is given in the next section.

In order to be able to conduct a fire risk assessment, for all scenarios to be investigated it is essential to:

- identify fire hazards that may lead to a fire scenario under certain conditions;
- associate with the scenario the probability of its occurrence (taking into account the preventive measures in place) as well as its consequences (influenced by the mitigation measures in place), in view of the objectives of the analysis, i.e. the type of vulnerability to be investigated.

The preliminary analysis of the fire hazards actually present is fundamental and can be conducted according to various methodologies such as checklists, classification, comparison with similar realities already analysed, expert judgement, etc., but it must necessarily lead to the identification of the fire scenarios to be analysed in terms of risk (estimation and evaluation).

Especially when developed using a semi-quantitative method, such as hazard classification, indexing, etc., fire hazard analysis is the main step in determining scenarios that are candidates for further investigation through deterministic consequence analysis (application of performance fire engineering concepts using specific algorithms or calculation codes). All fire scenarios that are identified as credible and appropriately representative must necessarily be precisely contextualised, indicating at the beginning of the analysis what are the main assumptions made for the purpose of conducting the analysis. These contextualising assumptions must be as measurable as possible so that the battery limits of the study can be clearly identified and subsequently developed in accordance with them.

The assumptions made constitute, together with the design specifications of both the building and the fire strategy (technical and organisational-management measures) and the definition of the objectives, the description of the reference context within which fire scenarios are developed and analysed. It is only through the explication of these concepts that the fire risk assessment becomes valid: the risk analyst exposes the boundaries of his or her work and consequently declares its scope of validity.

In view of the strong orientation (also for historical reasons) of the publication to the problems of fire in civil buildings, it states that an element that can significantly influence the estimation of the risk of a scenario can be the type and number of occupants. With reference to this subject, it is requested that for the analysis to be valid, precise behavioural scenarios ('behavioural scenarios') must be developed, better characterising the fire scenarios, which take into consideration the variability in terms of number of occupants, occupied areas, hypothesised behaviour in the event of an emergency, etc. The latter, with regard to the deterministic definition of the emergency evacuations scenarios, assumes that evaluations are carried out regarding exodus in emergencies, considering alternative configurations in terms of occupants, in all cases in which this performance objective plays a fundamental role, such as in hospital facilities, in places with high crowding such as amusement parks, railway stations, airports, etc.

The technical specification emphasises that for a risk analysis to be useful, both immediately and as a tool for planning improvement over time, the number of scenarios must be congruent with the degree of detail that the reality to be analysed requires.

In addition, it is required that the scenarios selected are actually such as to represent, as fully as possible, the fire risk so that it can be stated that the risk associated with the reality taken into consideration coincides with the risk re-composition of the individual scenarios analysed.

Obviously, it is unlikely to assume that all conceivable fire scenarios will be developed in a deterministic analysis phase.

The scenarios that it is not convenient to develop further, also for the purposes of optimising the time required to conduct a detailed deterministic analysis, clearly include all those that present a negligible risk with respect to the predefined risk acceptability criterion: these scenarios, in fact, do not contribute 'de facto' to determining the 'global' fire risk level of the compartment, unit or building/reality considered.

In order to facilitate the selection of scenarios, they can be grouped according to their own particularities and brought together within homogeneous ensembles for one or more characteristics (positioning, type of fuel, type of ignition, speed of development, probability of propagation and overcoming expected barriers, etc.). These ensembles can then be identified with a single reference scenario. In addition to this type of a priori grouping (i.e. carried out prior to conducting the risk assessment), it is also possible to think of a posteriori grouping to be implemented only after the assessment is complete (or after any deterministic investigation), on the basis of the homogeneity or affinity of the consequences associated with the various scenarios.

A fire hazard arises in all situations in which a given set of conditions can lead to the development of a fire, regardless of its magnitude and dynamics. A fire hazard may also lead to different fire scenarios in consideration of the variation of one of the conditions constituting the set.

The scenario is a qualitative and/or quantitative representation of a fire phenomenon that may occur in the event of a fire hazard if the underlying conditions are positively verified.

The fire scenario must be fully described for all the phases that compose it and that are essentially referable to four main moments: ignition, development (with reference to the mode and speed), maturity and decay. In addition, the scenario must be accompanied by the conceivable events that may affect one of the above phases (e.g. the presence of a mitigation measure, such as the activation or non-activation of active protection systems, the presence and guarantee of compartmentalisation, etc.) and by the boundary conditions (positioning, presence and characteristics of the occupants, specific vulnerabilities, etc.) that may influence the scenario.

In listing the scenarios, it is desirable for the analyst to precisely define the 'main' scenarios and those derived from them through variations in the above-mentioned conditions; furthermore, it is advisable that, for each scenario, he or she prepare as objective and quantitative a description as possible so that a possible third party can 'measure' the boundary conditions employed and verify not only 'how', but also 'how much' the scenarios vary from each other.

This third party may possibly coincide with the authorities with jurisdiction operating both in the authorisation phase and in the post-incident verification phase.

The specification emphasises that the analyst must produce an adequate justification of the selection of scenarios made and of their possible grouping. In particular, if he or she excludes from the analysis certain scenarios or groups of scenarios that are considered to be scarcely probable or in any case bearers of insignificant consequences (and therefore possess a negligible level of risk), he or she must provide evidence and justification for this in order to show how the preliminary fire hazard analysis has considered all possible cases.

Furthermore, the analyst, in developing the entire analysis, must make considerations regarding each scenario in such a way that they can be traced through his or her writings and that the scenarios considered can always be said to be representative and conservative. The suggestion to use appropriately conservative approaches in the risk assessment is formulated by the technical specification itself as the possible over-estimation of the risk could lead to compensate for the possible exclusion, from the group of those actually considered, of certain fire scenarios that would have contributed to a higher overall risk classification.

The use of a conservative approach for a portion of the analysis or for the analysis as a whole should also be explicitly stated, quantified and justified.

The ISO technical specification under review indicates the modalities for conducting both the estimates concerning the probability to be associated with each selected accident scenario and those concerning the expected consequences of the same. These methods are essentially attributable to the use of data from literature or related to operational experience, the use of specific models and calculation methods, and the use of expert engineering judgement.

It is up to the analyst to select the most appropriate method; however, as far as the alleged simplicity of expert judgement is concerned, it is important to emphasise that expert judgement cannot be a tout-court consideration: it is only valid if it is systematic and consistent, and therefore only viable, and therefore only if it possesses the typical characteristics of Delphi methods.

The ISO specification, in accordance with what has been indicated by all risk analysis publications (e.g. as seen for the Fire Safety Concepts Tree of NFPA 550), also emphasises, with regard to estimating the probabilities associated with an accident scenario, the independence between protective measures and causal events. It is essential for the risk analyst to be able to identify the 'common causes of failure' in order to always guarantee the independence of the treated events or of the preventive and protective measures: the violation of independence may lead, in fact, to a significant underestimation of the level of risk associated with a given scenario. It is therefore clear that the reliability of preventive and protective measures is therefore of great importance for the fire risk analyst and must be assessed, also using models, in a precise manner.

The estimation or characterisation of expected consequences is perhaps, despite the tools available today, a more complex activity than probability estimation as it is influenced by a wide variety of factors (some of which can only be determined probabilistically). Furthermore, it may happen, as is frequently the case, to overestimate or underestimate the fire risk in terms of expected consequences. For this reason, it may be useful not only to conduct in-depth or deterministic evaluations but also to define descriptive qualitative ranges of expected consequences to classify the outcomes of a possible fire scenario, in order to also organise the scenarios by means of a qualitative scale (as anticipated in the previous section on scenario grouping).

The ISO technical specification identifies the event tree technique and the fault tree technique as two methods suitable for defining a risk value associated with a fire event.

Where it is not possible to conduct a fire risk analysis with a degree of depth that can benefit from the use of well-established methodologies in risk analysis, it is always possible to conduct a simplified 'risk calculation' and its subsequent presentation simply by identifying reference categories for both the probabilities associated with each event and the expected consequences of that event. When such categorisation is possible, for both probabilities and consequences, risk matrices become a natural representation of summary fire risk: probabilities become the matrix rows and consequences become the columns (or vice versa) while the cells express the measure of risk. The risk acceptability criterion can easily be imposed on the category matrix by defining the combinations to determine which scenarios can be reasonably considered sustainable, given the measures present and the boundary conditions imposed.

Categorisation proves to be an excellent tool during the design phase of the fire strategy (i.e. during the selection and evaluation of prevention and protection measures) when insufficient time is available to conduct in-depth investigations (both with regard to probability and consequences), especially if these are to be carried out using models.

Since the fire risk analysis presented in the ISO technical specification implements the use of consolidated fire risk analysis techniques, it is desirable to use, in accordance with other guidelines, graphical tools to better communicate the estimated fire risk: in addition to the scenario matrix already mentioned, the available tools include risk curves, both F-N and EP types (as better defined by the guideline prepared by the SFPE).

As suggested by ISO 16732, fire risk presentation methods should then be used not only for the purpose of communicating the risk classifications associated with certain areas, compartments or zones (in consideration of the fire scenarios selected for them from the hazard analysis) but also for the purpose of better identifying the risk associated with the different existing protection alternatives. In this sense, the results of the risk analysis are directly used as an input for the selection of the best firefighting strategy in consideration of the hazards that are representative and appropriately conservative of the reality considered.

The technical specification defines the guidelines for conducting an assessment of the level of risk obtained as a result of the estimation phase against an acceptability criterion that must be predetermined and, as far as possible, quantified whether it is defined from operational experience, or by the authorities having jurisdiction, or simply from the application of 'ALARP' concepts (often coinciding with the highlighting of specific areas on a matrix representation of fire risk based on categories and essentially connected with an area of acceptability, an area of desirable intervention and a third area of unacceptable risk).

In the event that some scenarios highlight an unacceptable fire risk, it is implicit in the need for the analyst to identify suitable interventions to mitigate these scenarios. If appropriate technical and organisational/management measures cannot be implemented for some of these, the non-mitigatable fire scenarios should be compulsorily reconsidered, as they are, in fact, unacceptable and therefore underestimated when developing the corresponding fire risk analysis.

It is useful to highlight that the ISO technical specification foresees the possibility for the analyst to adopt safety factors (multiplicative) or safety margins (additive), compatible with the maximum uncertainty generated both in the estimation of the probability of occurrence and in the estimation of the expected consequences, within the risk classification activity. The use of such safety factors should be aimed at compensating for the uncertainty inherent in fire risk classification.

Paragraph 7 of the ISO standard discusses the phenomena of uncertainty, sensitivity, accuracy and distortion of results, referring to any potential difference between the risk value computed by the analysis and the actual risk value that the former is most likely to represent. Uncertainty derives not only from considerations of statistical variation in the results, but it is the result of errors or shortcomings in the representation of the risk scenario (and its magnitude/probability of occurrence), such as the failure to reproduce a phenomenon in the simulation of the fire event (e.g. the turbulent motion of flames) or the exclusion of human factors from the calculation of the evacuation time from a building. Uncertainties of these kinds that can be associated with the field of fire engineering are taken up in detail by another technical standard, namely ISO 16730.

Finally, ISO 16732 reports how, for the purpose of validating the results, i.e. for the resolution of the analysis on the uncertainty of the results, the criterion of expert judgement is, in the absence of observations (data), applicable, according to a peer-review approach, in the absence of a quantitative validation method.

The ISO/TR 16732-2 presents a complete example of fire risk analysis developed according to an index reflecting the numbering of the sections and subsections that make up the standard, thus allowing the development of the example to be directly related to the statements in each section.

The case study presented here, concerning a 40-storey office building with approximately 12 000 occupants, is extremely well known internationally. Developed in 1998 by Yung, Hadjisophocleous and Yager, on behalf of the Canadian National Resource Council, through the use of a major fire risk analysis tool (FiRECAM©), it represents, also in view of the year of publication, one of the high points of the application of fire risk analysis for the evaluation of alternative protection strategies with respect to an acceptable baseline, as well as the parallel conduct of a cost–benefit assessment expected in each situation.

In the case study, three different protection strategy alternatives are evaluated against a reference configuration determined by the application of all requirements derived from the prescriptive codes. These strategies were developed by varying the degree of compartmentalisation, the availability of automatic extinguishing systems, and the layout of the compartments.

Risk analysis and cost assessment are supported by the use of deterministic simulations of representative scenarios.

Two indices are calculated for each alternative option, known as the ERL and FCE, respectively. The first represents the 'Expected Risk for Life', the second, a function of ERL, represents the 'Fire Cost Expectation' and allows the evaluation of the economic benefits expected from the implementation of an alternative protection compared to full compliance with the prescriptive requirements. Therefore, in the case study presented in Technical Report 2 of Standard 16732, the ERL index corresponds to the estimated number of human losses per year as a result of fire in the building, while FCE is the total estimated value of loss as a result of fire, a value that includes the costs associated with the initial capital for the implementation of active and passive fire protection measures and the costs of maintaining these systems over time, such as inspection and periodic maintenance.

It is important to emphasise that separating the risks for occupants from the costs of protection directly eliminates the problem of assigning a monetary value to the loss of life (typical of some analysis approaches conducted for insurance purposes) and allows for a separate comparison of risks and costs.

The ERL index, with reference to the degree of safety associated with full regulatory compliance, is thus extremely suitable for determining the degree of safety with respect to fire of a protection strategy and evaluating equivalencies rather than the degree of achievement of individual predefined performance targets. The FCE index can instead be used for a parallel analysis of the costs associated with different proposed interventions.

Finally, it is extremely interesting to note that, among the alternative options that make up one of the proposed strategies, there is also the use of fixed extinguishing systems, with a higher degree of availability on demand: with the same discharge density (determined on the basis of the fire load and the type and layout of the fuels present in the building) an improvement in the degree of safety is defined, and therefore a decrease in the fire risk, by acting on a stochastic performance characteristic (an improvement that cannot be appreciated with a purely deterministic type of investigation).

Fire risk analysis, by its very nature, is the most suitable tool for assessing these aspects and for defining and selecting the input elements for deterministic investigation activities.

The ISO/TR 16732-3 section of the standard deals with the determination of the fire risk of an industrial production activity, specifically describing a propane gas storage facility and the associated operations of unloading the product from tank cars, storing the pressurised gas and loading the material for shipment by sea. The purpose of the application is to illustrate how the principles and evaluation logic of the 16732 standard can also be applied to industrial-type realities, limiting itself to the design phases of the facility and considering layout modifications at this stage prior to the actual production operations. The identification of fire scenarios and firefighting strategies, i.e. the risk reduction measures to be introduced, also refer to the design phase (so it does not consider, for example, a change in process operating conditions).

The analysis follows a precise workflow, firstly identifying the most probable scenarios that could lead to an accidental event (in the case study identified as the BLEVE of a pressurised propane tank).

In view of the fact that the number of fire scenarios distinguishable from one another is relatively too large for a precise analysis of each one, in the present case (but in a logic that can be extended to any analysis) the identification of the context, i.e. the influence of the layout of the propane tank loading/unloading area and the risk reduction measures to mitigate the frequency of occurrence of a pressurised storage tank BLEVE, is of paramount importance. The structure of the identified scenarios must be of a manageable size and at the same time be able to lead to a reasonable estimate of the fire risk related to the totality of the scenarios. The strategy proposed by the 16732 standard aims to achieve this by identifying potential hazards, combining scenarios into clusters and excluding scenarios associated with a negligible level of risk.

The 'Bow-Tie' representation, as shown in this book, is an effective way to identify the hazards related to the scenario examined; then the standard reports on several event tree structures, one for each fire strategy option examined, to estimate the value of fire risk, in terms of the frequency of occurrence of the top event (the BLEVE of the tank); finally, the definition of an acceptable risk level (value provided by relevant literature data) identifies which risk mitigation measures have a relevant effect for the scenarios examined.

7.1.2 ISO 16733

A critical step for a comprehensive, engineering approach to the discipline of fire safety is the identification of fire hazards and their underlying design scenarios.

This phase aims to provide a description of the level of risk that is an adequate and appropriately conservative and meaningful representation (also in terms of the credibility of assumptions and findings) of the fire risk present in the reality being analysed.

Given the critical nature of this activity, the ISO/TC 92/SC 4 commission considered it appropriate to develop a special publication defining the process of identifying and classifying fire risk for the purpose of selecting design fires; furthermore, with this publication it wished to emphasise the opportunity deriving from the use of risk analysis methods for the purpose of guaranteeing the significance and representativeness of scenarios that may be the subject of a subsequent deterministic type of investigation.

The technical specification ISO 16733-1, stating that the number of all possible fire scenarios in any given reality, such as a building, structure or means of transport, can be extremely high,[2] states that it is not possible to quantify all fire scenarios by analysis, unless it would consume a considerable amount of energy and possibly calculation time, as long as the analysis is also conducted in a deterministic manner for the assessment of the expected consequences.

For a deterministic type of assessment, scenarios must be selected in such a way that they can be considered representative and appropriately conservative. For a probabilistic risk assessment, on the other hand, scenarios must be grouped into structures so that, for each of these, the total expected risk can be calculated.

Since the scenario is characterised through the description of all the characteristic phases of a fire, both in terms of direct effects and indirect or secondary effects, in the sense of a 'domino effect', a risk assessment allows the outcomes resulting from all the possible accidental sequences that can be envisaged to be taken into account.

This ISO technical specification provides the key elements of a methodology for the selection of fire scenarios to achieve the set fire-safety objectives based on the risk assessment of scenarios (and

---

2 As stated in Section 5.1: *In reality, the number of possible fire scenarios in most building environments approaches infinity.*

thus the estimation of frequency and expected consequences) for different vulnerable elements, such as occupants, assets and property, the environment and business continuity, for which different scenarios may need to be assessed.

The fire scenarios selected, in accordance with the other ISO standards covered in this publication, must be able to be described quantitatively through design scenarios and characterised in terms of the variation over time of the quantities of interest characteristic of a fire (thermal energy released, development of CO, CO_2, other combustion products, etc.).

The technical specification supplements what is already defined in the Technical Report ISO/TR 13387 Part 2 'Fire safety engineering – Part 2: Fire scenarios and design fires'. In fact, since the ISO 13387 series publication is specifically addressed to approaches using deterministic analysis, the technical specification states that, when a deterministic investigation of the expected consequences is envisaged, a qualitative estimate of the probability and consequence, i.e. a minimum risk assessment, must also be carried out. If, on the other hand, it is necessary (or advisable) to carry out a full risk assessment, according to, for example, the requirements of ISO/TS 16732 and, above all, of the requirements that will most likely be introduced in the standard resulting from this publication, a quantitative estimation of the frequency and consequence must be carried out.

Based on the design scenarios, the firefighting strategy must be modified until a defined acceptable risk level is reached on the basis of an established criterion.

To achieve this, the analysis process must consider all scenarios (or their summary groupings) that are characterised by either a high degree of risk, a high frequency or a high expected consequence for one or more categories of vulnerable elements. In this way, it is possible to state that the most representative and appropriately conservative scenarios have been taken into account by the analyst.

The selection of scenarios may also take on an exclusively qualitative character. It is essential that there is a guarantee that the scenarios considered and subsequently excluded from the selection, in addition to not having a significant, or even unacceptable, level of risk, will not significantly alter the conclusions of the study performed. It would obviously follow from the possible alteration of the study that the selection made has erroneously excluded scenarios, perhaps because it excluded from consideration those events that, although they have low frequencies of occurrence, may lead to extremely significant expected consequences (which, as already widely reported, is not convenient in view of the possible perception of risk once the event has occurred).

The characterisation of scenarios and, subsequently, of design fire scenarios must not only take into account the typical information connected with a fire such as the mode of ignition, development, dynamics of effects, etc., but also all the information, especially stochastic information, that may influence accidental sequences, such as the effectiveness and efficiency of existing prevention and protection measures, both active and passive, their reliability and, therefore, probability of failure on demand, the distribution of occupants and the state of passage and ventilation openings.

Clearly, it is the appointed fire engineer who must define the risk analysis methods to be used for the selection of fire scenarios, in consideration of the reality to be examined, the degree of detail sought (possibly depending on the design status of the work in the case of assessments to be conducted as part of the design and construction process), the complexity arising from specific critical issues (layout, types of occupants, types of fire risk activities conducted, types of fuels present, etc.) and, of course, their expertise in carrying out these risk assessments.

Risk analysis is a discipline based on a set of structured, well-codified methodologies that have been widely used over the years, but only experience can validate the correctness of an approach.

It is advisable for the designer, if not familiar with risk analysis, to be assisted by an expert in this area who can support decisions by correctly prioritising fire risks based on a prior analysis of the expected frequency–consequence combinations for each scenario or cluster of scenarios and immediately verifying the findings against an acceptability criterion. Other possible approaches for identifying design fire scenarios that may be used include the following:

A list of predetermined scenarios relevant to the specific environment under consideration, which can be derived from standard codes or documents and defined as 'de minimis' scenarios: this simplistic and easily applicable approach may, however, lead to the underestimation of scenarios relevant to a single environment.

The application of a qualitative or semi-quantitative systematic approach in order to identify a set of credible scenarios for a deterministic analysis.

The selection of a cluster of scenarios to which defined probability and consequence values are attributed, based on techniques such as the construction of event trees, for a quantitative risk assessment. This strategy is effective when statistical data is available on historical fire events relating to environments similar to those examined although particular care is required when the probabilities of occurrence of scenarios are assigned in this way.

While for the first approach, reference is made to normative documents and codes, the reference standard for a purely quantitative approach is ISO 16732-1, discussed in this volume in the previous chapter, whose overall intent is to ensure that the chosen design fire scenarios include all credible scenarios, excluding those unanimously considered to be of acceptable risk.

The following section of ISO 16733 mainly focuses on the second approach.

The technical specification suggests the use of a method consisting of nine main steps; the numbers and types of steps identified, applied according to an extremely articulated but highly structured workflow, guarantee a complete examination of fire risks for the purpose of defining the design scenarios to be considered in subsequent in-depth studies:

Identification of specific safety issues.
Location of the fire (ignition point).
Type of fire.
Potential hazards generating other fire scenarios.
Fire risk control barriers.
Occupants' actions in case of fire.
Selection of project fire scenarios (or 'design fires').
Modification of scenarios based on availability/reliability of control measures.
Final scenario: Definition and documentation of the assessment.

Extremely useful for the practitioner who intends to apply the workflow proposed by ISO/TS 16733 are the informative annexes B and C that demonstrate, through practical examples, respectively related to a covered multi-activity stage and a production activity, how representative and credible scenarios of such realities can be determined through the 10-step methodology described in the specification.

The workflow envisages that both internal and external fire risks are taken into consideration. In the case of internal fires, emphasis is also placed on all those risks that are not normally taken into consideration, even though they are likely to have extremely significant consequences to fires that develop in certain situations such as maintenance, commissioning and decommissioning activities, or which take advantage in their evolution of particularities connected with the architectural layout and construction technologies that are to be introduced in the work (e.g. large vertical atria rather than double-walled curtain walls).

The identification of all the risks potentially carrying significant consequences is important because, for the purposes of characterising the type of fire, it is not possible to consider the description of the evolution of an outbreak in the first moments alone as sufficient: an in-depth investigation must be conducted into the accidental sequence that leads to the release of flammable product, the subsequent ignition (mode and probability of) and the evolutionary dynamics up to a stage that can be defined as 'stationary'.

A complete knowledge of the accidental sequences associated with a fire hypothesis is useful since the probability of occurrence, the states associated with the operation and malfunctioning of detection and mitigation systems, the probabilities associated with the time at which the ignition occurs, the probability of development in the various compartments, etc. (all considerations that can only be made through a probabilistic approach to fire risk) clearly define the types of accidental event arising and the expected consequences, both for the occupants, in terms of propagation, domino effects and secondary effects.

The potential fire hazard must be carefully assessed in consideration of both the characteristics of the substances likely to ignite and the boundary characteristics that may determine dynamics and/or consequences specific to the situation under consideration. An example of this is the presence of oxidants in such quantities that both the probability of ignition of a combustible and the dynamics of its development, as well as the probability of propagation, are significantly altered, rather than the presence of specific hazards associated with construction activities, in the case of a fire risk assessment on a building site, or maintenance activities.

In the systematic examination of the typologies, it is essential that the analyst identifies fire scenarios that, although apparently minor, may lead to undesirable indirect effects or have consequences for the safety of the occupants or property; suffice it to think of the unavailability of emergency systems for fire on cables in the industrial sector with the impossibility of securing the systems and the involvement in the fire of firefighting equipment and fixtures, resulting in their unserviceability.

It is clear that, downstream from the execution of the risk analysis, the planner must verify, with respect to each scenario identified, the secondary effects arising with particular reference to the possible involvement of means capable of coping with the event itself, thus proceeding to relocate rather than protect them.

Since knowledge of the accident sequence must fully represent all the possible evolutions and expected consequences of the fire phenomenon over time, considering the variability of certain elements, such as mitigation systems, the state of compartmentalisation, the state of ventilation and smoke management systems, the presence and response of occupants in the event of a fire emergency, etc., it is necessary to precisely identify these elements, assess their temporal positioning within the sequence as well as the variability and therefore the probability connected to different states. In this way, they can be properly taken into account when constructing fire scenarios.

Since both the variability of each element and the number of elements subject to variability are high, risk analysis, thanks to its methods, even the most classic ones, becomes a useful tool to select scenarios and their expected consequences on the basis of a risk classification criterion that operates on both representativeness and credibility of frequency and expected consequences for each scenario identified.

The selection of events for the purpose of scenario construction suggests two alternative methodologies as risk analysis techniques for ranking ('risk ranking'): the event tree and the fault tree.

Both the two methods are discussed in this book along with the 'Bow-Tie' diagram that, at a certain degree, can be considered the summary of both.

In particular, the fault tree is stated to be a valid method, used in combination with the event tree, for the definition of the conditional probabilities of the elements themselves considered: one thinks, for example, of the estimation of the probability of the failure to activate an active fire protection system from the data relating to malfunctions, failure rates and the time required for repair or periodic inspection of the system components.

The decision as to whether to employ a single method, rather than a combination of methods, is left, barring prescriptive constraints, to the discretion and expert judgement of the risk analyst who must make the choice in consideration of the appropriate degree of detail and the specificities characterising the reality under examination. Several methodologies may be jointly employed for the definition and subsequent selection of the risk level.

Combination is not as unusual an activity as one might think: taking the analyses used in industrial risk assessment as a reference, the methodologies that have been developed over the years to meet specific needs are increasingly being used in combination for better identification of accident scenarios associated with a given installation. An example of such a combination is shown in Figure 7.1, which shows the interaction between the Hazop/FMEA, fault tree, event tree and Layer of Protection Analysis (LOPA) methodologies.

Operability analysis (Hazop) in conjunction with FMEA analysis ensures that deviations from normal operation of a component-based system and the criticality associated with the failure (for different types of failure causes) of each component are identified, respectively.

The fault tree defines the frequencies associated with initiating incident events based on the reliability data of components identified as critical in relation to the concatenation of initiating causes and simultaneous failure to implement protective measures.

The LOPA analysis identifies, for each type of initiating event identified, the protection barriers present, which, called 'protection layers' in the original analysis, designate both protection devices

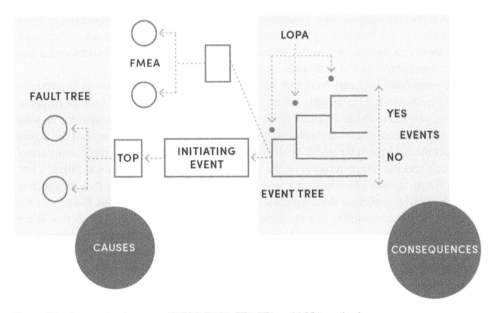

Figure 7.1   Interaction between HAZOP, FMEA, FTA, ETA and LOPA methods.

and administrative controls. In addition, it identifies the active and passive measures in place, their degree of availability (on demand) and the degree of independence existing between them, in order also to determine whether they are 'Independent Protection Layers', i.e. independent protection barriers, which are also necessarily distinguished by their reliability, ease of revision and the specificity of the danger they must contain.

The identification is conducted by LOPA with the aim of introducing into the event tree those key elements that are decisive for the evolution of the accident sequence. As a function of the probabilities associated with the states that the key elements may assume (generally functioning and malfunctioning), it then becomes possible to follow the evolution of the events up to the definition of the expected consequences for each alternative path.

The event tree, considering the time factor, analyses up to the expected consequences all conceivable accidental sequences concerning occupants, environment, property or business interruption. The differentiations inherent in the various incident scenarios that arise may also be used to classify them (by type of event, similar consequences, comparable degree of occurrence, etc.) in order to identify sets of scenarios based on one or more common characteristics.

The flexibility of analysis guaranteed by the combination of methods in order to better characterise accidental sequences has in fact become the strong point of some analysis techniques, which are nothing more than workflows aimed at exploiting the peculiarities and advantages of two or more 'traditional' analysis methods in a single evaluation procedure.

In this regard, it is useful to point out that the attempt to provide a better representation of the phenomena examined, also in order to be able to take into account, when describing accidental sequences, the peculiarities connected with specific initiating and promoting causes and with specific containment and mitigation actions, has led risk analysts to employ several methodologies at the same time and this, in turn, has led to the definition of real analysis tools based on the incorporation of these methodologies and the definition of specific original graphic notations to support these tools. One example of all is the Bow-Tie method.

In the application of one or more methodologies, the risk analyst is able, by means of a predefined acceptability criterion, to make possible simplifications, e.g. by not considering initiating or subsequent events characterised by a negligible frequency or by limited consequences, or by an overall classification of the risk that can be estimated as acceptable for the scenario considered. This takes the form of a process of selection of incident scenarios that must be documented and traceable by both stakeholders and authorities with jurisdiction.

It should be considered that certain design scenarios may be explicitly required by prescriptive codes or by authorities with jurisdiction to verify the building's fire performance response with respect to particular vulnerabilities (such as a certain category of occupants) or with respect to specific fire characteristics (related, for example, to its positioning within the layout rather than specific fire characteristics). With regard to vulnerability, for example, one cannot disregard certain specific characteristics of the occupants such as possible disabilities, a low degree of familiarity with the building, etc.; just think of healthcare facilities that, in addition to medical and nursing staff, host disabled patients as well as visitors of various kinds.

This attention to fire vulnerability and characteristics is emphasised in a special section of the technical specification where it is indicated how the factors such as the attitude to safety (from aspects related to education/information and training to the presence of a fire-safety management system), rather than the continuous presence of an internal emergency team or the proximity of a national fire brigade headquarters (understood as effectiveness and timeliness of intervention on call), are taken into account when conducting a fire risk assessment.

Following the identification of the fire scenarios that must be taken into account, the ISO/TS 16733 technical specification provides an extensive overview of the activities aimed at constructing the design scenarios representing the incident scenarios identified.

The construction of the design scenario is an activity based on the definition of the behaviour of the fire scenario through quantitative measurements of each of the phases characterising a fire, specifically incipient phase, development phase, stationary behaviour phase, decay and extinction phase, and based on the definition of the basic characteristics of each phase such as rate of heat release, rate of production of toxic and irritant species, smoke development, fire size, radiant energy and the resulting thermal regime.

These phases are discussed at length in a separate chapter, also in view of the fact that, in addition to the typical scenarios of civil buildings characterised by compartments, fire events that may occur in industrial-type installations are examined. Industrial scenarios, although different and slightly less subject to this classification of development phases, can be (and, in fact, are) treated using the same analytical approach for their risk classification.

7.1.3   ISO 23932

The standard ISO 23932 'Fire safety engineering – general principles' issued in its first edition in the year 2009 is a fundamental document for fire-safety experts and risk analysts. Prepared by the extremely active and oft-cited technical committee ISO/TC 92, Fire safety, subcommittee SC 4 'Fire safety engineering', it takes into consideration the recent incorporation of performance-oriented approaches to fire safety into the regulations of numerous countries.

Starting from the assumption that there are already reference ISO standards for conducting an engineering approach to the subject of fire safety based on the concept of 'equivalence' (suffice it to think of the corpulent ISO/TR 13387 standard developed by the same commission), this standard turns its attention to the unambiguous definition of the differences between a prescriptive approach and an approach oriented towards guaranteeing certain performances identified as objectives, and then indicating for the latter the requirements, methods and expected results.

Certainly, the ISO document is less 'descriptive'. However, it is not intended as a didactic or popularising tool but rather as a point of connection aimed at defining a workflow whose fundamental steps are detailed in other official ISO documents (standards, TS technical specifications and TR technical reports), such as the full-bodied and well-known ISO/TR 13387.

The contents of the ISO 23932 standard can be used by both the experts in charge of developing the analysis and those in charge of its verification, referred to as 'peer-reviewers'. With regard to these figures, the NFPA, limited to aspects related to fire risk assessment, decided to develop a real standard as a basis for conducting the verification of the analysis developed: the NFPA 551 document described in detail in a specific chapter of this book.

7.1.3.1   Scope and Principles of the Standard

The workflow defined in the ISO 23932 standard is intended to be of wide application: unlike most documents, guidelines and some prescriptive regulations, which are specifically designed for application to civil buildings characterised by 'compartments', this standard, as stated in its general principles, is intended to be a tool that can also be applied to industrial systems, means of transport and installations.

In analogy to what is also defined in ISO 16732, the ISO 23932 standard, in view of the critical issues connected with the definition of an acceptability criterion for fire risk, states that it is possible to use as an acceptability criterion complete compliance with the prescriptive type requirements defined by the standards applicable to the case to be studied.

It should be remembered that in other fields, such as industrial risk, the criterion of acceptability of the fire risk is either defined within standards or has been generally shared for some time between managers, experts and authorities in charge of verifying the analysis, or else it is defined in relation to the specific case through special methods, such as the one proposed by the AIChE in the well-known and very recent publication 'Guidelines for Developing Quantitative Safety Risk Criteria' of September 2009, developed by the Center for Chemical Process Safety. It is also worth remembering that ISO 16732, as reported in the specific paragraph, develops in Technical Report 1 a complete case study through a performance-oriented analysis that illustrates, according to different protection strategy alternatives, the risk, the expected consequences and the cost associated with the proposed measures, through a direct reference to the results that would be obtained from the complete application of what is foreseen by the standards related to the case study.

The 'equivalence', i.e. conformity to requirements, mentioned above must be 'measured' and this is not feasible by expert judgement alone. It is therefore necessary to define an acceptable level of risk on the basis of a documented criterion to be shared both with stakeholders, or interested parties, and with the authorities having jurisdiction, or to consider as acceptable risk the level of residual risk obtained by applying the requirements of a prescriptive code which, by definition, is congruent with the type of activity considered and whose requirements it expresses are defined with an appropriate margin of safety, according to the definition also taken from ISO/TS 16732.

It should be noted, however, that the parties concerned may implement 'voluntary objectives', i.e. define fire-safety objectives as more conservative than prescribed, possibly in consideration of historical analysis conducted on fires that have occurred in similar buildings and structures, historical experience and expert judgement of the analyst.

The engineering analysis, and consequently the assessment of risks arising from a fire (which is a fundamental step), must be developed so that the risks to the safety of people (occupants), the integrity of assets (the 'property'), the continuity of operations and the environmental risks determined by the direct effects of the fire are taken into account, such as the dispersion of fumes or the development of environmentally hazardous substances in combustion, rather than those of an indirect nature, such as the non-availability after fire of measures and installations aimed at protecting the environment and all those additional secondary events resulting from domino effects.

The ISO 23932 standard emphasises that the protection of historical and artistic assets ('heritage') is also fundamental. This is also in line with the historical experience that has always (but especially in recent years) seen a particular diffusion to buildings valuable for their art and history of both the tools and methods of firefighting engineering and the use of specific technical solutions for such environments such as finely atomised water technologies.

To each identified objective, the standard associates so-called 'functional requirements' (FR), i.e. definitions of how the predefined objectives are to be achieved, taking into account the specificities of the built environment.

The standard states the principle that a fire-safety analysis can only be fully developed and well documented if it is carried out as an integral element of the design and construction process and only if it is 'managed' over time, consisting of periodic inspections and specific procedures for ensuring fire safety, which are not to be understood as the procedures to be put in place for fire emergency planning but as the organisational and management activities, possibly codified through an appropriate fire-safety management system, aimed at maintaining the degree of safety achieved and the estimated acceptable fire risk for the reality.

In particular, the process must be put in place from the very beginning of the activities associated with the development of a project and must manage all aspects, including those relating to specific disciplines, that may have an impact on fire safety. These include, for example, security measures that may determine limitations on the safe exit of occupants in the event of a fire emergency or specific requirements for guaranteeing the energy performance of a building or for the management of a complex layout, consisting, for example, of curtain walls with ventilated walls or full-height atria crossing the buildings, elements that may determine conditions suitable for amplifying the consequences expected from a more strongly conveyed fire.

A perfect knowledge of the building or industrial installation that one intends to protect, as well as any modifications that may be necessary during the development of the project, is a fundamental requirement as an input to the fire risk assessment and the definition of one or more protection strategies.

Of course, risk assessment cannot disregard the identification of fire hazards, whether they are internal or external to the activity. Those of an internal type are characteristic of the reality being analysed and are more easily recognisable; those of an external type, on the other hand, are ascribable to natural events or neighbouring activities and have extreme relevance in the case of industrial-type risks, for the analysis of which it is always mandatory to consider the secondary effects on a unit, rather than on a complete installation, deriving from accidents that may originate outside, possibly in areas not owned (the so-called domino effect).

Existing risk analysis techniques (hazard identification and risk assessment) make it possible to develop activities that are congruent with the specific reality under study, characterised by a degree of detail appropriate to the state of the project (from Hazid or other preliminary methodology for hazard estimation to the development of non-negligible scenarios using deterministic methods) and representative of all the risks that may arise.

Hazard analysis is one of the fundamental steps in selecting fire scenarios, so it is crucial that it is conducted in a comprehensive, systematic and organised manner.

However, the use of risk analysis techniques should not be limited solely to the definition of the fire scenario but should be extended to the selection of design fire scenarios, whether the analysis is qualitative or quantitative. The choice of whether to carry out a quantification must lie primarily in the assessment of any mandatory requirements and must take into account the specifics of the case to be examined.

It is assumed, however, that even when strictly qualitative methods are used, the degree of risk is expressed by means of a classification so as to verify the change in the level of risk in view of alternative strategies or in comparison with a previous situation rather than desired by the prescriptive codes.

In more complex cases, the *fire-safety* design containing the analysis could be accompanied by a philosophy document (*trial fire-safety design plan*) which, on the basis of the risk assessment, constitutes an initial elaboration of the fire protection strategy, as well as being an input to the final study report; this document consists of a set of key fire-safety elements to be taken into account during the development of the analysis and fire-safety design.

The key elements of the philosophy document can be identified by employing the sub-systems ('SS') listed and developed in the ISO/TR 13387 Technical Report or by employing other tools which, in fact, also allow a more precise indication of the objectives and functional methods for achieving them (such as the fire concepts tree presented in NFPA 550).

A very important new element, as far as the ISO 23932 standard is concerned, is connected with the fact that risk assessment is not, as is often the case, exclusively cited as a method to provide evidence of fire hazards or to classify the fire risk in a qualitative or quantitative manner, but it is defined as one of the possible 'engineering' calculation methods for conducting the performance-oriented analysis. In this, it is lumped together with deterministic methods that are exclusively oriented towards defining, in more or less detail, the expected consequences and development of a fire phenomenon, methods that require the evaluation of aspects such as the availability of protective measures rather than scenario alternatives based on stochastic states.

In any case, the use of both deterministic and stochastic methods, rather than a combination of the two, must be subject to validation of the input data and findings.

With regard to the input data, it is necessary to verify the validity of the methodology used to define them, the confidence interval (uncertainty) and the degree of applicability to the specific case under consideration. Concerning the findings, it is necessary to validate them against similar cases already developed, information from full-scale tests or specialist literature, etc.

Even in the case of the use of engineering or deterministic calculation methods, the designer cannot neglect risk analysis and stochastic analyses in general: in fact, he or she must develop each scenario with and without the protection measures that may be unavailable for the purpose of characterising the expected consequences (for the purpose of estimating benefits) and must separately conduct an analysis of the reliability (in terms of availability on demand) of the protection systems.

Furthermore, by conducting a sensitivity analysis aimed at grouping scenarios, he or she must verify that the scenarios he or she has developed using a deterministic approach are not altered by a change in a state. If the check is successful for a given scenario, the latter can be said to be robust; if not, from the scenario under consideration, any changes that are likely to significantly alter the expected consequence must be investigated deterministically.

The standard provides that different engineering methods may be used for different scenarios considered.

At the end of the analysis, a technical report must be drawn up that illustrates the findings, clearly defines the conditions of use of the building and provides indications regarding the periodic inspection and maintenance procedures for managing the level of fire risk over time. These indications, including any restrictions to be maintained, guarantee to the operator that the level of risk deemed acceptable will be maintained throughout the life cycle of the building or industrial installation, barring any changes, whether of a technical or managerial organisational nature, which must be analysed in the same way in order to verify that the pre-existing estimated level of risk is not worsened.

Inspection and maintenance procedures, together with precise indications from, e.g. the risk analysis, must be found in the Fire Register.

In addition, a systemic management of the issue (with a management system sized according to actual needs) must be adopted, which includes the periodic review of the degree of safety in place and an inspection (audit) activity: these activities will be aimed at verifying that each key element of fire safety is kept effective and efficient in view of the analysis conducted (according to the requirements established) and, no less important, that the assumptions made are completely valid over time.

7.1.4   ISO 17776

The standard ISO 17776 describes the processes underlying the management of major-accident (MA) hazards during the design phases of offshore platforms and gas extraction installations. In particular, it is presented as a guideline for the development of strategies aimed at preventing the occurrence of detectable accidents while limiting their expected consequences.

Similarly, there is also a guide to assist the analyst in dealing with IR hazards in the operational phases.

Its scope of application can be extended to

fixed offshore structures;
floating systems with a production/storage and discharge function;
for the *oil and gas* industry in general.

The application of the line guides within the standard allows the analyst to include all non-negligible hazards related to major accidents, i.e. with the potential to generate physical harm to people, the environment and the continuity of operations.

The standard's guidelines cover different scales of installations, from the development of designs for new offshore facilities in their entirety to smaller-scale installations or the structural modification of existing facilities; moreover, its relevance can be extended to onshore production installations.

Mobile offshore structures are excluded from the scope of use of this ISO; however, most of the principles defined here can be adopted in the case as a guide to analysis.

Similarly, it is not possible to use the standard for the design of submarine structures, although the effects of major accidents on similar structures are to be considered, if they insist on offshore installations, i.e. falling within the scope. However, the construction, *commissioning* and *decommissioning* phases relating to offshore installations, as well as the risks associated with their security, are not considered in the ISO lines.

In any case, it is up to the analyst to decide whether to apply the requirements and guidelines of ISO, in full or only in part, depending on a preliminary assessment of the probability of occurrence and the expected consequences of potential scenarios associated with major accidents.

7.1.5   ISO 13702

ISO 13702 standard has been prepared to explain the main requirements and guidelines for the control and mitigation of fires and explosions on offshore production installations and in particular those plants that process petroleum (crude oil) and natural gas. Standard applies to fixed offshore structures, floating systems for production, storage and offloading (FPSOs) and petroleum and natural gas industries. This includes integrated installations, manned and unmanned plants.

This standard strongly suggests a structured approach to the management of fire and explosion issues in such areas, which necessarily takes into account the phases that characterise the entire life cycle of offshore installations. In particular, the standard emphasises that, on the occasion of each modification of existing installations, it is essential to proceed with a critical review of the existing protection strategy in order to make it congruent with both the intended changes and the available best practices.

Although the standard is preferentially applicable to offshore installations, the described approach and methodology for achieving fire-safety objectives are extremely general and widely applicable to even considerably different industrial contexts. The proposed approach is based on the selection of control and mitigation measures for fires and explosions determined by an initial and fundamental evaluation of hazards on the installation. The methodologies employed in this assessment and the resultant recommendations will differ depending on the complexity of the production process and facilities, type of facility, manning levels and environmental conditions associated with the area of operation. This results in a risk-based and performance-based approach where barriers are employed to control (limiting the extent or duration of a hazardous event) or to mitigate (reducing the severity of the resulting consequences) the fire/explosion risks, including any potential escalations that may pose threats to the people, to the environment and to the assets (including financial, business interruptions and consequential losses of fires and explosions). The escalation can be seen in the spread of impact from fires, explosion, flammable and toxic gas releases to equipment or other areas, thereby causing an increase in the consequences of a hazardous event.

Barriers, together with their specific functional requirements and performance criteria to be maintained over time, become the central element of the fire and explosion prevention and mitigation strategy resulting in being considered critical safety systems given their major role in the control and mitigation of fires and explosions and in any subsequent evacuation, escape and rescue activities of the occupants. Safety strategy is the result of the process that uses information from the fire and explosion evaluation to determine the control measures required to manage these hazardous events and the role of these measures, taking into account human factors.

This international standard assumes that risk assessment is performed within the principles and guidelines for risk management from ISO 31000 standards and states that risk assessment should be based on specific phases: risk identification, risk analysis and risk evaluation. In the context of fires and explosions, preliminary prevention measures such as inherently safer designs and ensuring asset integrity shall be emphasised wherever practicable.

Risks may arise from a number of different threats, and assessment should consider those elements, peculiar to the installations, that may pose a fire and/or explosion risk, considering:

a) nature of the fires and explosions which might occur;
b) risks of fires and explosions;
c) marine environment;
d) nature of the fluids to be handled;
e) anticipated ambient conditions;
f) temperature and pressure of fluids to be handled;
g) quantities of flammable materials to be processed and stored;
h) flammability and toxicity of materials in non-hazardous areas including accommodation and control station;
i) amount, complexity and layout of equipment on the installation;
j) location of the installation with respect to external assistance/support;
k) emergency response strategy;
l) production and manning philosophy;
m) human factors;
n) interaction with adjacent facilities and vessels.

A crucial document resulting from the application of the principles of the standard is the fire and explosion strategy (FES) that should be updated whenever there is a change to the installation that affects the management of the fire and explosion hazardous events. The FES level of detail depends upon the extension and the complexity of facility (considering that, in the offshore environment, large facilities also accommodate a great number of operators and third-party contractors in specific accommodations onboard).

FES is based on critical elements that should be considered in terms of

a) functional parameters of the particular system, e.g. essential duties that the system is expected to perform;
b) integrity, reliability and availability of the system;
c) survivability of the system under the emergency conditions that might be present when it is required to fulfil its role;
d) dependency on other systems or operational factors that might have an influence on the performance of the safety function when needed.

Those elements include a number of engineered controls and risk management activities:

- emergency shutdown and blowdown systems to initiate appropriate shutdown, isolation and blowdown actions to prevent escalation of abnormal conditions into a major hazardous event and to limit the extent and duration of any such events which do occur;
- specific activities such as control of ignition (to minimise the likelihood of ignition of flammable liquids and gases following a loss of containment) and control of spills (to provide measures for containment and proper disposal of flammable liquid spills);
- emergency power systems (to provide a reliable source of power in the event of failure of the supply from the main source of electrical power);
- fire and gas detection systems to provide continuous automatic monitoring functions to alert personnel of the presence of a hazardous fire or a flammable gas condition and to allow control actions to be initiated manually or automatically in order to minimise the likelihood of escalation;
- active fire protection systems to control fires and limit escalation, to reduce the effects of a fire to allow personnel to undertake emergency response activities including escape and evacuation and to extinguish the fire and to limit damage to structures and equipment;
- passive protections to limit escalation, to maintain functionality of critical safety systems and to allow emergency response;
- explosion mitigation and protection measures to reduce to an acceptable level the probability of an explosion affecting critical safety systems and other areas of the installation;
- inspection, testing and maintenance activities;
- activities to response to fires and explosions to reduce to an acceptable level the probability of an explosion affecting critical safety systems and other areas of the installation.

Standard underlines how the facility layout may result in safety issues and it underlines how the layout has direct effect on the explosion severity (considering the volume available, the blockage ratio and the number of obstacles, the obstruction position, the sideways venting availability, etc.).

Physical effects modelling (PEM) activities therefore play a fundamental role in the framework of a specific Fire and Explosion Risk Assessment (FERA), while at the same time FERA and engineered control management over time should be an integral part of the life-cycle management activities (Figure 7.2) of the facility.

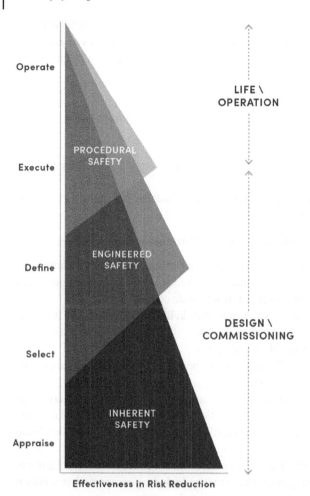

Figure 7.2 Risk management over time, from inherent safety principles to safe operation.

7.2 British Standards (BS) – UK

The British Standards Institute (BSI) is the body recognised by the Government of Great Britain as the 'National Standards Body' (NSB). In this capacity, the BSI actively collaborates with government agencies, companies, universities and public and private research institutes, as well as with end-user groups in order to promote the development, adoption and dissemination of standards, guidelines and norms of good practice.

The BSI develops guidelines, national standards and reference publications aimed at both industrial and civil sectors; many of the institute's proposals often become the basis for international standards, an example being the well-known ISO 9001, or other ISOs such as 14000, 27000, 20000, 10000 and 45000.

It is important to remember that Great Britain was the first to officially adopt the BSI standards relating to performance-based fire engineering, which, pending the promulgation of ISO standards, currently represents the most authoritative international benchmark.

The BSI also acts as a promoter of standards and technical publications worldwide, including through the development of specific cooperations with a large number of other national standardisation organisations.

Alongside the publication and promotion of technical standards, the institute is extremely well known internationally for an extensive series of technical manuals written by recognised experts and aimed at professionals and consultants working in the construction, manufacturing and service sectors.

The institute is more than a century old. In 1901, the associations 'Institutions of Civil Engineers', 'Institutions of Mechanical Engineers', 'Institutions of Naval Architects' and the 'Iron and Steel Institute' set up a committee called the 'Engineering Standards Committee' to standardise the sections of steel and iron manufactured by British manufacturers and used in the construction of bridges, railways and shipping. The committee succeeded in reducing the types of products on the market, thus saving an estimated £1 million a year.

In 1929, the originally established committee changed its structure and became an association, the 'British Engineering Standards Association'. In 1930, the association, having had its role in the field of technical standardisation as well as its moral stature recognised by the British government, took on the final name of 'British Standards Institution' (BSI). Furthermore, as was the case in many other countries at the same time, it took over the task of developing, disseminating and updating technical standards in the field of electrical engineering, taking over the task hitherto carried out by the 'British Electrotechnical Committee' (BEC).

The BSI is one of the founders of ISO (dating back to 1946), of the two European standardisation bodies CEN and CENELEC (founded in 1964), and has been involved in the IEC since its early days (1906).

Among the many standards developed in the field of fire engineering and fire safety, for the purposes of this publication, it is useful to present two extremely recent but little-known guidelines specifically dedicated to fire risk analysis and the subsequent preparation of an effective fire strategy.

7.2.1 PAS 911

The technical publication PAS 911 'Fire strategies – Guidance and framework for their formulation' is a British specification developed by the BSI institute that provides a formal method for the definition of a building's fire strategy and its documentation.

The term 'fire strategy' is intended to designate the set of fire precautions and basic principles that must guide the specification, design and implementation of fire-safety prevention, protection and management measures (whether technical or organisational/managerial).

For the purposes of this book, it is of fundamental importance to emphasise the close link between the fire risk analysis and the definition of the most appropriate fire-safety strategy, in consideration of the fire hazards identified and the level of risk associated with a specific activity. In fact, as already described, the ultimate aim of the fire risk assessment must be the definition of an effective firefighting strategy, which is sized in relation to the measured risks.

The technical publication PAS 911 formalises the ideal process underlying the definition of the fire strategy. In particular, it proposes a well-articulated and strongly structured workflow that ultimately leads to the preparation of a document illustrating both the general criteria underlying the firefighting strategy and the details of the technical and organisational-managerial measures for prevention and protection, as well as those related to the management of fire risk over time.

This document is also required to provide precise information about the findings of the fire risk analysis conducted, the fire philosophy adopted and the control measures put in place to ensure that an acceptable level of fire risk is maintained over time.

For these reasons, it is a candidate to become the point of reference for the subsequent in-depth study, design and implementation of fire safety, as well as its periodic review over time.

The publication, although issued in the United Kingdom, can also be used successfully and with undeniable advantages, given its wide applicability, in our country with non-significant modifications. In fact, it is released from both the specific regulatory requirements and the guidelines and standards taken as reference for fire design. The document only makes explicit the most suitable procedure for the formulation of a rigorous and justified fire prevention strategy, capable of giving it completeness and effectiveness independently of the specific body of applicable regulations.

The objectives of the publication in fact:

give uniformity to the workflow and the final document produced: since no standard establishes the binding content of the strategy, the form that the document describing it must take and the level of detail to be achieved, and firefighting strategies are difficult to compare;

ensure that all aspects relevant to the definition of the fire strategy have been taken into account in the above workflow: therefore, the working method presented by the BSI in the publication makes it possible to take into account all the fire risks present (assessed on the basis of the identified fire hazards) and to evaluate all the technical and organisational precautions to be taken into account in the fire strategy.

To support the proposed methodology, PAS 911 provides a number of 'tools' and 'templates' (presented in the form of graphs and tables) that can be used to facilitate decision-making processes; some of the tools are a reworking of those originally devised by the consulting firm Kingfell plc (which participated in the drafting of the BSI publication), starting in 1996.

The methodology proposed in PAS 911 operates according to a functional scheme that processes the input data (and thus the factors influencing the strategy) in order to subsequently define the various elements that must constitute fire protection.

The overall scheme can be summarised in five stages:

selection of the approach to be used, which may be prescriptive or performance-based;

identification and analysis of the elements influencing the choice of firefighting strategies; the elements can be traced back to the seven predetermined categories:
- system and management audit;
- the set of regulations;
- target setting;
- risk and hazard assessment;
- the characteristics of the building;
- the characteristics of the occupants;
- practical problems.

strategy formulation;

verification of the completeness of the formulated strategy by checking the following aspects on time:
- basic elements (principles of strategy);
- fire-safety management strategy;
- evacuation strategy;

– fire and smoke control strategy;
– firefighting strategy;
– fire protection strategy.

formalisation of the final version of the formulated strategy (including through the preparation of the summary document).

The first step is the preliminary decision-making phase, while the last is essentially a phase of consolidation, presentation and documentation of the work done.

At the heart of the methodology are the intermediate steps which, as can also be seen from this introductory schematic, divide both the data to be processed as input and the data to be produced as output into a well-defined subset of categories. The aim of the publication is to present a working method suitable for breaking down the problem of fire strategy formulation into a set of aspects, which can be more easily handled by the analyst, each to be examined individually.

The various stages of the workflow are discussed in the following sections.

The first stage of the process starts from the consideration that there are different ways of defining a fire strategy and the subsequent design or identification of control measures, both prevention and protection.

The selection of the strategy can be defined indirectly through full compliance with the prescriptive requirements of the normative body of reference or through the explicit use of fire engineering techniques aimed at performance assurance.

The publication does not exclude a mixed approach: in this case, it is extremely important that the summary document clearly states which aspects are subject to a prescriptive approach and those managed through engineering analysis, with appropriate justification of the reasons for a combined approach.

The second step suggested by the BSI involves identifying and analysing the input elements having an impact on strategy selection (see Figure 7.3).

If a prescriptive approach has been adopted, it is probably not necessary to conduct an analysis that considers all of the predefined categories of input elements, as it is the standards themselves that guide one towards the choice of certain 'certain compliance' solutions (intrinsic fire risk acceptability).

In any case, when conducting this phase, all elements must be assessed by examining available records, conducting interviews with personnel or occupants, and verifying any risk assessments already conducted on the same or similar realities.

This first element, which is, of course, only significant in the case of realities already in operation, draws attention to the prior conduct of an audit of the fire-safety management system.

Fire safety is, in fact, not only a function of the fire risk analyses and designs carried out, but also of the way in which they are conducted in order to maintain an acceptable level of fire risk over time (so-called operational control).

The purpose of periodic verification activities is to understand what roles and procedures are envisaged with regard to fire safety; what technical, organisational-managerial shortcomings and, if any, structural problems emerge from past history (in terms of accidents and near misses that have occurred, analysis of emergency response, etc., i.e. the organisation's operational experience); what peculiarities of the occupants or operations must necessarily be taken into account when formulating the new fire-safety strategy; etc.

Obtaining an overview of the criteria adopted for the definition of existing prevention and protection measures is essential in order to be able to assess whether these criteria are still applicable and justifiable (also in view of the above-mentioned operational experience gained).

Figure 7.3 Elements constituting a fire strategy: defining and resulting elements.

The information to be examined is that which can be gleaned from the firefighting strategies defined in the past, from existing codes and guidelines and from the regulations put in place over time (to control elements such as the conduct of work, the assignment of work to external suppliers, requirements on supplies, and education, information and training programmes).

Records of fire-safety meetings held, records of investigations conducted after incidents and near misses that occurred and records of false alarms that occurred are also important documentary evidence.

Where practicable, it is also useful to conduct an efficiency and effectiveness audit (e.g. comparing actual system performance against design performance) of detection, warning and extinguishing systems.

The strategies, with particular reference to all situations in which, even partially, the fire protection designer employs one or more prescriptive codes (including technical rules, standards and guidelines), must be drawn up taking into account the reference regulatory body and any other requirements deriving from corporate commitments, including requirements deriving from insurance companies, membership of industrial groups, professional and trade associations.

The primary objectives of the fire strategy must always include adherence to the relevant body of legislation and ensuring the safety and security of the occupants.

The primary objectives must be clearly spelled out and well defined as early as the start of the analysis. Primary objectives may include safeguarding of property, continuity of operations (activities) and protection of the environment (see Figure 7.4), as well as, of course, the safety of the occupants.

In view of this, the analyses carried out by fire analysts and fire planners must be developed with methodologies that are suitable for analysing the various aspects selected, even individually. The fire strategy developed must therefore be justified with respect to the various aspects considered in the defined objectives and the risks associated with them.

For each of the aspects considered, the BSI publication identifies a number of sub-objectives that can be taken into account for the evaluation of each primary objective.

The objectives of the fire strategy must coincide with the objectives used as the basis for the fire risk analysis in order to always ensure a perfect match between the hazard assessments and the associated fire risk measurement and control measures underlying the fire strategy.

The primary objectives of the strategy, and consequently all sub-objectives constituting each primary objective, must be clearly stated and specified in order not to become ambiguous. They must also be quantifiable to allow verification of actual achievement according to objective criteria, genuinely attainable, time-stamped and categorised in priority categories in order to unambiguously prioritise achievement in view of the primary objectives to be targeted.

Figure 7.4 Matrix of the objectives of the fire strategy.

The objective of occupant safety (referred to as 'Life' within Figure 7.4) must take into account a number of aspects including how occupants, visitors and external personnel are warned of the presence of a fire, their ability to safely exit the building or industrial installation, safe areas, etc. It is also essential to identify safety criteria for external rescue teams faced with a possible fire, i.e. how they can ensure the safety of the building or industrial installation. In addition, it is essential to identify criteria for guaranteeing the safety of external rescue teams who have to deal with a possible fire, i.e. how they will be provided with assistance and access to the resources needed to carry out the intervention, access to the building (including at height), etc.

The objective of property protection requires analysing factors such as the vulnerability to fire of structural elements and the value and fragility of particular interior elements (ceilings, doors, valuable decorations with historical-artistic value, etc.), furniture, machinery and equipment (including paper and computerised document archives), etc. For all the elements constituting the property and contained in the building, it is important to assess their value (also with respect to a possible resumption of activities in the short term in their absence) and to assess their degree of vulnerability with respect to a fire or with respect to the intervention of extinguishing systems.

The objective of protecting the continuity of business and operations ('Business') is assessed by analysing the impact that fire may have on operations, on the organisation's ability to achieve its business objectives and on the organisation's reliability in the eyes of suppliers, employees or neighbouring communities. Damage to the business thus includes damage to image.

In the analysis, it is extremely important to assess the time required for re-starting in the event of a fire, as the actual damage is a function of both direct damage and indirect damage, which includes charges for clearing away damaged material and equipment, charges for reconstruction, charges associated with lost production, charges associated with re-starting operations, etc.

The objective of environmental protection ('Environment') requires an assessment of the possible hazardous substances released into the environment as a result of a fire, an examination of how they may spread, a determination of the extent of the areas affected, an estimate of the cost of their removal and the long-term consequences affecting the usability of the building or industrial installation and the surrounding environment. As part of the definition of the firefighting strategy, it is also necessary to assess the environmental impact that may result from firefighting activities.

7.2.1.1 Risk and Hazard Assessment

The identification of fire hazards and the assessment of associated risks, the formulation of representative fire scenarios and the assignment of an overall risk profile can help to identify weaknesses in existing fire strategies and formulate cost-effective and beneficial solutions. It is not possible to define a fire strategy without detailed knowledge of the fire risks characteristic of a specific reality (also with reference to one's own or similar realities' operational experience).

The structure of a building, the materials used in its construction, its positioning, the arrangement of openings or structural elements to contain flames and smoke, the presence and safety of escape routes, and the availability of fixed and mobile firefighting equipment are all elements of a building that influence the development of an accidental event and must consequently be evaluated for the purposes of drawing up the strategy.

All of the elements mentioned (normally highlighted during a fire risk assessment) can alone determine the failure of the firefighting strategy: suffice it to think of the elements inherent to the positioning of the building (e.g. the accessibility of the surrounding areas or the distance from the fire brigade headquarters) that can become a significant obstacle for those who must carry out firefighting or rescue operations. Also consider the stability of the building's structure which, if compromised, can pose a risk to the lives of both firefighters and occupants in an emergency escape.

In fact, it is important to assess the possibilities of safe escape in the event of a fire for all occupants. For the purposes of defining a firefighting strategy including the evacuation plan, the layout of the building, the characteristics and availability of the escape routes in consideration of the reference fire scenarios are fundamental. Obviously, these assessments cannot disregard the characteristics of the occupants and the degree of crowding that can be found in the reality under consideration (even at particular times of the year, which is extremely significant for commercial establishments, for example).

The characteristics of the building must be taken into account in the light of its intended use as the activities conducted have a significant impact on the assessment of fire risks and the definition of the best firefighting strategy.

Many of the characteristics of the occupants, such as their estimated number, the type of occupants, their distribution within the building and their ability to reach the outside of the building or safe areas within the building, influence the process of determining the most appropriate evacuation strategy, whether they are regular occupants, visitors or external contract workers.

For the purposes of preparing an evacuation plan, it is essential to know, even for occasional occupants, their level of knowledge of the building and firefighting procedures, their ability to respond to the fire, any impediments to understanding dangerous situations or to reaching independently and in sufficient time the escape routes and areas of the building occupied by them (especially if they are not usually used to accommodate people), as well as the times of use of the building.

Furthermore, for certain types of buildings it is possible that, at different times of day, they are characterised by totally different risk and use situations: a peculiarity that must be taken into account in order to ensure an appropriate fire response strategy (or different strategies) for all intended use cases.

The fire risk analysis must be considered as an alienable input to the preparation of the internal emergency plan and evacuation plan.

By the term 'practical problems', the BSI publication intends to designate the set of elements that are typical and peculiar only to the reality under consideration and cannot fit into any of the categories already identified. These elements can be traced back to technical, logistical or economic constraints that may limit the solutions that can be adopted. They are often decisive in directing towards very precise choices which are the only ones that make the firefighting strategy effective or feasible, especially with reference to the relationship between costs and benefits obtained.

These peculiarities must be clear from the earliest stages of fire strategy design (just as they should have been clear when conducting the fire risk assessment), in order to avoid the selection of fire control measures that, due to specific conditions, although generally suitable, turn out to be for the specific application not to be implementable or not to be guaranteed over time.

The input data set out in this section is used to identify a firefighting strategy that aims to reduce the fire risks characteristic of the building or industrial installation while ensuring the overall compliance of the measures to be taken with the requirements of the existing regulations.

Each input element is appropriately evaluated to establish its adherence to the prescriptive or performance requirements of the protection, prevention and safety management measures in place (in the case of existing activities), or of the measures foreseen by the fire protection project (in the case of new activities). Should adherence to the requirements prove unsatisfactory, it is the fire protection designer's task to proceed to propose a series of additional or alternative measures (adequately assessing all available options and evaluating their equivalent degree of safety) that can lead to adherence to the requirements of the previous fire risk assessment.

The findings of the fire risk analysis cannot be ignored and, in the case of non-compliance, the risk assessment must be updated with the new considerations in order to verify that the changes resulting from the fire strategy definition activities do not lead to an aggravation of the previously estimated and deemed acceptable fire risk.

The BSI publication also provides an extremely useful and robust 'tool' for visualising the coverage provided, within the overall fire strategy, by the various fire-safety measures adopted (e.g. control of ignition sources, fuel control, room compartmentalisation, etc.). The graphical representation is based on a graph with eight variables (shown in Figure 7.5), in which each variable, which is a type of firefighting solution, can take on a value between 0 and 5, depending on how critical the solution is considered to be within the strategy.

As it is constructed, the graph immediately shows the priorities that have been assigned to the various solutions. It also makes it possible to assess the complexity and cost of the planned system: proportionally, it is possible to state that the greater the area covered by the set of solutions within the graph, the greater the complexity and cost of adopting all the planned measures comprising the strategy. The development of the same graphic representation can easily allow a comparison between alternative fire protection strategies.

In fact, it is possible to state that the representation mode introduced by BSI can be modified for the purpose of using the same graphic mode for the visualisation of a larger number of factors (or sub-factors), including those defined in NFPA 550.

In most cases, the objectives can be achieved by means of mutually equivalent solutions; in terms of the overall degree of safety, when there is this possibility, each acceptable solution should be weighed up to determine which one is preferable, taking into account both performance and ease of implementation and cost-effectiveness.

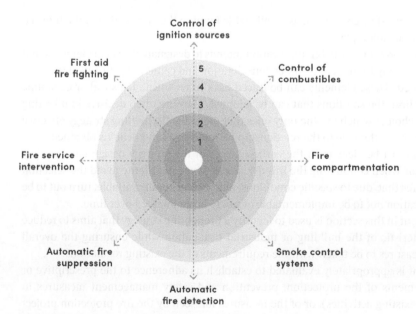

Figure 7.5 Strategy evaluation grid.

In this regard, PAS suggests adopting a formal method consisting of attributing a score (index) to each of the aforementioned characteristics and considering the product of the three scores attributed as the overall value of the solution. The ultimate aim is to implement a decision-making process based on objective and quantitative methods in order to make the analyses conducted and decisions taken easily traceable.

The definition of the strategy essentially aims at

producing a list of (existing or implementable) preventive, protective and fire-safety management measures over time;
assigning them a degree of importance, priority and responsibility for implementation;
identifying the relevant sub-category.

This list of measures will constitute the main element of the fire strategy document, which must also be accompanied by

a flow chart describing how existing or implementable processes and systems can interact with each other to achieve the basic objectives of fire extinguishment and evacuation of occupants;
a table showing the temporal sequence of events from the sighting of the fire to its control and the rescue of the occupants with the location of the barriers put in place to mitigate the accidental event (also useful to highlight any overlapping of elements).

The defined fire-safety strategy thus represents an overview of the fire precautions taken in consideration of the organisation's fire-safety policy, the applicable regulatory requirements, the fire risk assessment carried out and the fire-safety philosophy defined in the initial stages of the study.

The document containing the prepared fire strategy, approved by all parties involved, becomes the reference point for all subsequent fire-safety work.

As with the risk assessment document, it must be considered to be in continuous evolution and therefore susceptible to periodic review by the organisation in conjunction with the appointed designer. In particular, the institute advocates a review frequency of one year.

PAS publication 911 suggests a document organised in chapters or sections also in order to simplify review activities and possible amendment. In particular, it is requested that such a document be organised into at least six main sections, corresponding to the six 'sub-strategies'. The division into sections must not, however, coincide with a desire to deal with different aspects in a totally disjointed manner: the organisation's fire strategy must always be composed and periodically reviewed in its entirety, even if it is possible to update the contents of each individual aspect.

The first section in the PAS (referred to as the 'Principles of the Strategy') should provide an overview of the methodology followed in preparing the strategy.

This section is intended to present: the principles of the overall strategy, its foundations, the list of people involved in drafting it, the list of people responsible for implementing and verifying the strategy, the purpose for which the strategy is drafted, the primary and secondary objectives of the building, the limitations inherent in the strategy and the assumptions made during its development, and finally the findings of the fire risk analysis conducted.

The remaining sections identify very precise contents that must be present in the final document and specifically the following strategies:

The fire-safety management strategy: This includes elements such as fire prevention (policies, risk assessment and business process control measures), education, information and training of personnel (including third parties), maintaining the efficiency of firefighting systems, etc.

The evacuation strategy: This section must clearly identify and describe the suitability, signalling and activities for maintaining efficient escape routes and safety zones; it must contain the evacuation plan for the occupants (sized in consideration of the fire scenarios considered to be representative of all the use cases of the building or installation).

The fire and smoke control strategy: This section must clearly identify and describe the measures for controlling the spread of a fire (heat, smoke, combustion products, etc.) inside and outside the building.

The firefighting strategy: This describes how external *firefighting* support will be provided (including accessibility to the building or installation), the means available, logistical resources, etc.

The fire protection strategy: In this section, which must incorporate full fire emergency planning, fire identification procedures and systems, fire alarm systems, structural measures to control the spread of fire and smoke and fire suppression systems must be explained.

7.2.2   BS 9999

The standard BS 9999, 'Code for the practice of fire safety in the design, management and use of buildings', provides a structured guidance for the practice of fire safety, addressing various users such as architects, planners, surveyors, administrators, safety officers, consultants and engineers working in the field of fire safety.

BS 9999 takes a holistic view of fire safety (as summarised in Figure 7.6), allowing the introduction of compensatory measures to equalise and equate values of, e.g. the length of escape routes or the width of emergency exits.

The scope of the standard does not include building types that are considered in BS 9991 such as residential buildings, communal accommodation facilities (student halls of residence) and specialised living facilities, where, e.g. occupants may receive medical treatment; on the other hand, this standard can be applied to the design of new buildings and where existing buildings undergo alterations, such as alterations and modification of materials or extensions.

BS 9999 was revised in 2017 to better adhere to the latest good practice, new technology and to be consistent with other recent fire-safety standards, such as BS 9990 and BS 9991, which relate to the installation and design of non-automatic fire protection systems within buildings and the fire protection design and organisation for residential buildings, respectively.

Compared to the standards it replaces, BS 9999 introduces innovative elements, such as

a flow chart showing the sequential steps in the evaluation development process and providing user support for the application of the same document;

the revision of the fire-safety management clause, with reference to PAS 7:2013 (*fire-safety management system*);

the inclusion of fog extinguishing systems;

an expansion of the guideline on voice alarm systems and tables referring to standard fire growth curves.

In addition, the revised document presents updates with respect to recommendations concerning smoke and heat control, fire barrier systems, air conditioning and mechanical ventilation systems, and commercial complex buildings.

The standard emphasises the importance of the concept of maintaining fire-safety standards over time, through continuous fire-safety management over the entire life cycle of a building, suggesting line guides and recommendations, including measures for planners to ensure that the overall development of a building supports and promotes fire-safety management.

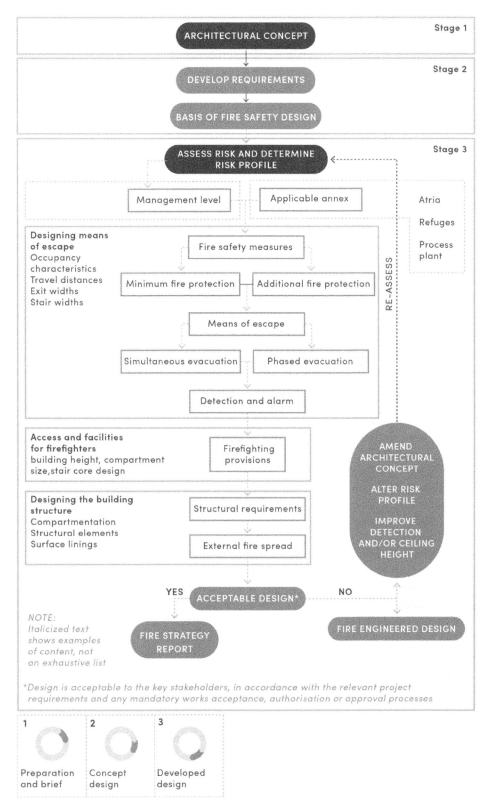

Figure 7.6 Flow chart about the BS 9999 workflow.

The objectives pursued by the recommendations and guidelines of BS 9999 are aimed at safe-guarding the occupants of the buildings under consideration and the rescue teams. At the same time, they can also support the risk analyst in pursuing other objectives of fire safety, such as the protection of property, the environment and the continuity of services, although some additional measures to this end fall outside the requirements of this document.

Finally, within the standard there is an assessment tool that ensures the effective robustness of the implemented firefighting strategies through testing methodologies related to elements such as facilitated access to emergency exits, the safety of people inside and outside buildings. The standard also covers the maintenance of detection systems as well as providing guidance on the fire training of workers through the organisation of efficient evacuation plans and the allocation of leadership responsibilities to staff in emergency management.

The standard identifies the key factors to be considered in determining the level of fire risk for new and existing buildings, namely:

- the expected probability of a fire occurring;
- the expected severity and spread potential of any fire;
- the ability of the structure to resist the spread of fire and smoke produced by the fire;
- consequential damage to persons in and around the building;
- the need to define a system of protection for facilities and their contents, the continuity of work activities and the surrounding natural environment.

In addition, safety factors to be considered in the risk assessment are defined, such as the adequacy of fire prevention measures, early detection of fire outbreaks by means of a detection and alarm system; the system of escape routes; measures to control the smoke produced by the fire; control of the rate of fire growth (the spread of fire on surfaces, cladding, etc.); the adequacy of the structure to withstand the fire; the degree of containment of the effects of the fire; fire separation within the building and between different buildings; active protection measures (extinguishing and controlling); the degree of fire containment; the degree of fire containment; the degree of fire safety; the adequacy of the structure to withstand the fire; the degree of containment of the effects of the fire; fire separation within the building and between different buildings; active protection measures (extinguishment and control) of the fire; support structures for the firefighting service; the quality of firefighting management measures; measures for the training of the emergency service and for ongoing checks; the characteristics of the occupants and the associated risk profile.

The allocation of a 'minimum package' of fire protection measures is determined, according to BS999, by the level of risk identified above.

Each building is assigned a risk profile so that the protective measures are appropriate to the specificity of the structure, in particular with regard to the usability and adequacy of the associated escape routes, as is detailed later in the standard in Sect. 5. E 7, regarding the adequacy of the complexity of the building's fire design.

The risk profile must reflect the characteristics of the occupants and the fire's growth rate, defined by the standard growth curves (slow, medium, fast and ultra-fast), also given in the SFPE publications and in a number of local regulations defined as risk-based.

All elements of a building that constitute its active and passive protection against fire must be properly designed, installed and regularly maintained and tested.

The effectiveness of fire protection is then defined in each of the three phases: design, construction and maintenance.

The fire protection strategy is defined in the start-up phase of the project, during which the fire protection systems are chosen, considering criteria such as their life-cycle assessment, the maintenance of their effectiveness through maintenance, their accessibility for inspection and maintenance throughout the service life of the building and the criticalities that may compromise the durability of the systems over time. The construction manager is responsible for the quality of the construction work and verifies that there is no interference between the different operations so that no compromise to the fire protection systems (active and passive) can result from installation or maintenance operations of structures, service technology systems, etc.

For example, the installation of additional ventilation systems on existing systems could cause the actual operation of existing ducts to be compromised or instructed if the operation is not properly planned; the figure of the person in charge of the work is crucial at this stage, where the various workers may not be trained in the requirements of the fire strategy and must be instructed through an integrated system of procedures to avoid such inconsistencies.

The standard indicates several recommendations to be followed during this phase: in particular, it refers to the installation of the materials constituting passive fire resistance and their specific and adequate installation, consistent with good engineering practice; the specified fire resistance must be guaranteed by the installation of several layers of material. Similarly, the same criteria are given for active fire resistance systems; in addition, the implementation equipment of the systems must be adequately installed and maintained.

With regard to the maintenance phase, the document indicates how fire protection systems may not guarantee their continuity of service if they are not properly inspected and maintained on a regular basis. In addition, maintenance operations that require structural changes to buildings may alter the effectiveness of passive fire protection systems (e.g. the installation of electrical cables or operations on the building's piping system); this must be checked before the premises resume serviceability and if necessary restored where a reduction in its effectiveness has been ascertained. In particular, this section recommends taking into consideration critical issues due to the ageing of buildings, ensuring that all the systems serving the building contribute to and do not hinder the firefighting strategies adopted, defining the possibility of permits for hot/cold work operations, monitoring and recording the status of all firefighting equipment, and ensuring an appropriate distribution of responsibilities in the management of the fire-safety strategy during maintenance.

7.3 Society of Fire Protection Engineers – USA (SFPE-USA)

The Society of Fire Protection Engineers (SFPE) was founded in 1950 and was then established as an independent organisation in 1971. It is a scientific association representing those working within the fire protection engineering sector; it has more than 4000 members worldwide, divided into around tens of local branches (national and international). The acquisition of 'Professional Member' status requires assessment by a specific technical committee of the candidate; therefore, the number of professionals who can boast the title is extremely limited compared to the total number of members, the list of which is officially published on the association's website.

The purpose of the association is to disseminate the science and practice of fire engineering with the aim of maintaining a high professional and ethical standard among its members and to encourage learning about the subject. To this end, the association is involved in the organisation of seminars, courses and conferences, the publication of books, journals and technical papers, and the dissemination of a regular newsletter. In addition, the SFPE is involved in the development and grading of the annual examinations for the professional training of fire engineers, in cooperation with the National Council of Examiners for Engineering and Surveying.

Various spontaneous committees and working groups collaborate within the association on technical projects that aim to further improve the state of the art in fire engineering.

In Italy, the objectives of the SFPE are pursued by the Italian Section founded in 1983 as Associazione Italiana di Ingegneria Antincendio (AIIA), now SFPE-Italy.

One of the association's most important works, related to the topics of this publication, is the 'SFPE Engineering Guide to Application of Risk Assessment in Fire Protection Design', currently one of the most significant and useful documents available internationally, which is discussed in the following section.

7.3.1 Engineering Guide to Fire Risk Assessment

The *SFPE Engineering Guide to Fire Risk Assessment* provides guidelines to be considered for the selection and use of fire risk analysis methodologies that should be adopted when designing fire safety or evaluating existing fire-safety measures within buildings, industrial facilities, etc., based on the fire risks identified. This guide is outmost important for all the fire professionals and it can be applied to integrate specific local regulation requirements.

A second edition of the guide, published in 2022, has been required to align its contents with the progress of research and practical experiences related to fire risk assessment, by providing broader discussion of identifying fire hazards and fire scenarios as well as a detailed quantitative and qualitative risk estimation methodology. The second edition of the SFPE Guide also includes a comprehensive and practical process to deal with uncertainty and sensitivity analysis and a more extensive discussion of additional items in the risk management process, such as risk communication, residual risk management and risk monitoring.

In particular, the SFPE Guide illustrates how to conduct a fire risk assessment according to a number of well-determined steps, starting with the definition of the aims of the assessment process and leading to the final decision on the acceptability of the identified risk.

All the phases that make up the process are included by the standard itself within a flow chart (Figure 7.7) that briefly illustrates the path to follow in order to carry out a correct and complete risk assessment, characterised by a degree of detail congruent with the complexity of the situation under examination.

The SFPE Guide examines, in great detail, each step, thus becoming a significant vade mecum for all practitioners. As it is also completely independent of mandatory regulatory requirements, its applicability can be found in every situation (in both civil and industrial settings).

The extensive bibliography listed in the guide and organised by topic allows for important insights.

At present, the SFPE publication can be said to be the main reference for all those who wish to adopt a well thought-out, strongly structured and disciplined workflow, which is widely used, easily understandable but at the same time coordinated with certain current issues (e.g. the use of fire engineering methods for a safety approach based on performance assurance, an issue that has itself always been closely followed by the SFPE).

Given the topics covered, the guide provides useful information on the choice of risk assessment methodologies, their use and the final assessment of the acceptability of the risk detected, always with the ultimate goal of conducting the 'Fire Risk Assessment' and the subsequent justification, documentation and presentation of the findings.

The following sections summarise the key aspects of the SFPE publication to which reference is made for further details.

First of all, the guide provides the definition of risk, fire risk and fire risk assessment. In particular, the risk is 'the potential for realising unwanted adverse conditions and considering scenarios and their associated likelihoods and consequences'. It becomes a way to measure

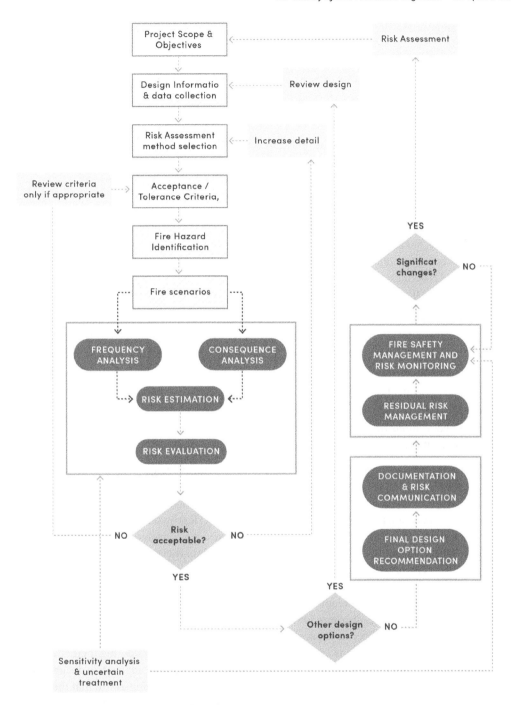

Figure 7.7 General overview of the fire risk assessment process.

(quantitatively or qualitatively) the fire-incident loss potential for fire protection engineering applications in event likelihood and aggregate consequences. According to this approach, fire risk assessment is the process to estimate and evaluate those risks associated with fires affecting buildings, facilities or processes, using one or more acceptance criteria. This process is used to:

- select an appropriate design, considering the risk (and the costs) associated with design alternatives;
- manage fire risk;
- inform resolutions of a regulatory process.

The general overview of a fire risk assessment process is shown in Figure 7.7, according to the revised Guide. The process covers four distinct phases of a fire risk assessment:

Phase 1 – Planning: The activities of this phase are intended to clearly define the scope and objectives, collect the necessary information to perform the analysis, identify the risk assessment method to be used and the define the tolerability criteria to evaluate the risks.

Phase 2 – Execution: It is the 'core' of the process, consisting in the 'risk assessment' phase, i.e. the steps required to identify the risk sources, analyse risks, characterise the fire scenarios and evaluate the risk. It is an iterative process as the analysis is expected to identify design improvements or modification that are necessary to further reduce the risks.

Phase 3 – Risk communication: Reporting it and using tools to better visualise the risk levels, e.g. Bow-Ties and risk matrices.

Phase 4 – In service: It is the residual risk management and monitoring phase.

A new section of the revised Guide is dedicated to the risk assessment method selection, which is related to the level of detail to which each scenario is described and quantified concerning the level of potential risk. The approach can be qualitative, semi-quantitative or fully quantitative depending of several factors, including:

- the information available;
- the complexity of the facility being analysed;
- the potential deviations from code requirements;
- the level of details necessary to make a fire risk-based decision.

Figure 7.2 shows how the revised Guide selects the fire risk assessment approach depending on the level of effort and the level of detail of the analysis.

The guide suggests that risk matrix should be used as a tool to evaluate and represent the outputs of the fire risk assessment, providing guidance on this topic. It clearly states that the risk of each of the scenarios included in the risk assessment is determined by considering the fire protection system and applicable features; in other terms, it recalls the role of the 'barriers', also known as risk control measures. The revised Guide also covers the fire risk estimation and evaluation, after having discussed how to establish acceptance/tolerance criteria, also introducing the possibility of conducting ALARP studies when required.

Moreover, the SFPE Guide also provides comprehensive steps that focus on sensitivity and uncertainty analysis, being necessary to evaluate the impact of the introduced hypothesis into the fire risk analysis, so that conclusions and recommendations are made considering their effects.

Once the evaluation process is complete, the risk assessment should be adequately documented and communicated. The guide provides useful insights into this topic, being aware that the documentation is the basis for the following, and last, step about residual risk management and monitoring.

The SFPE Guide dwells on the difference between the concept of 'probability' (which has a binary nature in that it represents the opportunity that a single event has to occur or not to occur) and that of 'frequency' (which allows multiple events to be characterised as it quantifies the possibility of a series of interrelated events occurring in a certain time interval) and emphasises that probability is not a suitable measure to describe the characteristics of uncertain events such as fire events and, consequently, justifies the use of frequency as a measure of the possibility of occurrence (or probability of occurrence) of a fire scenario.

The individual probabilities that contribute to the final calculation can be derived by experts in various ways depending on the meaning attributed to them: if they are considered a pure mathematical expression of the relationship existing between the possibility that an event has of occurring with respect to the totality of possible events (statistical view), they can only be derived from objective data (possibly through the application of appropriate formulas and in any case without limiting themselves to historical data of fire losses which may not be useful for predictive purposes) differently, and if they are understood as an expression of the expert's belief that an event may 'really' occur (subjective view), experience becomes prevalent over data and the expert's judgement will influence the probabilistic calculations in a decisive manner.

The estimation of consequences can be conducted using the following factors:

Historical experience: Historical data of consequences are used, provided they are appropriate for the case under consideration, according to a paradigm known as 'case-based reasoning'.

Expert judgement: This is based on the past experience of an expert or the application of a systematic and consistent process using procedures to enhance engineering judgement and operational experience such as the Delphi method, cf.

Consequence calculation models: Deterministic or statistical models that attempt to assess the consequences of a fire scenario developing in a building from a specific location.

The guide provides a detailed list of what must be included in the documentation:

names and roles of the participants in the analysis and the persons in charge of monitoring any subsequent changes;

objective of risk analysis;

scope of the analysis;

nature and list of identified and assessed hazards;

acceptability criteria and how they are constructed;

groups of scenarios chosen and why they can be considered representative of all potential scenarios;

types of scenarios excluded and their justification;

methods used in the course of the evaluation for the analysis of frequencies, consequences, uncertainty and overall risk, as well as justification of their appropriateness and possible limitations;

sources of the data used and justification of their appropriateness;

results of probabilistic and frequency calculations;

results of the uncertainty analysis and any considerations regarding their influence on the results obtained;

final results of calculated risks.

7.3.2 Engineering Guide to Performance-Based Fire Protection

Of all the professional associations with an international character, the SFPE is perhaps the one best known for its activities related to the in-depth study of issues aimed at defining a performance approach in fire-safety engineering, the Fire Safety Engineering also known by its acronym 'FSE'.

An up-to-date and highly significant summary of the association's work in this field is the technical publication 'SFPE Engineering Guide to Performance-Based Fire Protection'.

This publication, to all intents and purposes having the character of a guideline, represents one of the main points of reference for any safety expert interested in the search for useful criteria for the implementation of a workflow for the management of the fire performance engineering aspects of a building.

In addition to the aspects of the engineering approach, the guide provides a series of extremely up-to-date information on the deterministic investigation, carried out through the application of simulation models and calculation codes, of accidental consequences. Useful not only to experts in numerical simulation, the publication is an excellent resource for all professionals and people who, in various capacities, have to deal with fire projects. In fact, from the application of advanced techniques and procedures to identify the equivalence of alternative prevention or protection measures, it is also possible to optimise a design with respect to reference fire scenarios rather than optimising the very costs associated with the fire project.

The guide supports the designer in different aspects such as:

the definition and documentation of the achievement of specific fire-safety objectives to be maintained throughout the life cycle of the building, in accordance with the general objectives defined in the general fire-safety policy defined by the organisation;

the identification and management of the fire strategy within the limits specified as respecting the performance-based analysis;

the definition of fire protection measures, which ensure acceptable levels of safety and do not require the introduction of additional constraints either in the fire protection project or in the building project (or in any case in the subsequent management of the building).

In addition, the publication clearly defines the basic steps for managing a performance-based fire practice, specifically:

the definition of the scope of the fire project;
identification of objectives;
the definition of specific stakeholder objectives;
the identification and development of project fire scenarios;
the development and evaluation of protection alternatives;
documentation of the activities carried out (with particular reference to the outcomes, conclusive assessments, assumptions, limits and measures to guarantee the level of fire safety achieved over time).

For the purposes of understanding the authors' motivation for including the reference to this publication in this work, it is extremely important to note that the guide perfectly integrates the concepts of fire risk and fire risk assessment and clearly makes this link explicit within the workflow presented, in line with similar reference standards/specifications.[3]

In particular, the guide aims to indicate the methods and general workflow of analysis by which specialists can develop a fire-safety strategy based on measures to determine a level of fire safety that can be considered acceptable.

Acceptability itself cannot disregard the measure of the benefits expected from the implementation of the firefighting strategy and its individual measures, but, in fact, neither can it disregard the fire risk that can be considered as acceptable, as well as the representative and appropriately conservative fire risk of the reality being analysed.

---

3 See for example the existing relationship between ISO TS 16732 'FSE – Guidance on fire risk assessment' and ISO TS 16733 'FSE – Selection of design fire scenarios and design fires' discussed in this book.

Just as requirements derived from the application of prescriptive codes can be used as a term of comparison in a performance-type process, so fire risk classifications derived from norms, standards, previous experience, evaluations carried out using different methods can also be a valid point of reference for verifying the deviation (positive and negative) of the risk classification carried out within the project.

As part of the process underpinning the performance approach, the fire protection specialist, in addition to complying with all mandatory requirements, is responsible for identifying fire hazards and risks and informing all parties involved in various capacities about them. He or she must also assess and declare (in a manual defined as 'O&M', an acronym for 'Operations and Maintenance') the conditions that will satisfy over time the validity of the study carried out and thus the guarantee the level of safety achieved through the analysis, a level that coincides, by definition, with that of acceptable fire risk.

If the guideline is applied to the design of new buildings, it is desirable that the fire protection designer is involved from the start of the design phase because the performance-oriented engineering approach is itself a working model based on iteration and consideration of alternatives: the model works through successive steps that may have a different degree of depth depending on the results of the steps already taken and the information available at a given design stage. Consequently, the overall approach is more effective and efficient if such a model is applied from the initial design stages.

In this way, for example, fire risk analysis can be used extensively in order to identify the main fire hazards and risks at a preliminary stage, and consequently the alternative design strategies that can be implemented to reduce the degree of risk classified; or the number of fire scenarios to be subjected to deterministic investigation can be optimised; or fire risk analysis can be used in the various stages of building design to assess the influence of architectural, plant, structural, etc., changes on the risk classification carried out (for example, implementing a change management technique to prevent an increase in fire risk during construction or to identify which possible aggravations will need to be further investigated). It is also possible to use the fire risk analysis at the various stages of building design to assess the influence of architectural, plant, structural changes, etc., on the risk classification carried out (for example, implementing a change management technique to prevent an aggravation of the fire risk during construction or to identify the possible aggravations that will need to be investigated).

In general, the design team (or, if the work is already in progress, the construction management and execution supervision team) must inform the fire-safety designer of any changes that become necessary; in this way, within the scope of his powers and with his expert judgement, he or she will be able to assess the impact on the fire risk analysis already carried out, even in cases where there are apparently no direct implications.

Among the tasks that the fire planner may be tasked to perform is to assess whether the fire risk manifests equivalence to a prescriptive reference or whether alternative protection strategies should be examined. This task requires that the performance criteria have been suitably defined (in order to measure the degree of achievement of the objectives highlighted by the stakeholders), that possible fire scenarios have been identified and that the design fire scenarios on which the deterministic investigation is to be conducted have been selected from among them.

The identification and selection of scenarios cannot be conducted, except in the simplest cases for which expert judgement can lead to the same results, without the support of a risk analysis that enables the right prioritisation of the identified cases, the determination of which are to be considered representative, and the optimisation of the time and costs associated with subsequent investigations.

The fire risk analysis can be conducted in such a way as to take into due consideration all the characteristics that the SFPE guideline defines as essential to the description of a possible fire scenario and specifically those of the fire, the building and the occupants. An application of risk analysis conducted in these terms, since it cannot be reduced solely to an assessment of the fire load determined by the quantity of combustibles present, makes it possible to hypothesise the dynamics of a fire in relation to the aforementioned characteristics (of the building, the fire, etc.) and to assess the impact on the occupants and, in consideration of what is explicit in the standard, on the structures, the contents, the operations, the image of the organisation, etc.

Risk analysis can also be an excellent tool for conducting preliminary cost–benefit analyses,[4] analyses possibly based on historical data related to fire damage estimates.

Finally, with regard to fire scenarios, the risk analysis helps to select scenarios that are not too extreme (extremely improbable, too conservative or, on the contrary, excessively non-conservative) so as to ensure that the building's design remains congruent with the fire risk arising from the real dangers of the activity analysed: this means that it must not become excessively costly (improbable and extremely conservative scenarios), nor must it lead to an unacceptable level of risk for occupants (extremely non-conservative scenarios).

Design scenarios must be defined from a risk analysis as they are not a mere qualitative description of all the possible fires that may develop within a building, but rather a term of comparison suitable for verifying the degree of robustness of a given firefighting strategy with respect to representative, credible and appropriately conservative cases, also derived from historical and operational experience.

The guideline makes explicit, with particular reference to the identification of fire scenarios, some typical methods for conducting hazard identification. These include, in accordance with the contents of the specific guideline already examined and also published by the SFPE association, FMEA analysis, failure and malfunctioning analysis, typical and consolidated industrial risk analysis methods (Hazop operability analysis, PHA preliminary hazard analysis and failure tree and event tree) and 'What–if?' analysis; it also does not forget to mention the aforementioned historical and operational experience and the databases containing statistics and information on accidents that have actually occurred, even though these sources often have certain limitations connected with the impossibility of deriving information to derive probabilities of occurrence or to directly select scenarios on the basis of the number of occurrences.

The guide identifies two complementary phases for conducting risk assessment: a hazard analysis phase and a risk analysis phase. In the first phase, an attempt is made to identify ignition sources, fuels and development sequences; in the second phase, work is carried out to contextualise the hazards previously identified and associate them with a probability of occurrence and a series of expected consequences (parameters that can both be defined qualitatively and quantitatively).

The SFPE guidance suggests the use of the tree of fire safety concepts set out in NFPA 550 as a method for identifying general approaches and methods for achieving the intended objectives: traversing the NFPA tree from upper to lower levels ensures the systematic identification of protection strategies that can be developed into performance criteria for managing the fire risk identified through the analysis. The application of this methodology also serves to highlight possible shortcomings in the fire design since the NFPA tree brings together all the possible or foreseeable causes

---

4 It should be borne in mind here that the cost associated with the measures making up the strategy must include the costs of maintaining the measure in efficiency, which by their nature are spread over time.

and mitigation hypotheses in a coherent and effective manner; moreover, it allows the designer to identify project alternatives and impacts that have not been considered, simply by superimposing the NFPA 550 tree on the hypothesised strategies in a logical sense.

It is essential that each strategy is verified through the application of risk analysis and, in this sense, two alternative methods identified by the terms 'classical risk analysis' and 'risk binning analysis' are proposed.

The classical analysis is based on the assessment of the probability of occurrence and the expected consequences associated with each project fire scenario, leading to the re-composition of the overall risks.

The 'risk binning analysis' is based on quantifying the expected consequences of the most critical project scenarios in order to obtain a qualitative indication of the overall risk level.

The analysis can be conducted by classifying the scenarios into categories (one category for expected consequences and the other for expected frequencies), and then placing the combinations ('bins') obtained on a matrix; constructed in this manner, the matrix will graphically highlight the acceptability of the assumed project fire scenarios (with reference to the degree of acceptability of the fire risk predefined by the stakeholders).

Whichever analysis is conducted for the purpose of fire risk assessment, it is necessary for the final report of the performance-oriented project to indicate the purpose of the risk analysis, the assumptions, the boundary conditions, the findings (both as hazards and as risks), the acceptability criterion and the limits of validity; in this way, the risk analysis can be reviewed over time to decide, in consideration of changed conditions, whether it is applicable again or whether a new analysis is necessary to take into account the changes that have occurred.

7.4 Italian Fire Code

While it is a local regulation, given the origin of the authors of this book as well as their predominant professional environment, it is useful to quote the latest Italian Fire Code, available also in English, as a risk and performance-based tool for fire strategy design.

Fire safety of residential buildings and activities subjected to fire inspection are difficult tasks, especially when the safety targets have to be adopted in pre-existing buildings or in activities that are going to be modified into more complex ones. Usually, these circumstances show more constraints, and it can be challenging to achieve an acceptable level of fire residual risk by prescriptive fire regulations. Therefore, the Italian National Fire Rescue and Service, in charge of fire safety in Italy, in August 2015, issued a new Fire Safety Code whose design approach is more oriented to fire performance-based design rather than prescriptive fire codes. The flexibility of this novel fire design methodology offers a very completed tool for experts in order to design fire-safety measures for buildings and activities subjected to fire inspection. This section highlights the contents and the fire-safety strategy design methodology of the new Italian Fire Safety Code.

Fire codes and regulations play a fundamental role in reducing risks and achieving acceptable levels of fire safety, both in buildings and high hazard facilities.

Most of the Italian activities subjected to fire inspection, such as stores, malls, schools, hospitals, car parking facilities and industrial buildings, are subject to prescriptive fire-safety code. As well known, the prescriptive fire codes are consisting of specific requirements that attempt to specify all the different components and measures of the system to provide fire safety for a building or an industrial activity. Nevertheless, the contribution of each requirement to the level of safety provided by the system is not known and the interactions between the components are not generally known or taken into account. In addition, when the safety targets have to be satisfied in already built buildings or industrial activities that are going to be modified into more complex ones, there are more constraints to cope with and it could be difficult to achieve an acceptable level of fire risk using prescriptive-based fire codes and regulations. These inherent deficiencies lead to lack of flexibility, conservative outcomes and unnecessary cost burdens.

On 20 August 2015, the Italian Home Office released the Ministerial Decree 3rd August 2015 that contains a new approach to the fire-safety design of activities subjected to fire inspection. The technical Ministerial Decree is titled 'Approval of fire prevention technical standards, pursuant to Article 15 of Legislative Decree 139 of 8 March 2006', but it is commonly recognised among Italian fire officers and practitioners as the Italian Fire Prevention Code (IFC).

The IFC, following the fire-safety engineering principles, sees the process of fire-safety design considering the system as a whole by focusing on the safety targets whether they are life safety, property loss, business interruption, environmental damage or heritage preservation.

Furthermore, IFC gives a holistic approach to fire-safety design, guiding the practitioner during the fire design process in choosing the best fire provisions to reduce and mitigate the assessed fire risk to an acceptable level. The international state of the art of fire design codes, such as but not limited to the BS 9999:2008 'Code of practice for fire safety in the design, management and use of buildings', the NFPA 101 'Life Safety Code' and the International Fire Code 2009 have been considered during the developing stage of the IFC.

Nowadays, actual studies and experiments for understanding fire-related phenomena increase the capability of the fire engineering community to assess and predict the performance of structures and protection systems when exposed to a fire event. The use of analytical tools such as empirical models, finite element analysis and computational fluid dynamics, in conjunction with bench top and full-scale testing, has improved the ability of fire practitioners to develop performance-based solutions to challenging fire-safety design of high-risk industrial activities or complex buildings.

The fundamental assumptions of the IFC are as follows:

1) In ordinary conditions (no arson, no catastrophic situations), a breakout of a fire in an activity could happen only in one point of ignition.
2) In any safety design, the risk of fire cannot be reduced to zero; the fire-safety prevention, protection and management measures provided applying the IFC fire design process ensure a proper selection of the measure that minimise the risk of fire in terms both of occurrence and damages, at a level that could be considered as an acceptable level of safety. This paper highlights and discusses principles and the methods proposed by the IFC for the fire-safety design of activities subjected to fire approval and inspection.

7.4.1 IFC Fire-Safety Design Method

Advance structural engineering as well as material science innovation technologies satisfy the architectural demands to build up complex buildings that cannot comply with fire prescriptive codes. In order to assure an acceptable level of fire-safety, risk-based methods could provide an opportunity to determine the quantitative safety level. The main advantage of the risk-based method is the hazard versus safeguard determination both as probability and damage.

General design of the IFC fire risk assessment and mitigation strategy is based on the following principles. The first one is the overall applicability of the design procedure that should be specific for each activity subjected to fire inspection. The IFC method is oriented to 'Simplicity': given different choices to achieve the same safety level, the simpler one and the more easily achievable solution, by also taking into account the maintenance features, shall be preferred.

The design is 'module' oriented: the complexity of the fire design is split into easily accessible modules, which guide the designer towards the appropriate solutions for any specific activities.

Moreover, IFC has also been standardised and integrated to the fire-safety and fire protection language of international standards. The code is also fully inclusive: different disabilities (e.g. motor, sensory, cognitive, …), temporary or permanent, of the occupants shall be considered as an integral part of the fire-safety design. Lastly, the IFC was thought to be easily 'updatable': the document has been drafted in a format that can easily be kept up to date with the continuous improvements in terms of technologies and knowledge available in the fire-safety science.

The IFC design method is also 'Flexible': for each fire-safety project it gives design solutions that are semi-performance-based (the so-called 'deemed-to-satisfy solutions'). These compliant solutions contain prescriptive examples of materials, products, design factors, construction and installation methods, which – if adopted – comply with the performance requirements of the IFC.

If the deemed-to-satisfy solution could not be put in place, the IFC offers performance-based solutions called 'alternative solutions'. The alternative solution is any solution that can meet the IFC performance requirements, other than a deemed-to-satisfy solution, using the following allowed methods:

Fire-Safety Engineering (FSE).
Design solution based on innovative technologies products and systems.
Alternative, national or international, authoritative fire codes or regulations.
Experimental tests.

In any case, the alternative solution implies that the requested level of performance is, in any case, achieved.

The IFC method requires the scope definition describing the context of the activity pointing out the operating and environment peculiarity of the activity. The following step is choosing the safety objectives: life safety, property protection and safeguard of the environment. The third step is to identify and analyse the activity fire risks by means of a systematic evaluation based on fire hazard related factors. Then, to prioritise the risks in order to tackle them by means of fire preventions, protections and management measures, is required. The IFC final goal is aimed to achieve life safety, property protection and the safeguard of the environment in case of fire. According to the IFC document, the risk assessment is followed by the evaluation of the following simplified parameters:

R_{life}, risk profile concerning human life safety.
R_{prop}, risk profile concerning the property protection.
R_{env}, risk profile concerning the protection of the environment from the effects of the fire.

The fire risk profile connected to life safety – R_{life} – is evaluated as a function of the growth rate of a fire in a building compartment. The behaviour of building occupants in response to a fire is defined. The fire growth is a square type (typical parameter varies from 1, which represents a slow fire growth, to 4, which is an ultra-fast fire growth). The occupant characteristics are summarised into the following groups:

A occupants awake and familiar.
B occupants awake but unfamiliar.
C occupants that could be asleep.
D occupants receiving medical care.
E occupants in movement (stations, tunnels).

The occupancy characteristics chosen by the IFC are the same of those contained in BS 9999 [4]. R_{prop} is based on the building strategic nature or heritage, cultural, architectonic or artistic value coupled to the significance of the building contents, such as business continuities or property protection. The last simplified risk parameter R_{env} takes into consideration the risk of environmental damage or environmental contamination during and after the outbreak of a fire. R_{env} is assessed also taking into account the emergency response management. In Figure 7.8, the diagram highlights the general fire-safety design method proposed by the IFC.

The IFC structure is consistent with the international state of the art of the fire-safety science and engineering. Compared to the traditional prescriptive fire regulations, the performance-based approach implies a wide range of advantages such as flexibility in the choice of the most appropriate design solution consequent to a more realistic definition of fire scenarios. On the contrary, performance-based design requires more expertise and knowledge in this field, especially when the fire or evacuation numerical models are needed. In this case, in fact, the sensibility of the designer can strongly affect the results and therefore the proposed solution. The challenge of the next few years will be to guarantee the reliability of the integrated technology systems according to the RAMS approach (i.e. reliability, availability, maintainability and safety) in order to improve the safety performance.

The Italian Fire code is based on the following fundamental assumptions:

Under ordinary conditions, a fire in an activity will start at just one point of ignition.
The fire risk in a given activity cannot be reduced to zero.

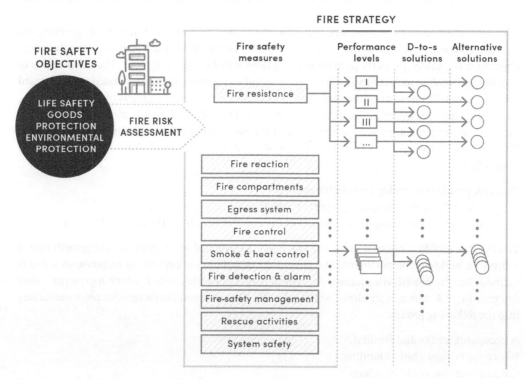

Figure 7.8 IFC fire-safety design method.

The *fire-safety* prevention, protection and management measures provided for in the Italian Fire Code have therefore been selected to minimise the risk of fire, in terms of probability and consequences, within limits *considered* acceptable.

Designing the fire safety of an activity means identifying the technical and management solutions aimed at achieving the *primary objectives* of fire prevention, which are

safety of human life;
protection of people;
protection of property and the environment.

The primary objectives of fire prevention are considered to be achieved if the activities have been designed, implemented and managed in such a way as to

minimise the causes of fire and explosion;
guarantee the stability of the load-bearing structures for a set period of time;
limit the production and propagation of fire within activities;
limit the spread of fire to adjacent activities;
limit the effects of an explosion;
ensure that the occupants can leave the activity safely or that they be rescued in another manner;
ensure that the firefighter/rescue squads are able to work under safe conditions;
protect historically or artistically significant buildings;
ensure business continuity in strategic works;
prevent environmental damage and limit any compromise of the environment in case of fire.

The design of fire safety for activities is an iterative process consisting of the following steps:

a) *Scope of the design*: The activity and its operation are described qualitatively and quantitatively in order to clarify the scope of the design.
b) *Safety objectives*: The safety objectives of the design, applicable to the activity, shall be specified.
c) *Risk assessment*: The fire risk assessment shall be carried out according to a defined process;
d) *Risk profiles*: Risk profiles shall be determined and assigned.
e) *Fire prevention strategy*: Risk mitigation is carried out through preventive, protective and management measures that remove hazards, reduce risks or protect against their consequences:
   – defining the overall *fire prevention strategy*;
   – Attributing *performance levels* for all fire prevention measures;
   – Identifying the *design solutions* that guarantee the achievement of the assigned performance levels;
f) if the *result* of the design is not considered compatible with the *purpose* defined in point a), the designer shall iterate the steps referred to in point (e) of this methodology.

The designer shall use an industry standard method for assessing fire risk depending on the complexity of the activity being handled. The fire risk assessment represents an analysis of the specific activity aimed at identifying the most severe but credible fire scenarios and the corresponding consequences for occupants, property and the environment. This analysis allows the designer to implement and, if necessary, integrate the standard design solutions.

In any case, the fire risk assessment shall cover at least the following topics:

a) Identification of fire hazards (for example, the following are assessed: ignition sources, combustible or flammable materials, fire loads, ignition–fuel interaction, any significant quantities of mixtures or hazardous substances, processes that may lead to fires or explosions and possible formation of explosive atmospheres).

b) Description of the context and environment in which the hazards may be found (for examples, accessibility and viability conditions, company layout, distances, separations, isolation, building characteristics, type of construction, geometrical complexity, volume, surface areas, height, underground levels, plan-volumetric articulation, compartmentalisation, ventilation and surface areas available for the extraction of smoke and heat, etc.).

c) Determination of the quantity and type of occupants exposed to the risk of fire.

d) Identification of property exposed to the risk of fire.

e) Qualitative or quantitative assessment of the consequences of the fire on occupants, property and the environment.

f) Identification of preventive measures which can remove or reduce the hazards that give rise to significant risks.

In areas of activity where *inflammable substances* are present in the form of combustible gases, vapours, mists or dusts, the fire risk assessment shall also include the risk assessment for *explosive atmospheres*.

After assessing the fire risk for the activity, the designer shall assign the *risk profiles*, as described next in this book. Basically, these are the following factors:

R_{life}, *risk profile* concerning *human life* safety.

R_{prop}, *risk profile* concerning the protection of *property*.

R_{env}, *risk profile* concerning the protection of the *environment* from the effects of fire.

The designer mitigates the risk of fire by application of an appropriate *fire prevention strategy* consisting of fire prevention, protection and management measures.

For each *fire prevention measure*, there are different *levels of performance*, graduated according to the increasing complexity of the expected performance and identified by Roman numerals (e.g. I, II, III, etc.). The designer shall apply all *fire prevention measures* to the activity, establishing for each the relevant *performance levels* in relation to the *safety objectives* to be achieved and the *risk assessment* of the activity.

Once the fire *risk assessment* for the activity has been carried out and the R_{life}, R_{prop} and R_{env} risk profiles are established in the relevant areas, the designer shall assign the relevant *performance levels* to the fire prevention measures. If the assigned levels are lower than those proposed, the designer shall be required to demonstrate that the fire-safety objectives have been achieved by using one of the *fire-safety design methods*.

For each *performance level* of each fire prevention measure, there are different *design solutions*. Application of one of the *design solutions* shall guarantee the achievement of the required *performance level*. Three types of *design solutions* have been defined:

a) Deemed-to-satisfy solutions.

b) Alternative solutions.

c) Solutions in derogation.

The designer using *deemed-to-satisfy solutions* is not required to provide further technical assessments to demonstrate the achievement of the related *performance level*.

Deemed-to-satisfy solutions are only those proposed in the relevant parts of the *fire prevention strategy* section of the standard.

Table 7.1 Additional fire-safety design methods.

Methods	Description and application limits
Analysis and design according to expert judgement	Analysis according to expert judgement is based on general fire prevention principles and on the breadth of knowledge and experience of the *fire-safety professional*, an expert in fire safety

The designer may make use of the *alternative solutions* proposed in the relevant parts of the *fire prevention strategy* section of the standard or may propose specific *alternative solutions*.

The designer using *alternative solutions* shall demonstrate the achievement of the related *performance level* by using one of the *fire-safety design methods* allowed for each fire prevention measure

If neither *deemed-to-satisfy* nor *alternative* solutions can be effectively applied, the designer may use the derogation procedure as provided for by the laws in force.

Designers who choose *solutions in derogation* must demonstrate the achievement of the pertinent fire prevention objectives using one of the *fire-safety design methods* (as those listed in Table 7.1)

The Italian Code also introduces the concepts of systems or installations with higher availability. They are systems or installations with a higher level of availability than the minimum required by the reference standards of the system or installation.

Higher availability for systems or installations may be achieved through

a) greater reliability (for example, by using components with lower failure rates, redundancy of power supply sources, extinguishers, critical components, introducing measures to reduce human error, specific protection measures against the effects of fire, etc.);

b) greater maintainability and maintenance support performance (for example, by reducing fault recovery times, scheduling maintenance for sectors of the system, periodic checks and testing, etc.)

In order to maintain the level of safety required for the activity, for systems or installations with higher availability, the management of degraded states or of the down states of the system must be anticipated (for example, by limiting the severity of degraded states, compensatory management measures, operating conditions or limitations, etc.).

8

Performance-Based Fire Engineering

The application of the principles of fire-safety engineering allows, similarly to other engineering disciplines, to define solutions suitable for the achievement of design goals through quantitative analysis, given an initial fire risk assessment. This is valid for both civil buildings and industrial assets.

The designer defines the purpose of design and then specifies the fire-safety goals that he/she intends to guarantee and translates them into quantitative performance thresholds. Subsequently, the designer identifies the design fire scenarios, the most serious events that can reasonably occur in the activity. Then, thanks to analytical or numerical modelling tools, the designer describes or calculates the effects of design fire scenarios in relation to the hypothesised design solution for the activity. If the effects thus calculated retain an adequate safety margin compared to the previously established performance thresholds, then the analysed design solution is considered acceptable (i.e. the performance requirements are not considering the selected fire-safety strategy). This is a performance-based approach that can be seen alternative or complementary to a design based on the compliance against pre-defined technical-specific requirements (prescriptive approach and associated solutions).

Note that it is not always necessary to use numerical models or even advanced numerical models like computer-aided simulation codes to assess the effects of fire scenarios and verify performances against acceptability defined thresholds.

This chapter summarises the design methodology of fire-safety engineering (or *performance-based fire design*) for alternative and equivalent solutions.

The performance-based design methodology usually consists of two phases:

a) *First phase – preliminary analysis*: The steps that lead to identifying the most representative conditions of the risk to which the activity is exposed are formalised together with the *performance thresholds* in relation to the safety targets to be pursued.

b) *Second phase – quantitative analysis*: Using calculation models, the qualitative–quantitative analysis of the effects of the fire is performed in relation to the goals assumed, comparing the results obtained with the *performance thresholds* already identified and defining the design to be submitted for the final approval.

The preliminary analysis phase consists of the following sub-phases necessary to define the risks to be counteracted and, consequently, the objective criteria for quantifying them, which are necessary for the subsequent numerical analysis.

Fire Risk Management: Principles and Strategies for Buildings and Industrial Assets, First Edition.
Luca Fiorentini and Fabio Dattilo.
© 2023 John Wiley & Sons, Inc. Published 2023 by John Wiley & Sons, Inc.
Companion Website: www.wiley.com/go/Fiorentini/FireRiskManagement

The *purpose* of the fire design is defined in this sub-phase. The designer identifies and documents at least the following aspects:

destination use and occupancy of the activity;

purpose of performance-based fire design;

any design constraints deriving from regulatory provisions or from specific requirements of the activity;

fire hazards associated with the intended use;

boundary conditions for the identification of the data necessary for the evaluation of the effects that could be produced;

characteristics of the occupants in relation to the type of building and the intended use destination.

After establishing the purpose of the project, in particular the destination and modes of use of the activity, the designer specifies the *fire-safety goals* in relation to the specific needs of the activity in question and to the design purposes.

Fire-safety goals should be based on the results of the fire risk assessment and should include any eventual requirements for operation phase.

With the fire-safety goals, for example, the level of protection of the occupants' safety, the maximum damage tolerable to the activity and its content, and the continuity of operation following an incidental event are qualitatively specified.

The next step is to transform the fire goals into *performance criteria*. These are quantitative and qualitative thresholds in relation to which the objective evaluation of fire safety can be carried out. By choosing the *performance thresholds*, the thermal effects on the structures, the propagation of the fire, the damage to the occupants, the goods and the environment become *quantitative*. These *performance thresholds* must be able to be used in the second phase of the design in order to objectively discriminate the design solutions that satisfy the fire-safety goals from those that do not reach the required performances instead. For the purposes of designing the safeguard of life, *the life safety criteria* are established. These are the thresholds used to define the *incapacitation* of occupants exposed to fire and its effects.

By definition, occupants reach *incapacitation* when they become unable to get secure themselves autonomously. The death of the subject follows in a short time.

Fire scenarios represent the schematisation of the most serious events that can reasonably occur in the activity (*credible worst-case scenarios*), in relation to the characteristics of the fire, of the building and the occupants.

The second phase is the quantitative analysis. This phase is made up of some sub-phases necessary to carry out the safety checks of the scenarios identified in the preliminary phase.

Firstly, the fire prevention professional elaborates one or more design solutions for the activity, congruent with the purposes already defined, to be submitted to the subsequent verification of satisfaction of the fire-safety goals.

Then, the designer calculates the effects (by using physical effects modelling) that the design fire scenarios would determine in the activity for each design solution developed in the previous phase. To this end, the designer uses an *analytical* or *numerical* model: the application of the model provides the quantitative results that allow for describing the evolution of the fire and its effects on the structures, occupants or the environment, according to the design purposes. The modelling of the effects of the fire makes it possible to calculate the effects of the single scenarios for each design solution. The modelling results are used to verify compliance with the performance thresholds for the design solutions for each design fire scenario. Design solutions that do not meet all the performance thresholds for each design fire scenario must be rejected.

The designer selects the final design solution from among those that have been positively verified with respect to the design fire scenarios, considering the costs and management future efforts connected with the operation phase.

Of course, a technical summary and a technical report must be provided as documentation supporting the design, including any eventual ALARP demonstration.

With the application of the performance-based methodology, the designer establishes the identification of the design fire prevention and protection measures on specific hypotheses and operating limitations: therefore, specific measures of *fire-safety management* must be foreseen so that the reduction in the safety level initially ensured cannot occur. The specific *fire-safety management* measures must *refer* to the aspects dealt with in performance-based design, with particular regard to the specific design solutions, to the fire prevention and protection measures adopted and to the maintenance of the operating conditions from which the values of the input parameters in performance-based design derive. On specific *fire-safety management* measures, they are subject to periodic checks by the activity manager according to the timing already defined in the design. As part of the programme for the implementation of fire-safety management, the measures taken in relation to the following points must be assessed and explained:

personnel organisation (including leadership);
identification and assessment of the hazards arising from the activity;
operational check;
modifications management;
emergency planning;
safety of rescue teams;
performance monitoring;
maintenance of protection systems;
control and revision.

These elements are crucial for a valid fire-safety management system (FSMS) since they control the fire risk over time, considering any eventual modifications during the life-cycle of the asset under consideration.

If the active protection systems are considered in order to reduce the thermal power released by the RHR(t) fire or otherwise contribute to mitigating the effects of the fire, *higher availability systems* must be installed.

The fire-safety professional can opt among the calculation models that the technical know-how of the sector makes available, based on evaluations concerning the complexity of the design. The designer who adopts sophisticated calculation models must possess a particular competence in their use as well as an in-depth knowledge of both the theoretical foundations underlying them and the dynamics of the fire. At present, the following are the most frequently used models (in the majority of cases, a combination of these is used):

Analytical models (including simple correlations).
Numerical models, including
  – fire zone simulation models for enclosed spaces;
  – field fire simulation models;
  – evacuation simulation models;
  – thermo-structural analysis models.

In their field of application, the analytical models guarantee accurate estimates of specific effects of the fire (e.g. the calculation of the *flashover* time in a compartment). For more complex analyses involving time-dependent interactions of multiple physical and chemical processes present in the development of a fire, numerical models are generally used. For the most relevant input parameters of the model, a *sensitivity* analysis of the results to the variation of the input parameter must be carried out. For example, the results of the analysis should not be significantly dependent on the size of the calculation grid. The simultaneous use of several types of models is allowed. For example,

specific models can be used for the evaluation of the activation time of a detection or extinguishing system and of the breaking of a glass as a function of temperature, to then insert the data obtained in a modelling carried out with field models;

a zone models can be used to initially evaluate the most critical conditions of the phenomenon, to then deepen the treatment of the effects with field models.

Specific models are available for physical effects modelling of industrial accidents.

In the following sections, this chapter describes the procedure for *identifying*, *selecting* and *quantifying* the *design fire scenarios* that are used in the quantitative analysis by the *designer* who uses fire-safety engineering. The *fire scenarios* represent the detailed description of the events that can reasonably occur in relation to three fundamental aspects:

fire characteristics;
activity (occupancy) characteristics;
occupant characteristics.

This procedure consists of the following steps:

identification of the possible *fire scenarios* that can develop in the activity, on which the outcome of the entire evaluation depends, according to the performance-based method;

selection of the *design fire scenarios* among all the possible identified fire scenarios;

quantitative description of selected design fire scenarios.

The first step of the procedure consists in *identifying all the possible fire scenarios* that can develop during the useful life of the activity. Related to this, all reasonably foreseeable *operating conditions* must be considered (e.g. temporary setups, different spatial configurations of combustible materials, modification of exit routes and crowding, etc.). To identify fire scenarios, the designer can develop a specific *event tree* starting from each relevant and credible initiator event. The process can be carried out in a *qualitative* way or in a *quantitative* way if statistical data are available from authoritative and shared sources. Each identified fire scenario must be fully and unequivocally described in relation to its three fundamental aspects: the characteristics of the fire, the characteristics of the activity and the characteristics of the occupants. In any case, the designer must specify whether the hypothesised fire scenario is related to a *pre-flashover* condition or a *post-flashover* condition, depending on the goal to be achieved.

In the phase of identification of the scenarios, the designer must take into account the fires that have affected buildings or activities similar to the one under examination by historical analysis and must describe

the ignition event characterised by a fire outbreak and by the surrounding environment conditions;

spread of fire and combustion products;

action of technological systems and active protection against the fire;

actions carried out by members of the company team dedicated to firefighting present in the environment;

distribution and behaviour of the occupants.

In the first step of the procedure, a large number of possible fire scenarios in the activity is generally identified. The purpose of this second step of the procedure is to reduce the number of fire scenarios to the minimum reasonable number, in order to lighten the subsequent work of verifying the design solutions. The designer selects the *fire scenarios* and extracts the subset of the *design fire scenarios*, making explicit in the design documentation the reasons that lead to exclude some from the subsequent quantitative analysis, referring to the tree of events already developed in the previous step or with other mode. The designer selects the *heaviest* among the *credible* fire scenarios. The *design fire scenarios* thus selected represent for the activity a fire risk level no less than that fully described by the set of all *fire scenarios*. The design solutions, respecting the *performance thresholds* required in the context of the *design fire scenarios*, therefore guarantee the same degree of safety even with respect to all of the other *fire scenarios*. The selection of fire scenarios is strongly influenced by the goal that the designer intends to achieve. For example, if safeguarding of occupants is the main pursual during the evacuation phase, scenarios such as those indicated below may be selected:

a fire of short duration and fast growth, which is accompanied by high production of smoke and combustion gas (for example, the burning of an upholstered furniture), is more critical than one that releases more thermal power, but which has a slow growth and lasts longer, even if the latter thermally stresses the present building elements more severely;

a fire of limited dimensions, which however develops in the vicinity of the evacuation routes of a room with a high density of crowding, may be more dangerous than one which emits greater thermal power, but which originates in a confined environment and which is located away from zones where occupants are expected to be present.

After the selection of the design fire scenarios, the designer must proceed with the *quantitative description* of each of them. The designer translates the qualitative description of the design fire scenarios, already elaborated in the first step, into numerical input data appropriate for the calculation methodology chosen for the verification of the design hypotheses. In relation to the purpose of the analysis, the designer specifies the input data for activities, occupants and fire, listed in detail in the following sections.

The characteristics of the activity influence the evacuation of the occupants, the development and dynamics of the fire and the spread of combustion products. Depending on the objective of the analysis, the quantitative description of the activity may include the following elements:

Architectural and structural features:
  i) location and geometry of the activity, dimensions and distribution of the internal environments;
 ii) structural description and characteristics of the relative load-bearing and separating construction elements;
iii) description of non-structural and finishing materials;
 iv) evacuation system: dimensions, distribution and emergency exits;
  v) size, location and state of effective opening/closing/breaking of design and potential ventilation openings, such as doors, windows, skylights and glazed surfaces;
 vi) barriers that influence the movement of combustion products.

Installations:
  i) active fire protection systems;
  ii) fire detection, signalling and alarm systems;
  iii) technological systems serving the activity, such as air conditioning, distribution or process installations.

Management and operational aspects:
  i) destination of use of the activity and the production process that takes place there;
  ii) organisation of the hosted activity;
  iii) possible actions implemented by the rescuers, foreseen in the emergency plan, able to alter the propagation of the products of combustion; these actions must be considered only exceptionally and evaluated on a case-by-case basis.

Environmental factors that influence the fire performance of the activity.

Depending on the objective of the analysis, the designer describes the characteristics of the occupants in detail, in relation to the impact they may have on the fire scenario. In particular, the description must take into account at least the following aspects where relevant for the purposes of the type of analysis:

overall crowding and distribution of occupants in the activity environments;
type of occupants (for example: workers, occasional visitors, the elderly, children, patients, etc.);
familiarity of the occupants with the activity and the system of escape routes;
occupant waking/sleeping status.

Depending on the objective of the analysis, the description of the fire consists in the quantitative characterisation of the fire outbreak, as a source of *thermal energy* and of *combustion products*, according to the relevance of the following parameters for the purposes of the type of analysis:

location of the fire outbreak;
type of outbreak: smouldering or with flame;
quantity, quality and spatial distribution of the combustible material;
ignition sources;
RHR curve (*rate of heat release and firepower*), as thermal power produced by the fire as time varies RHR(t);
generation of the combustion products taken into consideration (e.g. CO and particulate matter).

For purposes of quantitative characterisation of the fire, the designer can

employ experimental data obtained from direct measurement in the laboratory according to established scientific methodology;
use data published by authoritative and shared sources. The designer always quotes these sources with precision and verifies the correspondence of the experimental test sample (quantity, composition, geometry and test modalities) with the one foreseen in the design fire scenario, using a reasonably conservative approach;
use estimation methodologies.

Alternatively, the designer can use the pre-defined fires from technical standards.

The entire evolution sequence of the fire must be described, starting from the initiator event for a time interval that depends on the safety goals to be achieved as shown in Table 8.1.

The quantitative definition of the various phases of the fire reported here refers to the qualitative curve of the image in Figure 8.1. This methodology can be used

Table 8.1 Minimum duration of design fire scenarios.

Fire-safety goal	Minimum duration of design fire scenarios
Safeguarding the lives of the occupants (occupant life safety)	From the initiator event until the moment when all occupants of the activity reach or remain in a safe place. If the safe place is near or inside the construction work, any interactions between the maintenance of the load-bearing capacity of the construction work and the safe place must be evaluated.
Safeguarding the lives of rescuers (rescuer life safety)	From the initiator event up to five minutes after the end of the operations planned for the rescuers or the arrival of the fire brigade teams at the activity. The reference time for the arrival of the fire brigade can be assumed to be equal to the average of the arrival times taken from the statistics.
Maintaining load-bearing capacity in the event of a fire (structural safety)	From the initiator event until the structural analysis stop, in the cooling phase, to the moment in which the effects of the fire are considered not significant in terms of temporal variation of the characteristics of the stress and displacements.

to construct natural curves with an advanced numerical fire model, for the evaluation of the bearing capacity in fire conditions of the construction works;

evaluate the smoke flow rate during the fire for the design of SHES systems.

In applying the fire-safety performance-based method to life safety, the designer's goals can be

the direct and explicit demonstration of the possibility for all occupants of an activity to reach or stay in a safe place, without this being prevented by an excessive exposure to the effects of the fire;

the demonstration of the possibility for the rescuers to operate in safety.

The design must follow one of the internationally recognised procedures for assessing the position and condition of the occupants during the evolution of the fire scenarios foreseen for the activity.

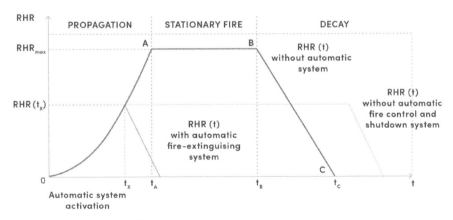

Figure 8.1 Phases of the fire.

The ideal design of an evacuation system should ensure that occupants can reach a safe place or stay put safely, without ever encountering the effects of the fire. This is therefore the first criterion to be used for most of the occupants of the activity. There are situations where the criterion of the first paragraph is not applicable, in particular for occupants who are in the first ignition compartment of the fire.

To resolve the provisions some standards introduce the ASET > RSET criterion exemplified in Figure 8.2.

The performance-based design of the escape route system consists essentially in the calculation and comparison of two time intervals defined as follows:

ASET (available safe escape time);
RSET (required safe escape time).

The evacuation system is considered to be effective if ASET > RSET, that is, if the time in which non-incapacitating environmental conditions remain for the occupants is greater than the time necessary for them to reach a safe place, and not be subject to such unfavourable environmental conditions due to the fire. The difference between ASET and RSET represents the *safety margin* of the performance-based design for life safety:

$$t_{marg} = ASET - RSET$$

In the comparison between different design solutions, the designer maximises the t_{marg} safety margin in relation to the hypotheses assumed in order to consider the uncertainty in the calculation of the ASET and RSET times.

Unless specific assessments are made, $t_{marg} \geq 100\%$ RSET is assumed. In the case of specific assessments on the reliability of input data used in performance-based design, it is permissible to assume $t_{marg} \geq 10\% \cdot$ RSET. In any case, it must be $t_{marg} \geq 30$ seconds.

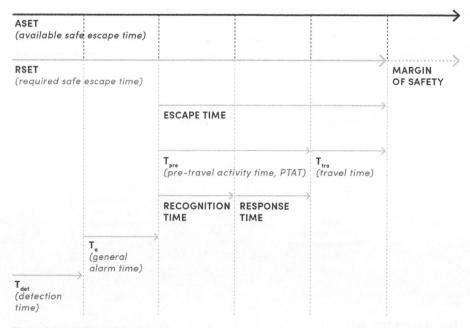

Figure 8.2 ASET > RSET criterion.

ASET, the time available to the occupants to save themselves, strictly depends on the interactions in the fire–building–occupant system: the fire ignites, propagates, and spreads its effects, smoke and heat in the building. The building resists fire by means of active and passive protection measures: fire prevention systems, compartmentalisation, smoke and heat control systems. Occupants are exposed to the effects of the fire in relation to the activities they carry out, their initial position, their path in the building and their physical and psychological condition. As a result, each occupant has an own ASET value. This complexity is solved by the designer with statistical considerations, with numerical calculation models or by assuming simplifying hypotheses. There are two *calculation methods* of ASET:

advanced calculation method;
simplified calculation method.

For the purpose of this book, only the simplified calculation method is discussed.

ISO/TR 16738 provides for the possibility of using the simplification hypothesis of *zero exposure*. According to this method, the designer uses the following very conservative performance thresholds:

minimum height of the stratified smoke from the walking surface equal to 2 m, below which the
  undisturbed layer of air remains and
average temperature of the hot smoke layer ≤200 °C.

These criteria allow the occupants to escape into undisturbed air, unpolluted by the products of combustion, and a value of the radiation from smoke to which they are exposed of less than 2.5 kW/m^2: the analysis is considerably simplified because it is not necessary to perform occupant exposure calculations to toxics, irritants, heat and obscured visibility. It is in fact sufficient to evaluate analytically or with numerical zone or field models the height of the pre-flashover smoke layer in the building.

The simplified calculation method is applicable only if the power of the fire related to the geometry of the environment is sufficient to guarantee the formation of the upper layer of hot smoke: the designer is obliged to check that this condition occurs.

RSET is calculated between the ignition of the fire and the moment when the occupants of the building reach a safe place. RSET also depends on the interactions of the fire–building–occupant system: the occupants' escape is strongly conditioned by the building's geometry and is slowed by the effects of the fire. The reference document for the calculation of RSET is ISO/TR 16738.

RSET is determined by various components, such as the *detection time* t_{det}, the *general alarm time* t_a, the *pre-travel activity time* t_{pre} and the *travel time* t_{tra}:

$$RSET = t_{det} + t_a + t_{pre} + t_{tra}$$

For the purpose of calculating RSET, the designer must develop the *most appropriate design behavioural scenario* for the specific case because the premovement activity and the evacuation speeds depend on the type of population considered and on the methods of use of the building. The parameters vary greatly if the occupants are awake and familiar with the building, such as in a school building, or sleeping and do not know the structure, as in a hotel.

As already indicated for ASET, each occupant also has an own value of RSET.

The *detection time* t_{det} is determined by the type of detection system and the fire scenario. It is the time required for the automatic detection system to notice the fire. It is calculated analytically or with specific numerical modelling of fire scenarios and of the detection system.

The *general alarm time* t_a is the time that elapses between the detection of the fire and the diffusion of information to the occupants, of the general alarm. The general alarm time will therefore be

equal to zero, when the detection directly activates the general alarm of the building;
equal to the delay assessed by the designer, if the detection alerts an emergency management centre that verifies the event and then activates the manual alarm.

In large and complex buildings, the alarm mode must be taken into account, which can be diversified, for example, in the case of a multi-phased evacuation.

The *pre-travel activity time* t_{pre} is the object of the most complex evaluation because it is the time necessary for the occupants to carry out a series of activities that precede the actual movement towards the safe place. The literature indicates that this phase often takes up most of the total evacuation time. The t_{pre} time is made of a *recognition* time and a *response* time.

During the recognition time, the occupants continue the activities they were carrying out before the general alarm, as long as they recognise the need to respond to the alarm.

In response time, the occupants cease their normal activities and dedicate themselves to activities related to the development of the emergency, such as gathering information on the event, stopping and safely setting the equipment, grouping their own group (work or family), fighting the fire, researching and determining the appropriate way of evacuation (*wayfinding*) and other activities, sometimes even incorrect and inappropriate.

Depending on the design behavioural scenario, these times can last even several tens of minutes. Table 8.2 shows some examples of evaluation taken from ISO/TR 16738.

The designer can use values different from those indicated in the literature as long as they are adequately justified, also with reference to evacuation drills reported in the control register.

The *travel time* t_{tra} is the time taken by the occupants to reach a safe place from the end of the pre-travel activities described above. The t_{tra} is calculated with reference to some variables:

a) the distance of the occupants or groups of them from exit routes;
b) the evacuation speeds, which depend on the types of occupants and their interactions with the built environment and the effects of the fire; it is demonstrated that the presence of smoke and heat considerably slows down the speed of evacuation as a function of visibility conditions;
c) the extent of evacuation routes due to geometry, dimensions, and sharp differences in height and obstacles.

In reality, when the occupants of densely crowded buildings flee along the evacuation routes, long lines are formed in the narrowing; furthermore, according to the development of the design fire scenarios examined, some paths may become impassable or blocked. The calculation of the t_{tra} must take these phenomena into account.

Currently, two groups of models are commonly used for the calculation of the movement time: *hydraulic models* and *agent-based models*. Hydraulic models predict with reasonable precision some aspects of occupant movement (e.g. flows through exits) but do not include important factors of human behaviour, such as familiarity with the building, person-to-person interactions and the effect of smoke on their movement.

Table 8.2 Examples of the evaluation of the premovement time, taken from ISO/TR 16738.

Parameters describing the activity taken from ISO/TR 16738	Pre-travel activity times ISO/TR 16738	
	$\Delta t_{pre(1st)}$ first fleeing occupants	$\Delta t_{pre(99th)}$ last fleeing occupants
Example 1: medium complexity hotel	20'	40'
• Occupants: *ciii, sleeping and unfamiliar*;		
• Alarm system: automatic detection and general alarm mediated by employee verification;		
• Building geometric complexity: *multi-floor building and simple layout*;		
• Safety management: *ordinary*.		
Example 2: large production activity	1'30"	3'30"
• Occupants: *A, awake and familiar*;		
• Alarm system: automatic detection and general alarm mediated by employee verification;		
• Building geometric complexity: *multi-floor building and complex layout*;		
• Safety management: *ordinary*.		
Example 3: nursing home care	5'	10'
• Occupants: *D, sleeping and unfamiliar*;		
• Alarm system: automatic detection and general alarm mediated by employee verification;		
• Building geometric complexity: *multi-floor building and simple layout*;		
• Safety management: *ordinary*;		
• Presence of employees in sufficient quantity to manage the evacuation of the disabled.		

Other types of models (e.g. *macroscopic/microscopic and coarse network/fine network/continuous models*) are the subject of intense scientific research and experimentation; currently, only partial validations of the results still exist. Therefore, the results must be evaluated with caution.

The performance thresholds for life safety determine the incapacitation of the occupants and rescuers when subjected to the effects of fire. The designer chooses suitable performance thresholds for the specific activity, in relation to the design fire scenarios, and in particular with reference to the characteristics of the occupants involved (e.g. elderly, children, with disabilities, ...). Compliance with the performance thresholds for life safety must be verified:

for the *occupants*: in all zones of the activity where there is a simultaneous presence of occupants, permanent or moving, and effects of the fire.

for the *rescuers*:

- only if they have a clearly defined role in the emergency planning of the activity,
- in all zones of the activity where there is a simultaneous presence of rescuers, stationary or moving, and effects of the fire.

Table 8.3 Example of performance thresholds that can be used with the advanced calculation method.

Model	Performance	Performance threshold	Reference
Obscuring visibility from smoke	Minimum visibility of reflective panels, not backlit, evaluated at a height of 1.80 m from the walking surface	Occupants: 10 m Occupants in premises of gross surface area <100 m^2: 5 m	ISO 13571:2012
		Rescuers: 5 m Local rescuers in premises of gross surface area <100 m^2: 2.5 m	[1]
Toxic gases	*Fractional effective dose* (FED) and *fractional effective concentration* (FEC) due to exposure to toxic gases and irritating gases, evaluated at a height of 1.80 m from the floor level	Occupants: 0.1	ISO 13571:2012, limiting to 1.1% the portion of occupants incapacitated at reaching the threshold
		Rescuers: no evaluation	–
Heat	Maximum exposure temperature	Occupants : 60 °C	ISO 13571:2012
		Rescuers: 80 °C	[1]
Heat	Maximum thermal radiation from all sources (fire, fire effluents and structure) of occupants' exposure	Occupants: 2.5 kW/m^2	ISO 13571:2012, for exposures of less than 30 minutes
		Rescuers: 3 kW/m^2	[1]

[1] For the purposes of this table, rescuers are understood that the members of the company teams are appropriately protected and trained in firefighting, in the use of airway protection devices, to operate in conditions of poor visibility. Further indications can be obtained for example from documents of the Australian Fire Authorities Council (AFAC) for hazardous conditions.

Table 8.4 Example of performance thresholds that can be used with the simplified calculation method.

Performance	Performance threshold	Reference
Minimum height of the stratified smoke from the walking surface below which the undisturbed layer of air remains	Occupants: 2 m	Reduced by ISO/TR 16738:2009, Section 11.2
	Rescuers: 1.5 m	[1]
Average temperature of the hot smoke layer	Occupants: 200 °C	ISO/TR 16738:2009, Section 11.2
	Rescuers: 250 °C	[1]

[1] For the purposes of this table, rescuers are understood that the members of the company teams are appropriately protected and trained in firefighting, in the use of airway protection devices, to operate in conditions of poor visibility. Further indications can be obtained for example from documents of the Australian Fire Authorities Council (AFAC) for hazardous conditions.

By way of example, in Tables 8.3 and 8.4, the performance thresholds for occupants and rescuers are reported with reference to advanced and simplified calculation methods.

A similar performance-based approach can be used for industrial assets with different and specific methods and tools but an identical workflow.

Similarly, the approach, where needed, should be considered the explosion risk together with the fire risks.

9

Fire Risk Assessment Methods

This volume is intended to be an introductory guide to fire risk management principles within organisations, not only industrial ones and across different sectors. Fire risk assessment is a fundamental element of the entire organisation divided into its process (main ones and auxiliary ones) and it needs special attention, including the aspects associated with the incoming elements and the findings to be made available for informed decisions. It is important to underline that several risk management standards are internationally available and recognised and those can be very useful to manage fire risk associated with assets in different domains. Among those a prominent role is played by ISO 31000 standard and associated guidelines.

As ISO 31000 outlines the principles and guidelines for risk management, the associated IEC/ISO 31010 technical standard reviews some techniques to be used for risk assessment. The focus, therefore, is on that phase of the risk management process that aims to answer the following questions:

- What can happen and why (risk identification phase)?
- What are the consequences and probabilities of the hypothesised events and what are the factors that mitigate their severity or reduce their probability (risk analysis phase)?
- What is the final risk level in relation to the tolerability and acceptability thresholds used (risk evaluation phase)?

These steps are the fundamental activities in a fire risk assessment as also stated by the *SFPE Engineering Guide* whose process (as depicted in Figure 9.1) covers four distinct phases of a fire risk assessment:

Phase 1 – Planning: The activities of this phase are intended to clearly define the scope and objectives, collect the necessary information to perform the analysis, identify the risk assessment method to be used and the define the tolerability criteria to evaluate the risks.

Phase 2 – Execution: It is the 'core' of the process, consisting in the 'risk assessment' phase, i.e. the steps required to identify the risk sources, analyse risks, characterise the fire scenarios and evaluate the risk. It is an iterative process as the analysis is expected to identify design improvements or modifications that are necessary to further reduce the risks.

Phase 3 – Risk Communication: Reporting it and using tools to better visualise the risk levels, e.g. Bow-Ties and risk matrices.

Phase 4 – In service: It is the residual risk management and monitoring phase.

Fire Risk Management: Principles and Strategies for Buildings and Industrial Assets, First Edition.
Luca Fiorentini and Fabio Dattilo.
© 2023 John Wiley & Sons, Inc. Published 2023 by John Wiley & Sons, Inc.
Companion Website: www.wiley.com/go/Fiorentini/FireRiskManagement

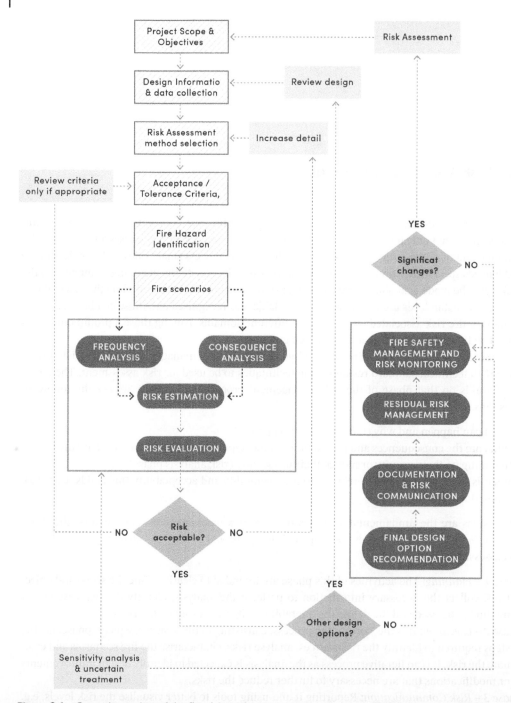

Figure 9.1 General overview of the fire risk assessment process as per the SFPE Engineering Guide.

ISO 31010 standard is general in its nature so it is a guide to be used for any type of organisation, regardless of the type of risks being assessed. It fully supports ISO 31000, meaning that a risk assessment conducted in accordance with ISO 31010 contributes to the effective and efficient satisfaction of the other risk management process activities specified in ISO 31000. The ISO 31010 standard introduces a number of techniques, with specific references to international standards where concepts and the application of techniques are described in greater detail.

It should be made clear that ISO 31010 does not provide specific criteria for identifying the need to perform a risk analysis, nor does it specify the type of risk analysis method to be adopted for a particular application.

The standard also does not list all available techniques, and excluding a technique from ISO 31010 does not mean that it is invalid. Moreover, the fact that a particular technique is applicable to a given circumstance does not imply that a particular technique must necessarily be applied.

The following will show some of the risk assessment techniques in ISO 31010. Their first classification concerns the possibility of being applied to a particular stage of the risk assessment process. Techniques applicable to the preliminary identification phase of hazards, those at the risk assessment phase and, finally, risk assessment techniques will be distinguished. The selection of 'risk assessment' techniques to be used can be influenced by several factors, such as

- the complexity of the problem and the methods required to analyse it;
- the nature and degree of risk assessment uncertainty, measured by taking into account the quantity and quality of the information available and the uncertainty requirements required to meet the objectives;
- the availability of required resources, in terms of time, level of experience, data and costs;
- the need to have quantitative output.

The purpose of this chapter is to offer the reader some insights about the methodologies that mostly use a net of specific tools related to particular aspects or types of risks. The reader, on his/her own, can then also conduct personal insights in relation to his/her own peculiar abilities and needs.

Since risk assessment is a fundamental (and critical) process for the organisation, where the methodology to be followed is not a prescriptive requirement, it must pay particular attention to selection in relation to the context, type of risks, skills and objectives to be pursued.

ISO 31000 standard apply to organisations having different complexities, sizes and resources and operating in different contexts. It is also the basis of modern high-level systems (HLS) for management. Therefore, the methodologies and eventually associated tools supporting the risk assessment process and the risk management methods should be selected considering the characteristics of the organisation but also some other key factors such as the maturity level to implement those methods, the competence and proficiency of internal resources and/or the availability of external consultants and advisors, the goals and the strategic objectives, and the past experience of undesired events and accidents (nature, number and main roots causes already identified).

9.1 Risk Assessment Method Selection

The selection of the risk assessment method(s) relates to the level of detail to which each scenario is described and quantified concerning the level of potential risk. In general terms, the analysis (i.e. the activities within the assessment where risk is evaluated) can range from qualitative to quantitative, including semi-quantitative approaches. This is primarily governed by the level of perceived risk, which may change as the overall assessment progresses and by regulatory bodies. In practice, the type of risk-based evaluation and level of detail should depend on the complexity of the risk and the decision-maker's needs. When selecting the type of analysis, it is necessary to consider several factors, including the information available, the complexity of the facility or process under analysis,

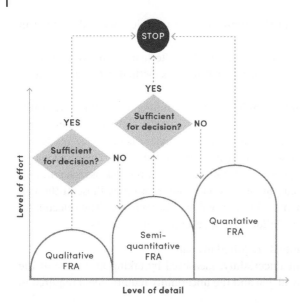

Figure 9.2 Iterative selection of fire risk assessment (FRA) approach.

the potential deviations from code requirements and best practices, and the level of detail necessary to make a substantiated decision about the tolerability of fire risk(s). Also the goal and objectives of the assessment should be carefully considered in the method or combination of methods selection.

Qualitative analysis refers to the evaluation of risk without explicit numerical quantification. In a qualitative assessment, fire risk is evaluated based on the merits of specific designs versus the postulated potential fire events. Qualitative risk methods may be appropriate for evaluating well-understood conditions associated with simple systems or configurations with established risk levels.

A semi-quantitative analysis refers to the evaluation of risk with simplified quantitative elements supporting assessment. This approach may be appropriate for evaluating configurations with minor deviations from code requirements or best practices and risk trade-off implications.

Finally, a quantitative analysis is a complete explicit quantification of frequencies and consequences to produce numerical risk levels. The need for a quantitative assessment often arises when evaluating novel, challenging or complex configurations with significant risk trade-offs. Additional factors influencing the need for a quantitative analysis include identifying significant uncertainties that need to be rigorously modelled and strong stakeholders' views and perceptions of potential risks. Alternatively, the relevant regulations may mandate a quantitative approach.

Figure 9.2 depicts an iterative approach, suggested also by SFPE, in which the level of detail increases as the level of quantification increases. Note that the level of effort increases as more quantification is required to support the conclusions. Also, a risk assessment may be developed with a combination of qualitative and quantitative approaches (i.e. semi-quantitative approach) while maintaining the rigour and analysis necessary to reach conclusions. The approach can be graded considering the current phase of the project where incipient stages may be faced with preliminary and qualitative methods.

9.2 Risk Identification

This section presents some methodologies generally used in the preliminary identification phase of hazards while remaining their applicability (more or less recommended) to the other phases of the 'risk assessment'.

9.2.1 Brainstorming

Brainstorming stimulates and encourages a free conversation between a group of people familiar with the subject for analysis in order to identify possible ways of failure and associated dangers, risks, criteria and options for the next stage of treatment. The term 'brainstorming' is often misused to denote any kind of group discussion. In fact, true brainstorming uses special techniques to try to get a participant's thoughts to be stimulated by other members of the group in order to support each other in the cognitive processing process. For this reason, effective facilitation is very important in this technique. It is achieved by stimulating the start phase of the discussion (the so-called kick-off), guiding the group at regular intervals through several relevant discussion areas and capturing the emerging aspects (typically in real time).

Brainstorming can be used in combination with other risk assessment methods, or it can be used alone as a technique to encourage thought at every stage of the risk management process and at every stage of a system's life cycle. It can also be used for high-profile discussions when issues are clearly identified or for detailed insights into particular issues.

Brainstorming places great emphasis on imagination. Therefore, it is especially useful when one wants to identify the risks associated with new technology, for which there is no historical data available or solutions already identified in other contexts.

What is needed for brainstorming is just a team of people with an appropriate level of knowledge about the organisation, system, process or application being evaluated.

The process can be formal or informal. Formal brainstorming is more structured, with participants preparing ahead of the session, whose purpose is defined and the ideas being questioned are evaluated. Informal brainstorming is less structured and often ad hoc. In particular, in a formal process the following steps are followed:

- The facilitator prepares suggestions in advance to stimulate and enrich the discussion.
- The objectives of the session are clearly defined, and the rules of the session are explained to all participants.
- The facilitator starts the discussion with a non-stop speech, and each explores their own ideas to identify as many aspects as possible. At this stage, there is not much discussion about whether a danger should or should not be on the list in order to avoid inhibiting the uninterrupted discussion. All stimuli are welcomed, and no one is criticised; the group's work advances rapidly to allow ideas to stimulate the so-called lateral thinking.
- The facilitator can direct participants to a new topic of discussion or change their perspective when the discussion deviates too much from the field they want to explore, or the thoughts are now devoid of added value. In any case, the idea is to gather as many ideas for the next phase of risk analysis.

The result of this process depends on the specific phase of the risk management process to which this method is applied. When applied to the preliminary hazard identification phase, which by far is the most frequent area of application, then the result will be a list of dangers or control measures put in place.

Among the advantages of this technique, it is possible to include its ability to stimulate creative thinking, thus helping to identify new risks and new solutions. It typically involves stakeholders with key roles and therefore supports communication between them. In addition, it is a fast and easy technique to implement.

However, brainstorming has some obvious limitations. The main ones are related to the quality of the discussion and therefore to the participants: they may not have the skills and knowledge required for their contribution to be effective. The methodology is also poorly structured, and it is, therefore, difficult to prove that all potential dangers have been taken into account with this process. Finally, particular group dynamics may arise where participants with valuable ideas remain silent while other participants dominate the discussion. This limit could be exceeded by using a computer forum, where participants speak on condition of anonymity, to avoid being subjected to personal or political pressure that could prevent the free exposure of their ideas. Similarly, the participants' ideas could be sent anonymously to a moderator and then discussed by the whole group.

9.2.2 Checklist

Checklists, or checklists, are a list of dangers (or risks or ways of failure of a control measure) generally developed from experience, both as a result of previous risk assessment and as operational experience in terms of past failures.

A checklist can be used to identify hazards, or risks, or to assess the effectiveness of control measures. This technique can be used at every stage of the life cycle of a product, process or system. Checklists are sometimes used as part of another risk assessment technique but are used more to ensure that the entire scope of assessment has been covered downstream from the use of a more imaginative technique capable of identifying new problems.

The application of this technique requires a clear identification of the scope. Only then is the checklist that adequately covers the entire scope selected. For example, a barrier checklist cannot be used to identify new hazards or new risks. The process ends with the use of the checklist by a person (or a team of people) who delves into every single element of the system that is being analysed and verifies that the items on the checklist are (or are not) present.

Checklists have the undoubted advantage of being able to be used even by non-experts. In fact, if well designed, they manage to enclose in a simple tool a wide range of skills, making them available to even the least experienced. This ensures that the most common issues are not forgotten.

In contrast, checklists tend to inhibit the imagination during the risk identification phase due to their highly schematic approach to the problem. In addition, they focus on the problems (in this case, the dangers) that are known, neglecting those knowingly and unknowingly unknown. By their nature, they tend to encourage a prescriptive attitude to risk management, passing on the idea that it is enough to tick a box to solve each problem. We must therefore be aware of these disadvantages.

9.2.3 What–If

The 'What–if?' technique is a systematic team study that sees a facilitator use a set of words or phrases during a session to stimulate participants to identify hazards. Typical phrases are used in combination with system/process/plant keywords being analysed to investigate how it is affected by deviations from normal operating and behavioural conditions. Normally, this technique is adopted at the system level and not at the individual elements that compose it, with a lower level of detail than – for example – a hazard and operability analysis (HAZOP).

This methodology was originally designed to identify hazards in the chemical and petrochemical industry; it is now widely adopted to systems, procedures, plants and organisations in general. In particular, it is used to analyse the consequences of the changes in the risk assessment subject and the risks associated with them.

The method allows the identification of hazards by recursively asking such structured questions: 'What would happen if ...? ' (what ... if) a certain system/process/equipment did not work as it should? The method also applies to procedures that govern processes.

Before starting the study, the system (or procedure, or plant, etc.) that is being analysed must be carefully defined, taking care to establish internal and external contexts through interviews with colleagues, documentation study, floor plans and drawings. It is also important that the level of expertise and experience in the working group is carefully selected in order to avoid poor results.

The methodology applies according to the following process:

- Before the study begins, the facilitator prepares a list of guiding words or phrases that can be based on pre-defined sets or be created on ad hoc to ensure extensive coverage of hazards.
- During the session, the internal and external contexts of the system (or process, or equipment, etc.) that is being analysed are discussed and shared.
- The facilitator asks participants to speak and discuss:
  - known risks and dangers;
  - past experiences and incidents;
  - known and existing control measures;
  - any legal requirements and other constraints.
- The discussion is facilitated by means of questions that use the structure 'What ...' and a key-word/subject. Phrases like 'What ... if' can also be 'What would happen if ... ', 'Could someone or something ... ' or 'Someone/something has ever ... '. The aim is to stimulate the study group to explore potential scenarios, their causes and consequences.
- The risks are therefore summarised, and the group takes into account the control measures put in place (barriers).
- The group confirms the description of the risk, its causes, consequences and expected barriers and then records the data.
- The working group considers whether the control measures are adequate and effective, and agrees on a measure to monitor their effectiveness. If it is not satisfactory, the group considers the possibility of further dealing with the risk, possibly defining potential additional barriers.
- During the discussion, further questions 'What ... if?' are placed in order to identify further possible risks.
- The facilitator consults the list he/she has previously created to monitor the status of the discussion and suggests any scenarios not considered by the team.

The output of this technique is generally a register of dangers (and, if desired, the risks associated with them). If the assessment has gone so far as to identify corrective actions, then they will form the basis for the risk treatment plan. In its objectives, the technique is similar to HAZOP and failure modes and effects analysis (FMEA). The substantial difference is that it is less structured.

The advantages of this technique lie in its broad applicability to all forms of systems, installations, situations, circumstances, organisations or activities. It allows relatively quickly to identify the dangers and risks associated with them. This technique is also oriented to the 'system', allowing participants to observe the system's response to deviations rather than simply examining the consequences of a component's failure. The what–if study can also be used to identify opportunities for improvement, identifying those actions that lead to an increased probability of success of processes and systems. In addition, the involvement of colours that are required to ensure and monitor the effectiveness and efficiency of existing barriers reinforces their sense of responsibility. Although the technique lends itself to a qualitative or semi-quantitative approach, it is often also used to identify those hazards and risks that will later be analysed through a quantitative study.

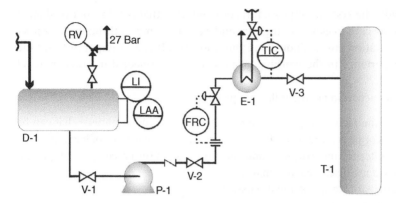

Figure 9.3 Feed line propane–butane separation column (RV: relief valve; LI: level indicator; LLA: low-level alarm; FRC: flow recorder controller; TIC: temperature indicator controller).

Table 9.1 Example of 'what–if' analysis.

Question 'what–if'	Consequences	Recommendations
... the pump P-1 shuts down? ... valve V-1 is accidentally closed?	The liquid level rises in D-1 Feeding T-1 interrupted, causing operational upset	RV will open if LI fails
... the FRC valve is leaking?	Possible fire due to flammable mixture	Schedule a more frequent maintenance Substitute it with a double-seal system

However, the 'What–if?' analysis requires that the facilitator be a competent and experienced person in the scope of the evaluation, and this could be a limit. It also requires careful preliminary preparation so as to avoid unnecessary waste of time during the course of the session. It is also necessary to be aware that the high-level application of this technique may not bring out complex causes, which can only be found with a detailed analysis. Finally, not only the facilitator but the entire working group must have a broad knowledge and experience base in order to avoid the failure to identify certain dangers.

In order to have a comparison between 'What–if?' and HAZOP, the following scheme will be analysed with both the methodologies to underline pros and cons of each. The example takes inspiration from Assael and Kakosimos (2010). Let us consider the flow diagram in Figure 9.3. It is a feed line of the propane–butane separation column. Initially, the mixture enters in vessel D-1; then it is pumped (through P-1) towards the column T-1. A flow recorder controller (FRC) valve controls the flow rate, and the heat exchanger E-1 pre-heats the mixture before entering in T-1.

In Table 9.1, there are two very easy examples of 'what–if' scenarios.

9.2.4 HAZOP

The HAZOP was invented in the United Kingdom during the 1970s. It is a very structured technique allowing to identify those hazards related to process deviations of parameters with respect to the normal range of activity. A HAZOP analysis can be used for every kind of process. The structured path to identify the possible deviations follows the below steps:

Table 9.2 Guidewords for the HAZOP analysis.

Guideword	Meaning
NO	Complete negation, fully absence of
LESS	Quantitative decrease
MORE	Quantitative increase
REVERSE	Logical opposite
OTHER THAN	Complete substitution

Establish a list of keywords, intended as parameter's modifier. A typical set is shown in Table 9.2.

Establish a list of parameters, intended as those physical dimensions whose setting affects the process (like pressure, level, temperature, flow, composition and so on).

Combine each keyword with all the parameters to identify all possible deviations. Obviously, some resulting deviations might have no sense or might be not applicable in the specific context being analysed.

Determine all the possible causes for each deviation.

Determine all the possible consequences for each cause.

Develop the necessary corrective actions to face (avoiding or mitigating) the hazardous scenarios being identified.

When performing a HAZOP analysis, balanced team composition is crucial to obtain good results. The HAZOP team requires the following personnel:

A team leader to guide and help the team in reaching the objectives of the analysis. It is not necessary for the team leader to known technically the specific process being investigated since other members are required to bring that knowledge.

A scribe (eventual) to write down the results emerging from the brainstorming activities of the team.

Experienced operator(s) with the process being analysed and its standard and emergency procedures.

Technical specialists, like instrumental, electronic, mechanical or plant operator(s), depending on the specific process/plant (this category may include technologist engineers).

It should be well kept in mind that the primary goal of a HAZOP analysis is the hazard identification: this means that engineered solutions must not be found during a HAZOP session, thus avoiding to waste time. Clearly, if the corrective action is obvious, then the HAZOP team may recommend it; instead, when the solution is not so immediate to be reached, the task must be left to the engineering team.

Even if it is preferable that the HAZOP analysis is carried out at the earlier stages of the design, so to positively influence it, on the other hand, an already complete design is required to perform an exhaustive HAZOP. The compromise could be carrying out the HAZOP as a final check, once the detailed design is ready.

A HAZOP may also concern an existing facility, and it is generally used to identify hazards due to plant modifications or to propose modifications in order to reduce risks. Being a structured method, HAZOP is widely used in the process industry.

Taking inspiration from the already discussed example in Figure 9.3 about the feed line propane–butane separation column, Table 9.3 presents the scenarios corresponding to the outcomes

Table 9.3 Extract of an example of the HAZOP analysis.

Guideword	Deviation	Causes	Consequences	Recommendations
NO	No flow	The pump P-1 shuts down (failure or power loss)	The liquid level rises in D-1 Feeding T-1 interrupted, causing operational upset	There is already an RV. Place a high-level alarm (HLA) on D-1
		The valve V-1 is accidentally closed by an operator	Pump overheats: possible mechanical damage of the seal, leakage and fire feeding T-1 interrupted, causing operational upset	Point out the error in the operating procedures
LESS	Less flow	The FRC valve has a minor leak	Hydrocarbons in the air, possible fire	Schedule a more frequent maintenance Install a double-seal system

of the 'what–if' analysis in the previous paragraph to make the difference immediately visible. The example concerns the parameter 'flow', and one more time is restricted to the subset of causes found in the 'what–if' analysis. A full HAZOP is more extended to what presented in Table 9.3.

The primary goal of a HAZOP analysis is to identify hazards, which means that engineering solutions should not be found during a HAZOP session, thus avoiding wasting time. Clearly, if corrective action is obvious, then the HAZOP working group can recommend it. If, on the other hand, the solution is not immediate to be reached, this activity should be left to the engineering department after the conclusion of the HAZOP session.

Although it is preferable to conduct the HAZOP session in the early stages of the project, so as to positively influence it, it should be noted that a complete project is also necessary to perform a comprehensive HAZOP analysis. The trade-off is, therefore, to use a HAZOP study when a complete view of the process is available, but detailed design changes are still allowed.

A HAZOP can also cover an existing plant and is generally used to identify hazards associated with plant changes or to promote changes in order to reduce risks. Being so strongly structured, HAZOP is widely used in the process industry.

To perform a HAZOP study, one must have drawings, datasheets, logical and process control diagrams, know the layout, maintenance and operational procedures, emergency response procedures and so on.

The objectives of the HAZOP analysis are met thanks to a systematic examination of how every single part of the system, process or procedure will respond to changes in key parameters using guiding words.

This technique offers significant advantages, such as

a systematic and thorough examination of the system, process or procedure (with ample opportunity to conduct an expert assessment of the human factor by expert judgement);

the involvement of a multidisciplinary group that also includes those people with direct operational experience on what is being discussed;

any ability to generate solutions (risk treatment actions);

applicability to a broad spectrum of systems, processes and procedures;

the explicit consideration of the causes and consequences of human error;

the allocation, as output, of a written register containing on time all the information analysed and organised by criteria (e.g. by process or phase, by deviation, by consequent scenario, by the need of integration, etc.).

On the other side, HAZOP is an extremely time-consuming technique, and detailed analysis requires excellent documentation from which to extrapolate the information needed to start the risk identification process. Often, there is also a risk that the discussion will focus on the details of design, neglecting the context, and even the generality of the problem analysed. For this reason, it is essential to set the scope of analysis and objectives together with the working group before the session. The whole process is heavily based on the expertise of those present and designers in particular, for whom it may be difficult to remain objective enough to find any problems in their project. For this reason, it is recommended that the working group include people not directly involved in the project (system, process, procedure, etc.) under review.

The HAZOP analysis is similar to FMEA, in that both techniques identify ways of failure of a process, system or procedure, with their causes and consequences. It differs, however, taking into account the unseeded effects and deviations from the expected operating conditions, working backwards to find out the possible causes and ways of failure, while the FMEA takes its cue precisely from the identification of ways of failure.

9.2.5 HAZID

A different tool that can be used in order to perform a preliminary hazard analysis is hazard identification (HAZID). The HAZOP is generally used late in the design phase; therefore, the identified safety and environmental issues can cause project delays or costly design changes. Instead, HAZID is a structured brainstorming (guideword based) that is generally carried out during early design so that hazards can easily be avoided or reduced. The objective of HAZID is to identify all hazards associated with a particular concept, design, operation or activity (as stated in ISO 17776). Typically, the structured brainstorming technique involves designers, project management, commissioning and operation personnel. Like a HAZOP, HAZID is based on inductive reasoning so it is necessary that the analyst has sufficient experience in 'safety', to think as widely as possible in order to ensure that predictable major accident hazards are not overlooked, including low-frequency events. To do so, the analyst focuses his/her attention on

the hazardous substances used as inputs, as intermediates and as outputs;
the used chemical processes;
the equipment, components and materials being used;
the plant layout;
the environment surrounding the plant;
the safety systems;
the inspection, control and maintenance activities.

To conduct a HAZID analysis, it is suggested to subdivide the scope of the analysis into homogeneous areas or functional groups, as from the process schemes (Table 9.4).

Then, the hazards associated with the performed activities are taken into account (fires, explosions, toxicity and so on). It is not necessary to list all the possible causes for each incident; indeed, it is sufficient to identify a significant number of them to determine the probability of its occurrence. Pre-defined lists of hazards are generally used, like the ones provided in ISO 17776 related to the hazards that can be encountered in the petroleum and natural gas industries. The approach should be applied to each area and hazard guideword, asking the following questions:

Is the guideword relevant?

Is there something similar that should be identified?

What are the causes that could lead to a major accident?

What are the credible potential consequences?

What are the preventive and mitigating barriers already specified (or expected)?

Are there any additional barriers that could be proposed?

Are human barriers (if any) reasonable?

Is further (quantitative) analysis required to understand better the consequence of the hazard?

What recommendations can be made?

An example of a list that can be used during the preliminary analysis is shown in Table 9.5. Finally, the analyst identifies the consequences of each assumed event, considering the most conservative scenario. The consequences can be pre-defined too, as shown in Table 9.6, as well as the preventive barriers and the mitigating measures.

Table 9.4 Subdivision of the analysed system into areas.

Area	Designation	Details	Flammable inventory	Toxic inventory	Comments	PDF
1	1st stage separator	1st separator	Hydrocarbons	–	–	#
2	Crude booster pumps	Crude booster pumps and process area	Hydrocarbons	–	–	#
3	...	...	...	...	...	...

Table 9.5 Hazards and assumed event in HAZID.

Hazards	Assumed event
Hydrocarbons under pressure	Leakages
Toxic substances	Leakages
Lifting facilities	Falling parts
Transportation/traffic	Collision
Utility facilities	Loss of function

Table 9.6 List of typical consequences.

Consequences
Pool fire
Jet fire
Toxic gas cloud formation
BLEVE
VCE
Others

The results of a HAZID analysis are arranged using HAZID worksheets, showing clear linkages between hazardous events, hazards, underlying causes and control measures/safeguards (if any) as well as the corrective actions. An example of HAZID worksheet is shown in Table 9.7.

This methodology is particularly suitable to identify the dangers to be analysed later by Bow-Tie. In fact, the results of a HAZID session can be superimposed on the input data required by a Bow-Tie analysis that, as best illustrated in the dedicated section, precisely includes the danger (associated with the guiding word of HAZID), the top event (associated with the deviation analysed), the causes, consequences and barriers. It is therefore convenient to set up a HAZID session prior to a Bow-Tie analysis. Note that this 'tandem' coupling of these methodologies is also referenced in some regulatory contexts, such as the regulations related to major accident prevention such as the 'Seveso' and the 'Offshore' European Directives managing the industrial risk respectively associated with onshore and offshore industrial chemical plants.

It should be noted that, like HAZOP, the HAZID method, with the appropriate changes in the taxonomy used and in the way homogeneous areas are identified, can well be applied in areas other than that of originality coinciding with the oil and gas world and for the risks different from those characteristics of the installations of the chemical process industry.

9.2.6 FMEA/FMEDA/FMECA

FMEA is a technique used to identify ways in which components, systems or processes can fail to achieve their project intent.

The FMEA identifies

- all potential ways of failure of the various parts of a system (a way of failure is what is observed to fail or perform incorrectly, including reduced efficiency compared to what is expected);
- the effects that these failures can have on the system that contains them or related systems;
- failure mechanisms;
- how to avoid bankruptcies and/or mitigate the effects (or possibility) of failures on the system.

Failure modes, effects and criticality analysis (FMECA) adds to an FMEA the ability to assign a score to each identified failure, in relation to its own importance or criticality. This critical analysis is often qualitative or semi-quantitative, but can also be quantitative if real failure instalments are used.

Failure modes, effects and diagnostic analysis (FMEDA) also essentially takes into account the operation of diagnostic systems compared to the simplest methodologies from which it derives,

Table 9.7 HAZID worksheet.

Node: Process and instrumentation diagram (P&ID) #:				Date: Revision	
No.	Deviation	Cause	Consequences	Safeguards	Recommendation

thus offering itself as a tool for the definition of system 'target' fault instalments, the creation of the reliability block diagram (RBD) and, therefore, valid support for functional safety analyses related to the determination, allocation and verification of safety integrity levels (SILs) defined under the IC 61508 and IEC 61511. Like FMEA and FMECA, FMEDA is often used in reliability engineering and is not required to give a detail description here but to provide only an overall overview, as is the case with the other methodologies described in this chapter.

There are several FMEA applications: for components and products, for systems, for production and assembly processes, and for services or software.

FMEA, FMECA and FMEDA can be applied during the design, implementation or exercise of a physical system. Techniques can also be applied to processes or procedures, for example, to identify potential medical errors in the health system or ways of failure in maintenance procedures.

In general, methodologies can be used to

support in choosing high-dependency project alternatives;
ensure that all systems and process failure modes and their effects on operational success are taken
    into account;
identify human errors and their effects;
provide a basis for planning maintenance and testing operations of physical systems;
improve the design of procedures and processes;
provide qualitative or quantitative information for risk analysis techniques, such as the fault tree.

For their application, one needs enough data to understand how failure can occur. This information can be derived from drawings, analysed process flow diagrams, environment details and other parameters that can affect the process, e.g. historical–statistical information such as failure rate, where available.

The FMEA runs according to the following process:

- Definition of the scope of study and expected objectives.
- Training of the working group.
- Understanding of the system/process subject to the FMEA analysis.
- System/process breakdown into its components/phases.
- Definition of the function satisfied by each component/phase.
- For each component or phase listed, identify the following questions:
  - How can any party reasonably fail?
  - What mechanisms could produce these ways of failure?
  - What effects could result from failure? With what consequences?
  - How is the fault diagnosed?
- Identify actions to change the design to compensate for detected failures.

For the FMECA, the working group continues further by classifying each failure identified in accordance with its own criticality. There are several ways to do this, including the following:

- The mode criticality index.
- The level of risk.
- The risk priority number (RPN).

The mode criticality index is a measure of the probability that the failure mode considered will actually cause a system-leading failure. It is defined as the product between the frequency of occurrence of the specific mode of failure for the probability that it will follow a system failure for the time the system is operational. Typically, this index is applied to equipment failures, for which each term of the product can be defined quantitatively, and the failure modes all have the same consequence.

The level of risk is achieved by combining the consequences of a failure mode with the likelihood of failure. Usually, it is used when the consequences of different fault modes are different from each other. The level of risk can be expressed qualitatively, semi-quantitatively or quantitatively.

RPN is a semi-quantitative measure of criticality obtained by multiplying numbers of scale graduates (typically from 1 to 10) chosen for the consequences of failure, the probability of failure and the ability to detect the problem (a mode of failure acquires priority if it is difficult to detect). This method is widely used in the quality studies of a system/product.

Once the failure modes are identified, corrective actions are identified, implementing them from the most critical fault modes.

The FMEDA, on the other hand, pushes the analysis towards the proper times of functional safety. The interested reader will be able to consult the IEC 61508 to find out the details.

The FMEA analysis is documented in a report that contains all relevant information such as

- details of the analysed system;
- how the analysis is performed;
- preliminary recruitment;
- data sources;
- results;
- criticality (if FMECA) and methodology used to define it;
- any recommendations.

The advantages of using these methodologies are as follows:

- They apply widely to the ways of failure of equipment, systems and even humans, hardware, software or procedures.
- They identify the ways in which components fail, their causes and effects on the system and their presentation in the form of reports that are easily readable even by non-experts.
- They avoid the need for costly plant modifications, identifying problems in the early stages of the project.
- They identify individual modes of failure and the consequent need for redundant security systems.
- They provide input data for the development of monitoring programmes, highlighting key features to monitor.

On the other hand, these methodologies can only be used to identify individual ways of failure and not their combinations. Finally, their application often takes a long time and can be difficult and tedious for particularly complex systems.

An example of the methodology selection of a new LNG site is provided in Tables 9.8 and 9.9, including an application of a screening methodology.

Engineering phases of new installations request for the selection of the risk/analysis approach considering the complexity (and inherent risks), project LOD (and available documentation) as well as external requirements (e.g. requirements from local authorities or regulations, such as Seveso Directive in the EU or Control of Major Accident Hazards [COMAH] in the United Kingdom). There are many available risk assessment tools, as those listed in IEC 31010. For assets posing industrial risk, some of the most common methods are given in Crawley (2020), where they are discussed considering engineering phases. In some cases, local regulations allow the use of a

Table 9.8 Phases and HAZID studies.

Project phase	HAZID studies	Resources (documentation/people)	Deliverables	Comments on method selection
Concept	HAZID (coarse)	Concept Material safety data sheets (MSDS) Preliminary plot plan and operation philosophy Hold-Up Conditions (process/design) RAGAGEPS Accident history for similar units Design team Operation team	HAZID report ('hazard/scenario/control' structure). TOR	HAZID is maintained during the various phases of the project to feed the top events identification/quantification methods (Bow-Tie and layer of protection analysis [LOPA]). Across different stages, the HAZID is conducted using a different degree of detail. HAZID can initially be applied using process-facility-related guidewords (as those in Appendix to ISO (2016b)) and then refined using specific requirements as indicated below. Site layout will help in initial considerations for traffic management. RAGAGEPS include European Committee for Standardization (2016) and National Fire Protection Association (2019).
	Inherent safety assessment	See above Facilitator Design team	Risk ranking based on alternatives (hold-ups, distances, separations, etc.).	Inherent safety study conducted according to the principles described in AIChE (2008) has proven to be an efficient tool to assess the safety level of LNG facilities at initial stages of design (Health Safety Executive 1996) to develop indicators deriving from severity of consequences and credibility of loss of containment events, as well as evaluate alternatives.
Preliminary design	HAZID	Front-end engineering design (FEED) documentation Facilitator Design team Operation team	HAZID report and preliminary risk register with actions prioritised using a consequence-based criteria. TOR	HAZID guidewords can be integrated with specific requirements to be verified during preliminary design as those suggested by best available techniques (BATs) (e.g. ISO 2016a) that give some indications about design requirements, site layout and equipment location, environmental considerations, security, LNG operations, LNG tank design, emergency shutdown (ESD) and F&G.
	F&E index	Main equipment datasheets MSDS People number in onsite and offsite buildings Risk acceptability criteria and prescriptive requirements from LA Design team	PHA (safety indices calculated for the main areas: tank, pump and loading bay) and preliminary damage areas.	In some methods (as the Dow-Mond described in Dow Chemical Company (1987) or the Safety Weighted Hazard Index (SWeHI) described in Khan, Husain, and Abbasi (2001)), there is also a preliminary estimation of damage areas. People number in buildings is required to verify prescriptive separation distances as per Annex B of ISO (2016a).

		Input	Output	Description
	WHAT-IF?	Procedures (LNG tanker offloading/fuelling) Facilitator Design team Operation team	Deviations and associated consequences from the intended operation.	What-if? Method will address behavioural issues in order to give suggestions about the operating procedure as well as the human–machine interface (HMI) of the card reader and the dispenser. Task analysis could also be used due to the simplicity of the procedure. A brainstorming method is deemed to be sufficient for the installation.
	Bow-Tie	HAZID outcomes Design team Operation team	Top events with threats, consequences and barriers (preventive, mitigative and technical/behavioural).	Bow-Tie has been selected to support a barrier-based risk management approach during both project and operation phases as suggested in several publications (AICh6E 2018), also considering the communicative power (better than fault tree analysis [FTA]/event tree analysis [ETA]).
	Consequences estimation	HAZID + Bow-Tie outcomes FEED documentation LA prescriptive requirements Design team HSE experts	Scenario damage areas defined for effects thresholds layered on the maps with vulnerable receptors inside and outside the facility. Verification of separation distances against requirements in Tables B.1/2/3 of ISO (2016a).	Considering safety issues connected with the new layout of the organisation and considering the concerns of the LA for effects outside the fence, it is important to carry out, during the preliminary design phase, the consequence estimation, for identified top events on vulnerable receptors (carriageway and housing estate). Prescriptive separation distances from BATs can be reduced with this specific assessment and considering eventual fire walls (Barry 2002).
Detailed design	HAZID (detailed) – checklist based	Engineering documentation (P&IDs, component data sheets, operation philosophy, plot plans and hazardous area classification [HAC]) Failure rates and frequencies, enabling factors as defined in CCPS (2015) and CCPS (2013).	HAZID report with barriers, risk ranking, MAE's, actions, comments and TOR.	
	HAZOP (on process and on procedures)		HAZOP report (deviations, causes, protections, actions and risk ranking).	HAZOP is a method that can be used to assess the process and also the operation procedures. In this phase, HAZOP can be extended with a matrix to estimate and prioritise risks.

(Continued)

Table 9.8 (Continued)

Project phase	HAZID studies	Resources (documentation/people)	Deliverables	Comments on method selection
	LOPA		LOPA report for the top events	LOPA will quantify the top events from the Bow-Tie method. It is a straightforward method to apply and can give an estimation of the risk for both process deviations and random failures. Due to the simplicity of the plant, a full quantitative risk assessment (QRA) is not required. LOPA could be completed with the information about consequences effects distances. LOPA is preferred also to the combination FTA + ETA that requests for more data and resources.
Construction	Checklist	Engineering documentation Design team Construction team	Safety plan for construction activities Qualification requirements of people for components joining operations as per (ISO 2016a).	

Table 9.9 HAZID considerations applied to a typical industrial asset.

Sections	Equipment	Event	Potential causes	Potential consequences	Controls		Priority
					Action	Preventive/mitigative	
						P M	
LNG loading bay (from truck to tank)	Hoses	Loss of containment with consequent LNG release	Leakage or rupture due to • ageing; • pull-away; • seal leak or couplings not tightened on hose.	Pool fire • Flash fire • Explosion • Cryogenic hazards; • Rapid phase transition in case of contact with water, as shown in Gaz de France, CERMAP, Cryogenic Studies Section. 2021.	Anti-driveaway and breakaway systems Pressure and level interlocks to stop transfer Resilient material		

(Continued)

Table 9.9 (Continued)

Sections	Equipment	Event	Potential causes	Potential consequences	Controls		Preventive/ mitigative		Priority
					Action		P	M	
	LNG pump		Leakage or rupture due to • ageing; • cavitation while liquid level is too low; • loss from seals; • air inlet and formation of flammable mixture; • impacts.		Seal-less pumps Minimisation of seals O$_2$ analysers in pump Equipment layout in order to reduce collisions with vehicles coupled with collision protections and a traffic management procedure				
	Connection pipes		Leakage or rupture due to • ageing; • overpressure; • mechanical constraints; • loss from seals; • low temperature; • impacts.		F&G detection, with alarm and ESD				

Cryogenic liquid detection on the foundation, with alarm and ESD

Piping and flexible hoses' (ISO 21012) ratings suitable for maximum operating conditions and LNG.

Equipment and instrumentation compliant with ATEX

Coverage to avoid the contact with meteoric water

Suitable slope to avoid the accumulation of LNG, above the truck

Continuous presence of a trained operator/truck driver and TVCC

Dry-powder fire extinguishers and fixed foam systems

Wheeled foam monitor

Emergency button to stop tank loading, in safe position (connected to ESD system)

Availability of hydrants for cooling or for feeding foam system

Pressure and level alarms/interlocks

PSV and vent at safe location

Use of cryogenic tank, with double walled and solid insulation used in the anulus to retard heat transfer from the outer vessel to the inner

Vacuum-loss alarm in the anulus

Use of a loading pump (from truck to tank) with a maximum pressure lower than tank design pressure

Resilient material

(Continued)

Table 9.9 (Continued)

Sections	Equipment	Event	Potential causes	Potential consequences	Controls Action	Preventive/ mitigative P	M	Priority
					Extruded and non-welded connection pipes to the tank			
					Minimisation of seals/couplings			
					Tank layout in order to reduce and/or eliminate the risk of collisions with vehicles			
					Tank horizontal configuration, to limit seismic and thermal stresses			
					Calculation of storage tank as per seismic regulation			
					Seismic alarm (eventually EWS)			
					Denial of access without personnel (preferably with fence)			
LNG storage tank	Tank	Loss of containment with consequent LNG release	Ageing, overpressure, random leaks and tank-overfilling due to control system/human error	Pool fire, flash fire, explosion and RPT	Disposition of wall, especially on the sides facing the truck and the dispensers			
					TVCC, preferably with thermal imaging cameras for cryogenic releases and flames			
					F&G, with alarm and ESD			
					Cryogenic liquid detection on the foundation, with alarm and ESD			

Distribution systems	LNG pump from tank to dispenser(s)	Loss of containment with consequent LNG release	Leakage or rupture due to - ageing; - cavitation while liquid level is too low; - loss from seals; - air inlet and formation of flammable mixture; - impacts.	Pool fire, flash fire, explosion, RPT	Piping ratings suitable for maximum operating conditions and LNG Equipment and instrumentation compliant with ATEX Coverage to avoid the contact with meteoric water Suitable slope to avoid the accumulation of LNG, with displacement towards safe area with coverage Dry-powder fire extinguishers Wheeled foam monitor Availability of hydrants for cooling or for feeding foam system Pressure interlocks to stop transfer Resilient material Seal-less pumps	Pressure interlocks to stop transfer Resilient material Seal-less pumps

Table 9.9 (Continued)

Sections	Equipment	Event	Potential causes	Potential consequences	Controls		Preventive/ mitigative		Priority
					Action		P	M	
	Connection pipes		Leakage or rupture due to: - ageing; - overpressure; - mechanical constraints; - loss from seals; - low temperature; - impacts.		Minimisation of seals				
					O2 analysers in pump				
					Equipment layout to reduce the risk of collisions with vehicles coupled with collision protections and a traffic management procedure				
					F&G detection, with alarm and ESD				
	Dispenser(s)		Leakage or rupture due to: • ageing; • loss from seals; • low temperature; • - security loss.		Cryogenic liquid detection on the foundation, with alarm and ESD				
					Ratings suitable for maximum operating conditions				
					Equipment and instrumentation compliant with ATEX				

Disposition of wall on the sides facing the tank

Coverage to avoid the contact with meteoric water

Suitable slope to avoid LNG accumulation, especially under vehicles

Trained operator in place

Denial of access without personnel (preferably with fence/gate)

Fire extinguishers

Wheeled foam monitor

Emergency button to stop tank loading in safe position

Availability of hydrants for cooling or for feeding foam system panel of different approaches to demonstrate control of risk. There is no single 'best' method for each phase while each step may require a pool. Some of them are included for 'FEED' in Tables 9.10 and 9.11 for 'detailed design', along with their references (international standards, RAGAGEPs or references), applicability to the risk assessment phases of ISO 31000 and identification (yellow)/ analysis (green)/evaluation (cyan).

Table 9.10 'Feed' methods.

Inherent safety confirmation study	Rahman, Heikkilä, and Hurme (2005)	In FEED phase, a plot plan is to be prepared; therefore, the inherent safety study prepared during the CONCEPT phase should be revalidated, as suggested in CCPS (2009).
HAZID (coarse)	ISO (2016a)	It is a very robust tool to identify threats having high impact/ low likelihood, even considering emerging threats (e.g. Na-Tech, cyber and physical security) as input for layer of protection analysis (LOPA)/random failures + PEM studies. HAZID is preferable to what–if? (described in chapter 11 of Crawley (2020)) since it is based on a structured approach starting from a checklist that can be customised on the basis of the knowledge (including technical literature on ethylene plants to derive TEs from Olivo (1994)).
Relative risk ranking	Dow Chemical Company (1987), ICI – Imperial Chemical Industries (1985) and Khan, Husain, and Abbasi (2001)	In a complex plant, it is fundamental to carry out a relative risk ranking among the process units to consider layout/ height/operating conditions/hold-ups and chemicals as penalties and available measures. The Safety Weighted Hazard Index (SWeHI) method could give some insights about ethylene plant risks, also in terms of consequence distances: this is useful for preliminary judgement about domino effects among plant units and offsite consequences. Application of relative risk ranking on multiple configurations is possible to optimise engineering and define requirements for following phases.
LOPA	CCPS (2011), CCPS (2013), and CCPS (2015)	Given HAZID, LOPA will consider top events (cause–consequence pairs) with a quantitative assessment based on the number and probability of failure on demand (PFD) of IPLs, as well as on some PEM calculations. LOPA could be useful for the SIF SIL classification. It could be replaced by FTA–ETA, both described in Crawley (2020), to be developed for TEs; this approach would be more time consuming and would require more expertise, data and the use of two methods.
Random failure + PEM	Benintendi (2021), TNO (2005) and (TNO (1997)	Study of PEs considering random failures (flanges, pipes, etc.) would complement threats from process deviations, completing the TEs and would avoid the need of a (full) quantitative risk assessment (QRA). Due to the characteristics of the ethylene, some specific PE studies could be considered (to be refined in the detail design phase): dispersion, flare and facility siting studies.
Risk matrix + 'as low as reasonably practicable' (ALARP)	Risk management: Expert Guidance (2021), HSE (2001) and ISO (2016b)	RMA is applied to the risks to have an overview of those identified with a specific plant configuration and define (given a risk criterion) the acceptability. The ALARP study should be carried out to demonstrate that the residual risk is reduced as far as reasonably practicable. In the FEED phase, a qualitative ALARP model can be applied.

Table 9.11 'Detailed design' methods.

HAZID (detailed)	See Table 1	See Table 1
HAZOP	IEC (2016b)	Consolidated method to consider hazards and operating problems in oil and gas (O&G)/petrochemical plants. HAZOPs of ethylene plants main equipment are available as technical literature for reference. Operating part could be complemented by HRA. It is also a good starting point to define SECEs.
HRA	Chapter 6 in Crawley (2020)	HAZOP results can be integrated by HRA at different degrees of complexity to verify the human factors (HF) associated with critical tasks (as suggested in chapter 6 of Crawley (2020)). Among the available methods, the SPAR-H (Idaho National Laboratory 2005) resulted to be also applicable in the O&G sector.
Layer of protection analysis (LOPA)	See Table 1	See Table 1. In detail design, the SIL study should also complement the SIL allocation. LOPA has been selected for feed and detailed design but could also become the basis for an operation phase barrier-based risk management built on IPLs (and SECEs/SCTs), eventually to be combined with Bow-Tie (CCPS 2018) diagrams for risk monitoring, communication and training.
Quantitative risk assessment (QRA)	TNO (1999), CCPS (1999), and TNO (1992)	QRA resulting indices as location-specific individual risk (LSIR), individual risk per annum (IRPA), PLL and fatal accident rate (FAR), given regulation acceptability criteria, can be used in the risk evaluation phase too.
PEM	See Table 1	See Table 1.
Risk matrix + ALARP	See Table 1	See Table 1. In this phase, a more sophisticated ALARP model can be applied: for the considered ethylene plant identify, control, assess and follow-up (ICAF) model, as described in DNV (2002), can be effectively employed. Populated risk matrix and ALARP studies should be made available via a risk register for the plant operation phase to manage the management of change (MOC) process and periodic revalidation.

9.3 Risk Analysis

This section presents some methodologies generally used in the risk analysis phase while remaining their applicability (more or less recommended) to the other phases of the risk assessment. Since risk analysis activities generally require a 'quantitative' approach, a number of methods, cited by ISO 31010, have been selected, which allow for numerical insights, including, possibly as a deepening of initial qualitative studies such as HAZID and HAZOP.

9.3.1 Fault Tree Analysis (FTA)

This method, created in the Bell Telephone Laboratories in the early 1960s, intends to reconstruct the exact sequence of primary and intermediate events leading to a top event failure. It is useful to recognise those situations that may give rise to undesired consequences when combined with specifically identified events. The main structure of a fault tree is shown in Figure 9.4. An alternative way is to draw it vertically, as shown in Figure 9.5, whose basic events are shown in Figure 9.6. They are as follows:

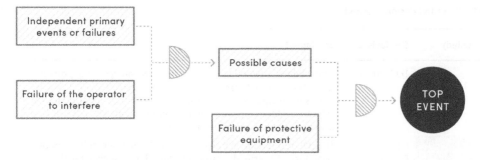

Figure 9.4   Basic structure of a fault tree (horizontal).

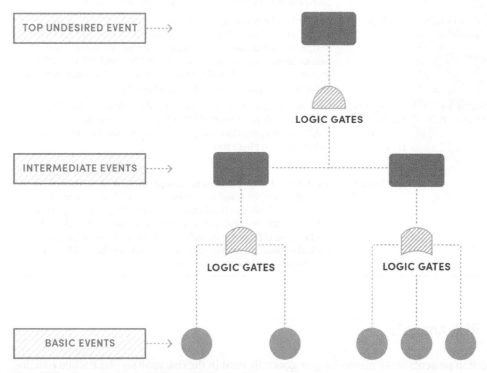

Figure 9.5   Basic structure of a fault tree (vertical).

- Base events – further analysis not useful.
- Undeveloped event – events not analysed further at this time.
- Top or intermediate event – events that are further analysed.

The FTA is an analytical deductive technique to analyse failures. It focuses on a particular unde-sired event and attempts to determine its causes. The undesired event is known as 'top event' in a fault tree diagram: it is generally a complete, catastrophic failure of the system under investigation. The top event is the final effect, but it is also the starting point of a fault tree analysis. This explains why the formulation of the top event must be punctual and exhaustive: this will ensure the good-ness of the outcomes provided by the FTA. The FTA diagram is a graphical representation of both parallel and sequential chains of failures that take the pre-defined undesired event (i.e. the top

BASE EVENT   UNDEVELOPED   TOP OR INTERMEDIATE
                EVENT   EVENT

Figure 9.6 Basic events.

event) to occur. Usually, each fault in an FTA is the combination of system failures (mechanical failures or human error), and the ineffective/missing/failed safeguards put on to stop the chain of failures, but that revealed incapable of doing so, for a determined reason.

It is important to underline that a fault tree does not take into account all possible system failures or all possible causes potentially at the base of the event. Indeed, every fault tree is designed for a particular top event, which represents the starting point that will define the development of the rest of the tree: only the contributing failure modes are indeed considered. Every single fault in an FTA is combined with the others through AND/OR logic gates. Other logic Boolean gates can be used, but, generally, they are not required. The usage of logic connectors is useful when a single event could be caused by one or more factors that must act at the same time.

After the top event is identified, the analysis of the faults proceeds level by level: firstly, the possible and most general immediate causes are considered, always finding support in the collected evidence. Potential failures that are eliminated by the matching with the evidence are then further investigated, thus finding the second level of causes. The iterative approach continues until the found causes are considered sufficiently detailed to stop the investigation.

It could also happen that more than one path between the same faults at the origin and top event is found. When this happens, and the tree is fully drawn in a flow chart, the more realistic path between the final failure and a specific set of causes is called 'minimum cut-set' and represents the shortest path between the two.

Fault trees, which may be considered as reversed FMEAs, are used to guide the investigative resources in the most probable causes. Up to now, the description of the FTA seems to define a qualitative analysis. Even if it is possible to leave the fault tree without any number, significant advantages are taken if it is used in a quantitative approach. Data about the failure probabilities are taken from historical databases (when available) or the guides provided by the manufacturers or independent publication. Obviously, there is software that performs a computer-based fault tree analysis. The numerical data about the probabilities of human errors, component failures or environmental factors are combined using the mathematical rules for probability, depending on the logic gates on which multiple causes converge. Usually, the logic gates are those that are shown in Figure 9.7, where

- OR GATE – the output event occurs if ANY of the input events occur;
- AND GATE – the output event occurs if ALL of the input events occur;
- TRANSFER GATE – transfers to or from another part of the fault tree.

OR GATE   AND GATE   TRANSFER GATE

Figure 9.7 Gates.

For instance, an 'AND' gate means that all the previous factors must be fulfilled to generate the subsequent event. From a probabilistic point of view, this is translated into a single probability of the subsequent event, obtained by multiplying all the probabilities of the single causes each other (according to the combined probability rule). Instead, if the connection is through an 'OR' gate, the likelihood of the resulting event is equal to the sum of the probabilities of the single causes. Carrying on in this way, the probability of occurrence of the top event is found. In risk assessment, this information is combined with the severity of the event (for instance, obtained through numerical simulation, in a quantitative approach) to assess the final risk of the top event.

For example, the well-known fire triangle can be transferred into an FTA using the diagram in Figure 9.8.

It clearly appears how FTA is an analytical tool for establishing relations; it does not provide any direct information about how to gather evidence. The strengths of the fault tree, even when used in a qualitative approach, are its ability to break down an accident into root cause.

Here, an outline of how to conduct a fault tree analysis is presented:

Develop the problem statement (i.e. the top event and the reason for the investigation).

Identify the first layer of inputs for the incident, considering basic components or procedures. Remember to consider all the possible inputs of failures for the considered equipment or procedures.

Define the relationship between the top event and the first layer inputs through a logic gate.

Evaluate each first layer input by identifying their second layer inputs.

Define the relationship between a single first layer input and its second layer inputs through a logic gate.

Continue with other layer inputs until the required level of investigation is reached (typically, when the root causes are found, the iterative procedure stops).

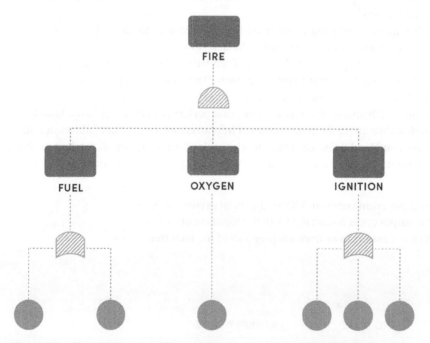

Figure 9.8 Fire triangle using FTA to combine fuel, oxygen and ignition components.

If required, gather additional information to complete, support or eliminate some branches or single inputs of the tree.

Document and report the result of the FTA, also highlighting the minimum cut-set path and the probabilities related to it.

A further example is now shown. Consider the flammable liquid storage system in Figure 9.9: it is kept under pressure by nitrogen, and a pressure controller is used to maintain the pressure between certain limits otherwise an alarm is sent to the control room. The relief valve RV-1 opens to the atmosphere in case of emergency. Considering the tank rupture due to overpressure as the 'top event': the corresponding fault tree is shown in Figure 9.10.

Another example of a fault tree is shown in Figure 9.11.

This technique can be used to calculate the availability on demand of the active fire protection systems, as shown in Figure 9.12.

9.3.2 Event Tree Analysis (ETA)

The event tree analysis (ETA) determines the potential consequences in terms of undesired incident outcomes, starting from an initiating event (i.e. equipment or process failure). The aim of the ETA is therefore complementary to an FTA goal. Indeed, an FTA explains how an undesired event can result from previous failures (allowing to find also the root causes), while an ETA examines all the possible consequences of the undesired event.

The structure of a typical ETA diagram, from initiating events to outcomes and consequences, is shown in Figure 9.13.

This technique is among the most difficult to apply in practice due to the number of intermediate events that results in a very large number of different outcomes. Indeed, meaningful results are obtained only if the undesired (or even desired) events, from which branches are created, are fully anticipated and well defined. It is therefore clear that the application of the method requires strong

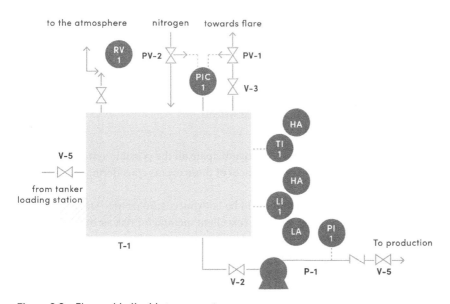

Figure 9.9 Flammable liquid storage system.

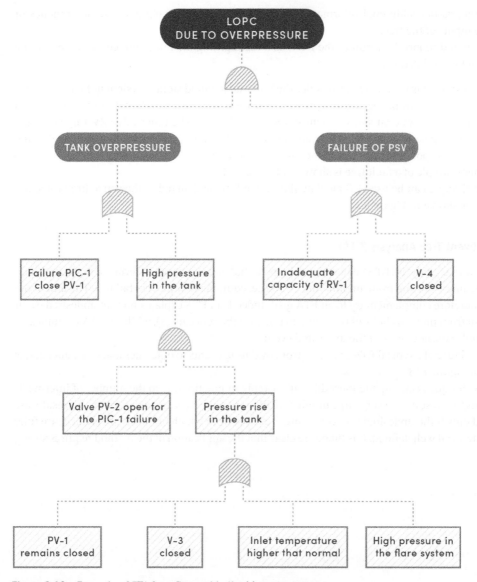

Figure 9.10 Example of FTA for a flammable liquid storage system.

practical experience in the specific domain in order to anticipate all the possible system (intermediate) events and to explore all the possible consequences of those events considering all the available combinations.

The event sequence is defined by barriers that could be both successful or not. The tree underlines how likely was the occurred event. The ETA is an excellent method for risk assessment, being
used to identify possible event scenarios. In the incident investigation, the actual incident path
may be underlined among all the possible ones (Figure 9.14).

From a quantitative point of view, the frequency of occurrence of each scenario is determined
starting from the likelihood of the initial event and combining it with the probabilities of failure of
the barriers put in the position to create the nodes for diverging branches of the tree. As usual,

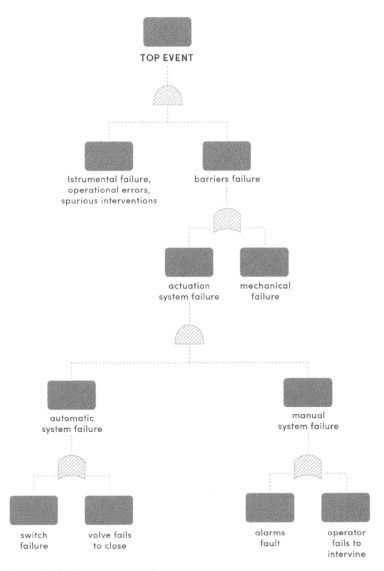

Figure 9.11 Fault tree example.

combined probability rules are followed. Often, the probability of occurrence of the initial event is obtained from a fault tree analysis: in other words, the top event of an FTA is the starting point for an ETA. The combination of the two methods represents the Bow-Tie, the further risk assessment that aims in the full comprehension of an undesired event both looking for its causes (FTA) and all its possible consequences (ETA). Bow-Ties are powerful tools to view, at a glance, both preventive and mitigating measures.

Once the probability of failure of a specific barrier is known (from the historical databases, experience and so on), the likelihood of being successful is complementary to the unit (i.e. the probability that the barrier will fail or will not fail is one).

Let us consider a pipe connected to a vessel, as shown in Figure 9.15. The possible consequences of a rupture of the pipe in the point 'P' need to be found. The system is equipped with an excessive

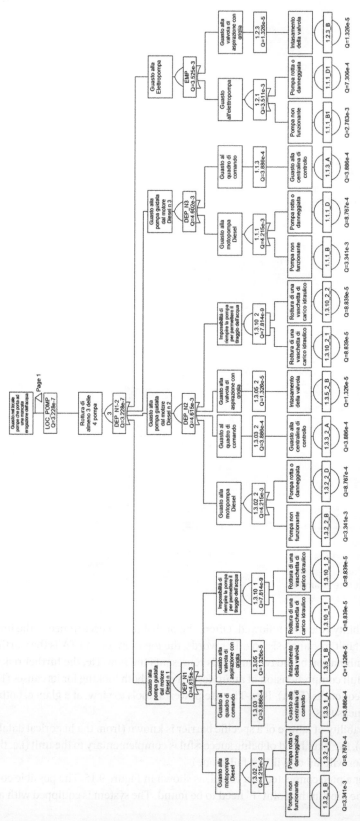

Figure 9.12 Representative fault tree developed for the subsets of the analysis.

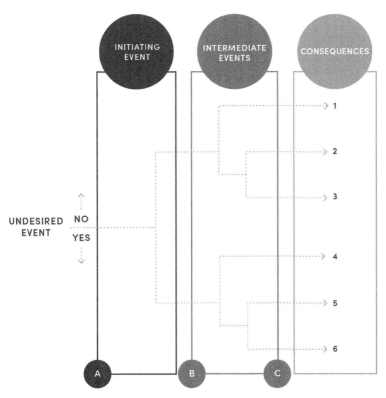

Figure 9.13 The structure of a typical ETA diagram.

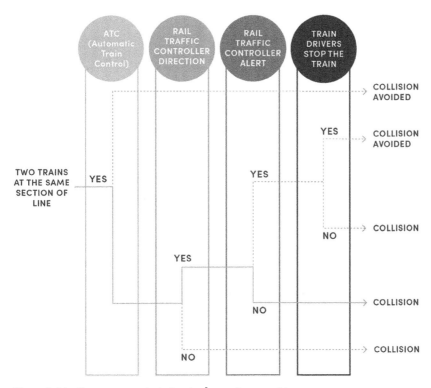

Figure 9.14 Event tree analysis for the Åsta railway accident.

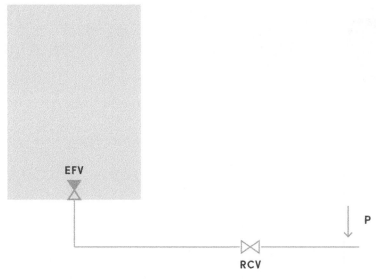

Figure 9.15 Pipe connected to a vessel.

flow valve (EFV) and a remote controller isolation valve (RCV). The resulting event tree is shown in Figure 9.16. In the event tree, the pipe rupture is the undesired event (A) with a probability PA, while the EFV failure is an intermediate event (B) with the probability PB, like the RCV failure (C) that has a probability PC.

According to the resulting event tree, the probability of a continuous leak is given by PA × PB × PC; the likelihood of leakage until RCV is closed is given by PA × PB × (1 − PC); finally, the likelihood of a minor leak is given by PA × (1 − PB).

An example of ETA is shown in Figure 9.17.

9.3.3 Bow-Tie and LOPA

Risk management, once context analysis and risk assessment are completed, results almost entirely in the management of risk control measures or barriers. Managing risk over time, ensuring that it remains within certain acceptability values, effectively means ensuring that the measures put in place by the organisation to contain the risk remain intact, effective and efficient over time. Starting from a simple reflection, two risk analysis methods based on the concept of barriers are presented here. They, therefore, offer a different perspective on risk management, positioning themselves as tools to support an entire risk management system that intends to adopt the barrier-based perspective that is emerging in the technical standards of voluntary adoption, such as those mentioned previously in this book. This approach is completely compliant with the framework of the ISO 31000 standard for risk management and can be adopted for the implementation of modern risk-based SMSs. The barrier is, therefore, a measure of control or grouping of control elements that, in itself, can prevent the development of a cause in a top event (preventive barrier) or can mitigate the consequences of the top event once it has manifested itself (mitigating barrier).

The Bow-Tie, thanks to its powerful ability to communicate information graphically, has established itself as the main technique of risk analysis based on the concept of barriers. The entire chapter, as a preferred methodology by the author of the book for the illustration of the principles of ISO 31000, is dedicated to it. In this section, we provide a simple introduction to the method.

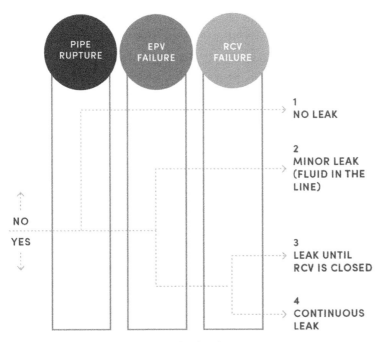

Figure 9.16 Example of event tree for the pipe rupture.

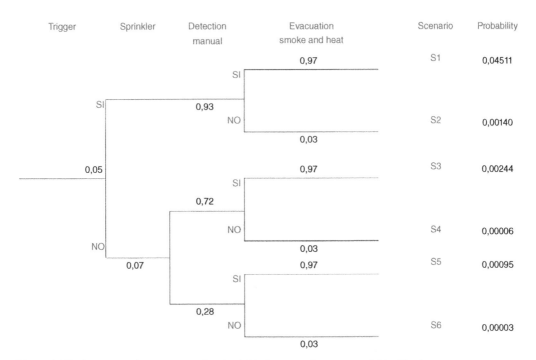

Figure 9.17 Representative event tree for the initiating event 'paper material ignition'.

The Bow-Tie technique involves the development of logical flow diagrams developed in three distinct zones. An example of a Bow-Tie diagram is shown in Figure 9.18:

- Zone 1 (Prevention) is represented on the left side of the diagram; identifies all causes (blue rectangles) that can be associated with the unwanted event and, for each of them, highlights all the specific protection systems (both plant and operational control) that help prevent the unwanted event. Zone 1 can be considered equivalent to a simplified fault tree.
- Top event is represented in the centre of the diagram and uniquely identifies the danger considered (yellow and black striped rectangle) the primary incidental event called top event; this event can, in turn, evolve based on the dynamics of the incident in alternative incidental scenarios.
- Zone 3 (protection) identifies all potentially generated incidental scenarios (e.g. burning gas jet, explosion, flash fire, etc. ...) and the combination of all the elements that allow its development, including all protection systems that can mitigate its effects. Zone 3 can, for all intents and purposes, be considered equivalent to a simplified event tree.

The Bow-Tie technique allows to identify and evaluate the frequencies and consequences associated with scenarios and to quantify the contribution of protective and mitigating systems (barriers) under normal production conditions. This aspect is also discussed next in this book and requires the introduction to the methodology of analysis that underlies the frequency quantification of Bow-Ties: layer of protection analysis (LOPA).

9.3.3.1 Description of the Method

The Bow-Tie risk assessment methodology owes its name to the typical shape that the Bow-Tie diagram generally takes: that of a *papillon* (Figure 9.19). Its strong ability to immediately transmit complex information through its powerful (though simple) notation and graphic design has made it one of the most widely used and appreciated risk analysis methods worldwide, regardless of type, size and complexity.

Its applicability is extremely wide and allows the analyst to deal with different types of risks with the same method (possibly combined with other methods), standardising their treatment, documentation, communication and discussion within the organisation, which are peculiar aspects of the risk management process required by the ISO 31000 standard.

However, reducing its potential in the (albeit powerful) graphic notation alone is very reductive and risks diminish its value within a larger overall framework that is risk management.

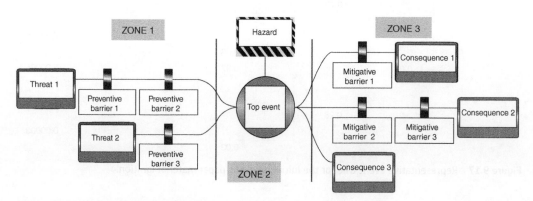

Figure 9.18 Bow-Tie diagram structure.

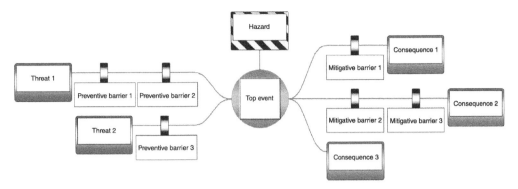

Figure 9.19 A typical Bow-Tie.

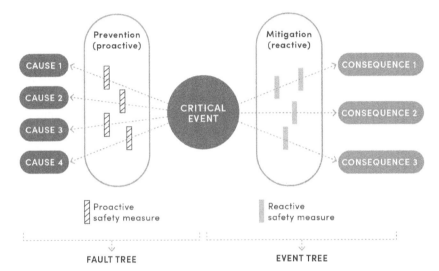

Figure 9.20 Bow-Tie as the combination of an FTA and an ETA.

In order to offer a first definition, the Bow-Tie diagram can be understood as the fusion between a fault tree (FTA) and an event tree (ETA) (Figure 9.20). The junction point between the two trees, i.e. the centre of the Bow-Tie, is the so-called top event. In one fell swoop, the Bow-Tie therefore provides an exhaustive overview of everything that could cause an unintended event (the top event), through the branches to the left comparable to those of a cause tree, and the possible consequences of the top event, through the branches to the right superimposed on those typical of an event tree.

However, it should be noted that in the 'simplified' definition of Bow-Tie as the mere union of a cause tree and an event tree, most of the inconsistent and incorrect uses of the Bow-Tie methodology fall. There are in fact several and also important differences between the logic behind a cause tree and the left side of a Bow-Tie, first of all the independence between the causes and their direct ability to determine the top event. In other words, there is no possibility of considering the top event as the result of two or more causes that must occur at the same time; from a logical point of view, it would seem that the causes in Bow-Tie are connected only by OR doors, unlike the cause trees where both OR and AND doors are used. As far as the right-hand side of the Bow-Tie is concerned, it can be observed that the representation of the typical intermediate events of an event tree does not translate so immediately into a BT diagram, which does not explicitly take intermediate events into account although the experienced risk analyst knows that a reasoned and careful use of barriers could bridge these differences. What is discussed here represents only some of the reasons that led the authors, with a strong sense of self-criticism, to present the parallelism between Bow-Tie and the union between FTA and ETA as a 'simplified' definition, useful for novice analysts to understand the basic concepts of Bow-Tie. In general, this definition is not to be considered fully accurate and that is why the Center for Chemical Process Safety (CCPS) of the American Institute of Chemical Engineers (AIChE) has decided to publish the first official guidelines for a correct use of the Bow-Tie method, in order to gain more important advantages and reduce cases of misuse.

The Bow-Tie provides a comprehensive answer to the three basic questions that any risk manager knows and considers the basis of their work:

- Do you know what could go wrong?
- Do you know the systems (i.e. barriers) that are in place to prevent this from happening?
- Do you have enough information to say that these systems are working effectively?

These questions will be presented several times to the reader due to their importance, which could be referred both to the initial and periodic risk assessment phase and the continuous management phase. They make a summary about the principal issues an organisation should face if it wants to start dealing with undesired events that may arise during business operations.

The first question is therefore related to hazard identification and risk analysis, the second to barriers and the third to their effectiveness. It is clear that from a barrier-based perspective (i.e. the IPLs of other methodologies), these are the questions that need to be answered in order to assess whether or not business risks are properly controlled.

Like other methods, the Bow-Tie diagram – with its explicit, complete and intuitive graphical notation – is nothing more than a model used to interpret the complexity of reality. It is therefore not the tool that every organisation must necessarily adopt, but rather the model that better than many others makes it possible to read and interpret the complex reality of risks and their management in the opinion of the authors of this volume who have used it in different situations, with different types of risks and with the need to illustrate the results of consulting activities to various types of stakeholders, both internal and external to the organisation. The method has also played a leading role in teaching activities on risk analysis and management, with particular reference to courses based on the principles of ISO 31000.

It, like other methods based on the concept of barriers, is based on the well-known Swiss Cheese Model by James Reason. According to this model (Figure 9.21), which is widely used every day to explain the malfunctioning of complex technical and/or organisational-management systems, the measures to control a hazard (i.e. barriers) can be metaphorically compared to slices of Swiss cheese, which are placed between the hazard and the accident.

The holes in Swiss Cheese slices represent their never being 100% effective (or, if you prefer, reliable): some holes are due to latent conditions, and others due to active failures and in general to random failures. Generally, the barriers (at the design stage) are placed in such a way as to avoid the alignment of such weaknesses, thus ensuring that, although each barrier is not 100% effective, the overall system is still safe. However, under certain conditions (with a specific frequency), it may happen that the holes, i.e. the intrinsic weaknesses of each barrier, align with each other, thus allowing the danger to become a real accident.

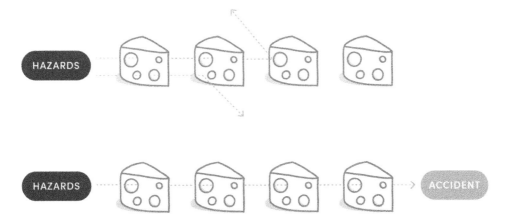

Figure 9.21 The 'Swiss Cheese' Model by James Reason.

As it is made clear in the course of reading this book, the Bow-Tie method is used in the following cases:

- Reasoning in a structured way with respect to complex systems.
- Making risk-based decisions.
- Communicating and conducting risk awareness, information and training activities.
- Monitoring barriers with periodic audits to record their status.

When compared to other risk analysis methodologies, the Bow-Tie method has a number of advantages that are explored in more detail next in this book:

- It is easier to interpret due to effective and intuitive notation as well as strong graphics.
- It reduces the complexity of information.
- It makes barriers (preventive or mitigative) immediately visible.
- It shows the effectiveness of barriers.
- It is combined with incident analysis (reactively).
- It combines with audit plans (proactively).
- It provides a well-implemented structure to the management system consistent with a risk management process developed according to ISO 31000 principles.

On the other hand, one of the disadvantages is the impossibility to hold against multiple simultaneous causes (normally connected with AND logical doors in a fault tree). It is however possible to overcome this limitation by grouping within the same Bow-Tie case all those causes that must occur simultaneously in order to reach the top event.

Given this evident limitation, it is possible to say that in any case Bow-Tie method is the most intuitive approach to risk assessment and to risk management for all those organisations, operating in different contexts and sectors, that have the goal to approach an initial risk-based thinking built around the evolution of the barriers they have already in place, initially to understand the need of an increase in number, quality, definition and robustness of those. Subsequent further assessments are then performed with more sophisticated methods.

9.3.3.2 Building a Bow-Tie

In this section, the constituent elements of a Bow-Tie diagram are presented. The following eight elements constitute the Bow-Tie diagram, as shown in Figure 9.22:

- Danger;
- Top events;
- Causes;
- Consequences;
- Preventive barriers;
- Mitigative barriers;
- Escalation factors;
- Escalation factors controls (preventive and mitigative barriers to escalation factors).

Each informative element making up the diagram is univocally associated with a graphic representation: this makes the model elaborated, immediately readable and comprehensible.

It is clear from the outset that the names used are not the only ones that can be encountered in the various sectors of industry. The same elements in fact are also indicated through some synonyms, more or less widespread depending on the application context. It is worth remembering that the Bow-Tie method, born in the bedrock of the practices and methodologies for guaranteeing the safety characteristics of industrial risk in the 'oil and gas' sector and in particular in that of 'offshore' industrial assets, is now widely used also for completely different types of risk. Therefore, we can often hear about proactive and reactive barriers, indicating those placed on the left (preventive) or right (mitigative) sides of the Bow-Tie; or we could use the term 'threats' instead of 'causes', or 'incidental scenarios' instead of 'consequences', or 'degrading factors' for escalation factors, or 'secondary barriers' for escalation factor controls. The same 'barrier' may be referred to as 'control measure', 'risk reduction measure', 'IPL', 'individual protection layer' or 'independent protection layer', etc. depending on the organisational context and how certain terms have already been used (and established) in this context. What is important is to understand that these are synonyms and that, within the methodology illustrated in this book, they all assume the meaning that is clarified in the following sections.

In any case, while fully respecting the principles set out in ISO 31000, the user of the method is free to integrate further elements and specific taxonomies aimed at better describing, with respect

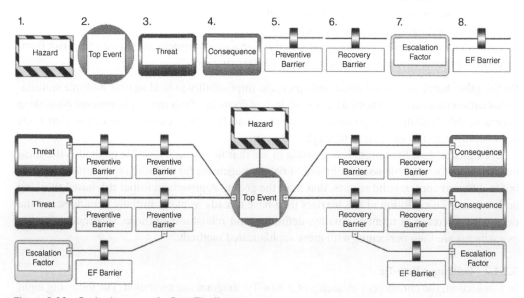

Figure 9.22   Basic elements of a Bow-Tie diagram.

to the use of the 'basic' system, the risk scenarios typical of the context in which it operates. In the same way, he/she can extend the basic information tools of the notation with the elements, also of quantitative type (discrete or stochastic), as a result of the application of other techniques in parallel specialist in-depth studies.

It is therefore very important to underline that, if ISO 31000 standard does not require the use of a specific risk assessment method (or even the barrier-based approach), multiple methodologies could be combined together in order to achieve better and more useful results for the organisation goals. In this book, we give emphasis to the combined use of the Bow-Tie method and LOPA for the risk quantitative assessment, as well as for the use of the Bow-Tie and barrier failure analysis (BFA) methods to optimise the resources in the active and reactive phases of risk management; but there are other infinite combinations. FTA and ETA could be used to calculate the probability of failure on demand (PFD) of technical barriers, HRA could be used to assess the probabilities associated with human factors (errors as threats, responses as behavioural mitigation barriers) and so on, with a number of new possibilities.

For example, Figure 9.23 shows a simplified Bow-Tie with the barriers prescribed by the Italian law to manage fire risk in a tunnel. In order to visualise the alternative solutions, the Bow-Tie in Figure 9.24 has been created, showing the equivalent measures. As the analysis revealed some weaknesses due to a real fire occurred in road tunnel, the analyst suggested additional risk control measures and communicated them using a third Bow-Tie, the one shown in Figure 9.25.

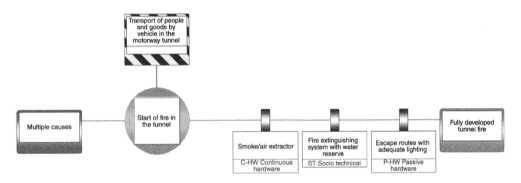

Figure 9.23 Barriers prescribed by Italian standard to manage fire risk in a tunnel.

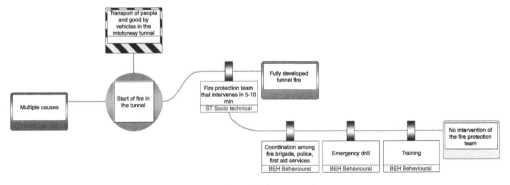

Figure 9.24 Equivalent measures to manage fire risk in a tunnel.

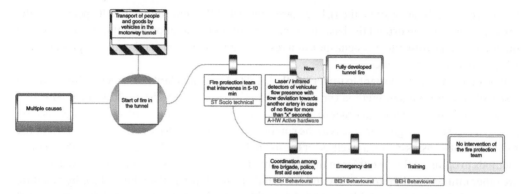

Figure 9.25 Suggested additional risk control measures to manage fire risk in a tunnel.

The causes must be able to cause the top event independently of each other (they are therefore in the 'OR' logic between them). This is a fundamental assurance placed on the basis of the implementation of the method.

As a prompt reference, the Bow-Tie guiding principles are summarised in Figures 9.26 and 9.27.

9.3.3.3 Barriers

There has been extensive discussion of how risk management is focused in its control. This is done through control measures, i.e. barriers. This book differentiates between primary and secondary barriers. The former are those found on the main branches of a Bow-Tie diagram, directly interposed between causes and top events or between top events and consequences. The latter, on the other hand, only appear if escalation factors are also defined and serve as measures to support primary barriers from the degrading threat of escalation factors. Therefore, secondary barriers do not directly prevent or mitigate the sequence of events: this is the task of primary barriers.

A barrier can be defined as any measure taken against an unwanted 'force' in order to maintain a desired state. They can be both physical and intangible. For example, in the case of a Bow-Tie diagram having

- top event = 'loss of containment' and
- consequence = 'fire',

the 'automatic deluge active fire protection system' barrier clearly represents a mitigation barrier, i.e. to be placed on the right side of the Bow-Tie diagram, as its effectiveness is only expressed after the top event has occurred. The analyst then asks the following question for each barrier: 'does the barrier act before or after the top event has occurred?'. If before, then it should be placed on the left side; if after, then it should be placed on the right side. By way of an example, a containment basin intervenes after the accidental release of material has occurred: it is therefore a mitigating barrier. On the other hand, the execution of the activities foreseen in the inspection and maintenance plan can be clearly classified as a preventive barrier.

In reality, the placement of barriers in the Bow-Tie graph is not always so immediate. In fact, barriers generally have one of the five functions as shown in Figure 9.28.

In general, a barrier performs its function by executing the following three actions:

- Detecting the danger.
- Deciding how to act.
- Acting as decided.

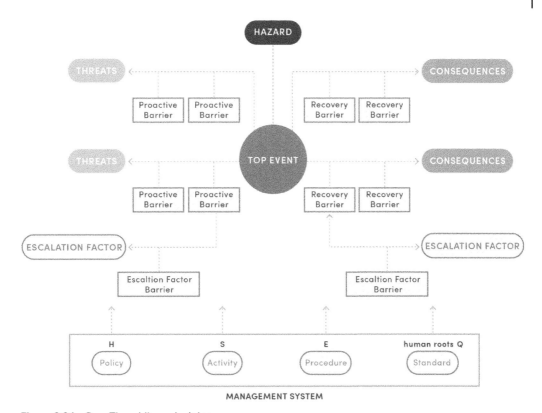

Figure 9.26 Bow-Tie guiding principles.

It is therefore essential to identify the subjects that carry out these actions and any sub-systems of the barrier that can be traced back to them. In the example in Figure 9.29, it can be observed that the three functions have been differentiated and the subjects responsible for the various sub-systems are also different. If, on the other hand, the responsible subject is common to all three subjects carrying out the three actions (detect, decide and act), or in any case one prefers to observe the problem in terms of systems and not functions, then it is recommended to use a single macro-barrier: the one at system level.

Given the centrality of barriers in the development of barrier-based risk management, it is deemed necessary to offer the reader some insights that also recall appropriate scientific reference literature.

Due to the origins and the widespread use of the Bow-Tie method in the chemical sector, most of the literature focuses on safety and process safety. But definitions given in those contexts are pretty useful in a number of difficult cases.

Safety barriers can be technical or organisational and perform one or more safety functions, thus determining their purpose. Generally, the function of a barrier is defined by a verb and a noun, but what is important is that it is clearly identified. The classification proposed by the AIChE CCPS guidelines is shown in Figure 9.30.

Adopting the barrier-based perspective, risk management is therefore barrier management and barrier management is an integrated part of risk and safety management, where after identifying hazards, estimating frequency and assessing consequences, barriers are analysed too. According to the theory of the safety management developed by Li and Guldenmund, barriers are the input

HAZARD

▶◀ Describes the desired state or activity
▶◀ Is part of normal business
▶◀ Has the potential to cause harm if control's lost
▶◀ Defines the context and scope of the bowtie diagram

e.g.: Driving a car, hydrocarbons in containment, landing an aircraft

TOP EVENT

▶◀ Is a deviation from the desired state or activity
▶◀ Happens before major damage has occurred
▶◀ It's still possible to recover
▶◀ Hazards can have multiple top events

e.g.: Losing control over the car, loss of (hydrocarbons) containment, deviation from intented flight part

THREATS

▶◀ Are credible causes for the top event
▶◀ Are not barrier failures
▶◀ Should lead directly to the top event
▶◀ Should be able to lead independently to the top event

e.g.: Driving on a slippery road, pipeline corrorsion, loss of positional awareness

CONSEQUENCES

▶◀ Are the hazardous outcomes arising from the top event
▶◀ Describe the direct cause for loss or damage
▶◀ Describe how the damage occurs

e.g.: Car rollover, ignition of vapor cloud, mid-air collision

ESCALATION FACTORS

▶◀ Are factors that reduce the effectiveness of a barrier
▶◀ Should be used sparingly to highlight real issues
▶◀ Tip: focus on critical barriers
▶◀ Tip: avoid repetition and duplication

e.g.: Forgetting to wear the seatbelt, no manteinance done, person not trained

BARRIERS

▶◀ Prevent, control or mitigate undesired events or accidents
▶◀ Can be (a combination of) behaviour and hardware
▶◀ A barrier system contains a detect, decide & act component

e.g.: Wearing a seatbelt, Blow-Out Preventer, Ground Proximity Warning System

PREVENTIVE BARRIER ▶◀ Eliminates the Threat or prevents the Top Event

RECOVERY BARRIER ▶◀ Avoids or mitigates the Consequence

ESCALATION FACTOR ▶◀ Reduces the effect of the escalation Factor

Figure 9.27 Bow-Tie guiding principles.

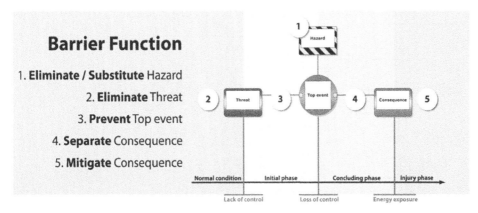

Figure 9.28 Barrier functions.

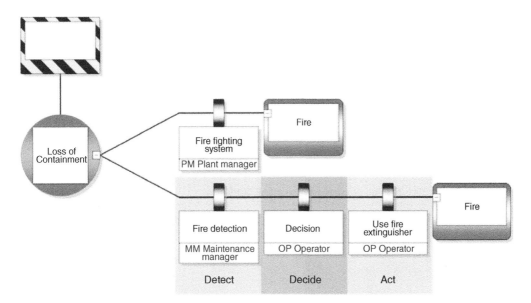

Figure 9.29 Barrier systems.

of the model, while safety performance is the output. Of course, barrier management includes managements on processes, systems, solutions and measures.

Verification is an important step in the barrier management since the barrier's performance have to remain suitable over time, maintaining adequate condition to meet the requirements.

The concept behind a safety barrier is usually related to the so-called energy model, widely used as an accident model. According to it, a barrier is seen as a measure to protect the vulnerable target from the unwanted release of energy from a hazardous source as shown in Figure 9.31.

Some authors proposed the generic safety functions related to a process model as shown in Figure 9.32.

A common way to define the performance standards is the functionality, availability, reliability, survivability and interaction (FARSI) approach, according to whom barriers are described in terms of functionality, availability, reliability, survivability and independence.

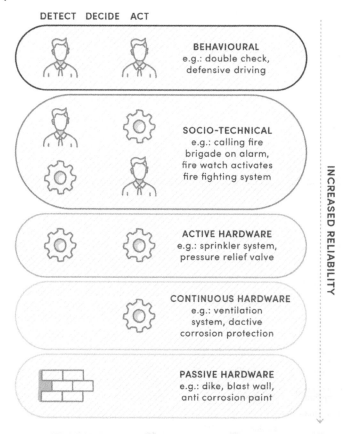

Figure 9.30 Barrier classification promoted by the AIChE CCPS Guidelines.

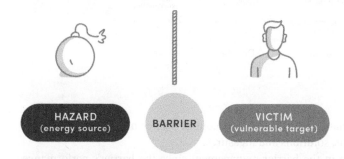

Figure 9.31 The energy model by Haddon.

Confusing a cause with the failure of a barrier is a very common mistake that must be avoided. Identifying the lack of maintenance on the level switch as a cause is an error that, if not removed, cancels any possibility of improvement because it hides an opportunity for the organisation to pursue continuous improvement. Similarly, identifying antilock braking system (ABS) failure as a cause of a loss of control of a vehicle is a mistake: ABS is a preventive barrier, and failure to intervene is a failed barrier. The cause is 'abruptly braking' or 'slippery road pavement' or 'abruptly braking on slippery road pavement', depending on the context and scope of the Bow-Tie.

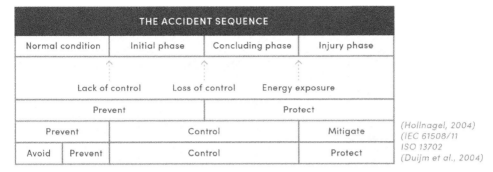

THE ACCIDENT SEQUENCE			
Normal condition	Initial phase	Concluding phase	Injury phase
	Lack of control	Loss of control	Energy exposure
Prevent		Protect	
Prevent	Control		Mitigate
Avoid	Prevent	Control	Protect

(Hollnagel, 2004)
(IEC 61508/11
ISO 13702
(Duijm et al., 2004)

Figure 9.32 Generic safety functions related to a process model.

It is clear that a consistent use of the methodology cannot ignore the knowledge of terms, references and definitions, otherwise the benefits of applying the method will not be exploited and, more strictly, significant errors of assessment will be made.

It is possible to identify three verbs to describe the functioning of any barrier present in the system under consideration: those are 'detect', 'decide' and 'act', as summarised in Figure 9.33

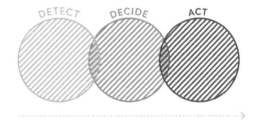

Figure 9.33 Actions of a barrier.

A barrier, as discussed in the previous sections, represents both the fulcrum of the risk assessment, and the key element on whose operation the process of ensuring the maintenance of the level of risk over time is grafted, to be considered valid within the Bow-Tie methodology, must meet at least the following three requirements:

- *Effectiveness*: Barrier must be able to prevent the top event or mitigate a consequence by acting as and when expected. 'Training' or 'competence' are not effective barriers: they are secondary barriers. Likewise, the 'fire and gas detection system' is not an effective barrier because although it is an important system it alone is not capable of mitigating the expected consequence, merely detecting a danger (i.e. 'decide' and 'act' actions are not envisaged). An effective barrier is instead 'operating intervention on fire & gas system alarm with emergency shut down activation'.
- *Independence*: Barrier must have a direct and independent impact on the cause, top event or consequence. This criterion therefore excludes those systems that share common causes of failure or failure modes.
- *Evaluability*: Barrier must be capable of being evaluated for efficiency and effectiveness (i.e. 'auditable') to verify expected performance. Formally, this criterion is met by assigning requirements or performance standards to the functionality of the barrier, which can be periodically verified by comparison with minimum performance criteria judged to be acceptable (e.g. the technical barrier consisting of an active fire protection system operating with foam can be considered suitable and functioning not only if present, not only if activated, it supplies the extinguishing agent, but only if tests carried out at a suitable frequency, if necessary in accordance with specific regulatory requirements, show compliance with the performance defined in the design activities of the barrier itself with respect to the risk scenario for which it was designed, such as the expansion ratio and the specific discharge density).

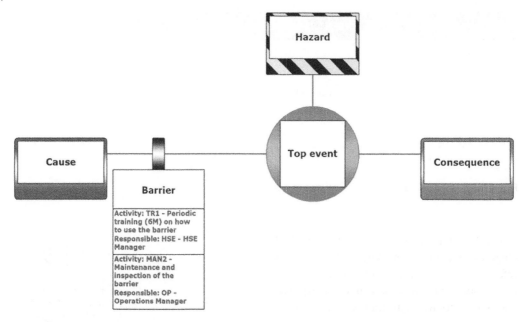

Figure 9.34 Defining 'activities' for a barrier.

Escalation factors, which are not mandatory within a Bow-Tie diagram, must be used with the utmost care in order to avoid improper use. The use of escalation factors, if not controlled, can lead to a nesting of escalation factors and secondary barriers, if these are used recursively.

The alternative is to transform the nested structure of secondary barriers into 'activities', as shown in Figure 9.34.

The Bow-Tie method is not only the approach to initial context definition and risk assessment but also a valuable aid for the periodic verification of the state of barriers in a systemic perspective. In fact, there is clearly a direct correlation between the level of risk of an organisation and the instantaneous state of the barriers, i.e. all those preventive and mitigative controls designed and implemented to obtain a degree of risk reduction congruent with the criteria of acceptability and tolerability adopted by the organisation.

Thanks to the introduction of the 'activities', the possibility offered by the Bow-Tie method to support the organisation in barrier management activities is even clearer. In fact, not only will it be possible to identify a responsible person within the organisation for each barrier to ensure its integrity and functionality, but also the individual activities in support of a barrier will have a responsible person, a certain frequency, a priority, one or more supporting documents and so on. In fact, the risk management system is being implemented in a 'barrier-based' perspective.

9.3.3.4 LOPA Analysis in Bow-Tie

The Bow-Tie method can be used for both qualitative and quantitative risk assessment, overcoming in fact the use of the method for the only immediate and intuitive graphic notation that, given the knowledge of the fundamental elements, can be used in a short time in real applications. In this second case, through some very simple rules, the Bow-Tie allows to calculate the expected frequency of consequences once the following input data are known:

- Frequency of all independent causes.
- Probability of failure on demand (PFD) of all barriers.

Obviously, other frequency reduction/amplification parameters will also have to be assessed, which the analyst will have to evaluate, based on experience and objective data defined by the context and scope, in relation to time to risk, exposure and other factors, such as the degree of vulnerability of the targets to a given potential impact.

In the following sections, the operations to be done for the frequency quantification of Bow-Tie will be presented, one step at a time. Once the frequency of the consequences is known and combined with the magnitude level that the analyst will obtain on the basis of other methods that are not dealt with here, then the level of risk associated with the consequences of the Bow-Tie diagram, for example through the use of a risk matrix, will be immediately determined.

The tool that allows the frequency quantification of Bow-Tie is the LOPA that represents, according to ISO 31010, a proper methodology.

Thanks to this tool, deliberately illustrated here in simplified form, it is possible to combine the frequencies of the causes with the probability of failure of the barriers in order to obtain the frequency of occurrence of the consequences.

First of all, the difference between frequency and probability must be clear to the reader. While frequency is a dimensional quantity, measured in occasions/year (occ/year), probability is instead an adimensional measure (expressed from 0 to 1) used here as the possibility that a barrier does not perform its function satisfactorily.

Therefore, a barrier with PFD = 0 is obviously an ideal barrier since it never fails, i.e. it has 100% reliability and availability. In reality, no barrier has a PFD = 0 because there are always latent failure mechanisms (in the case of hardware barriers) or human factors (in the case of software barriers) that make the PFD, however small, different from zero. To admit, as the opposite extreme, that PFD = 1 means that the barrier in question always fails whenever it is called upon to intervene, so in fact it offers no reduction in risk.

The LOPA analysis therefore foresees that the frequency of each independent cause is multiplied by the PFDs of the barriers placed along that specific 'cause–top event' branch. This multiplication is carried out for all independent causes. If the contribution of all independent causes is added together, the frequency of the 'top event' is obtained. In mathematical terms,

$$f_{top\ event} = \sum_{i=1}^{N} f_i \cdot \prod_{j=1}^{M(i)} PFD(j(i))$$

- $f_{top\ event}$ is the frequency of the top event;
- $i = 1, ..., N$ indicates the ith cause;
- f_i is the frequency of occurrence of the ith cause;
- $j = 1, ..., M(i)$, where i indicates the jth preventive barrier (whose maximum number M depends on the branch under consideration, i.e. the ith case under examination);
- $PFD(j(i))$ indicates the PFD of the jth preventive barrier placed on the ith cause.

At this point, the frequency of the consequences is obtained by multiplying the frequency of the top event by the PFD of the barriers placed on the branch of the consequence considered. In mathematical terms,

$$f_k = f_{top\ event} \cdot \prod_{j=1}^{M(k)} PFD(j(k))$$

- f_k is the frequency of occurrence of the kth consequence;
- $j = 1, ..., M(k)$ indicates the jth mitigation barrier (whose maximum number M depends on the branch under consideration, i.e. the kth consequence under consideration);
- $PFD(j(k))$ indicates the PFD of the jth mitigation barrier placed on the kth consequence.

The method can be enriched, as mentioned above, by conditional modifiers, i.e. multipliers that aim to also include, within the quantitative analysis carried out, evaluations on the time of exposure to risk or other factors such as temporal or spatial constraints that are necessary to satisfy in order to be exposed to risk.

The application of LOPA analysis to the Bow-Tie method used preliminarily to construct the risk model to be assessed, with the mathematical formulas presented above, is based on two important hypotheses:

- Independence of causes (the frequency of occurrence of a cause is independent of whether or not all other causes occur).
- Independence of barriers (there are no common causes of failure between barriers identified on the same branch).

Summing up:

- Causes are quantified by assigning each cause an occurrence frequency (occ/year), which can be obtained either through technical–scientific literature data (database) or through more detailed evaluations such as a fault tree or human factor assessments.
- The quantification of preventive and mitigative primary barriers is performed by assigning to each of them a probability of failure on demand, which can be obtained either through technical–scientific literature data (database) or through more detailed assessments such as a failure tree or human factor assessments.
- The quantification of the top event is performed by combining the frequencies of the causes with the PFD of the preventive barriers according to the logic discussed above.
- The quantification of the consequences is done by combining the frequency of the top event with the PFD of the mitigation barriers according to the logic discussed above.
- There is no explicit quantification of escalation factors and secondary barriers. Any reduction in the reliability/availability of the primary barrier by the escalation factor and secondary barriers must be considered when determining the PFD of the primary barrier to which the escalation factor and secondary barriers refer. Generally, this results in an increase in the PFD of the primary barrier.

Finally, it may also be useful to distinguish three types of occurrence frequencies:

- Those unmitigated calculated without barriers, i.e. assuming that PFD = 1 for all barriers. In this way, the risk of incidental scenarios is assessed if the organisation has not installed any barriers.
- The current ones calculated assuming the PFD of preventive and mitigative barriers installed by the organisation to lower the risk.
- The mitigated ones calculated considering the contribution of any additional barriers that the organisation should implement to further mitigate the level of risk.

This distinction makes it possible to immediately appreciate the contribution made by barriers (whether they are actually present or only assumed) to the reduction of the level of risk.

These assessments are particularly useful for the identification of any judgement of importance or criticality of the barriers, for the conduct of cost–benefit analyses or for the prioritisation of risk-oriented adaptation or improvement measures.

Finally, it is sometimes useful to define a correlation factor between the PFD of a barrier and its effectiveness in order to correct the value of the PFD according to a set of performance factors recognised as driving factors for the integrity and functionality of that specific barrier. This is extremely useful when it is necessary to perform a relative risk ranking on a pool of similar assets, for which it is not important to know the 'absolute' risk level, but the relative risk level of one asset over the others, in order to properly prioritise possible barrier improvement actions on the basis of risk-based decisions.

For example, the effectiveness of a barrier can be measured on a normalised scale from 0% (zero effectiveness) to 100% (maximum effectiveness). The numerical scores obtained can also be associated with a qualitative judgement, as summarised in Table 9.12.

The criterion for assigning the numerical value to the effectiveness of a barrier is as follows:

EFFECTIVENESS = EXISTANCE * AVERAGE(PS)

In other words, first the existence of each individual barrier is assessed, assigning, for each individual asset:

- Yes, if the barrier exists (EXISTANCE = 1).
- No, if the barrier does not exist but should exist (EXISTANCE = 0, which forces the effectiveness to 0%).
- NA, if the barrier does not exist and it is not necessary to foresee it.

In the last two cases, no further evaluation is necessary.

In the first case (existing barrier), it is instead necessary to assign qualitative scores to the performance standards (PS) identified for each individual barrier, such as the existence of an updated procedure, the execution and frequency of maintenance and inspection activities, and so on.

Table 9.13 can be used to assign scores to the PS.

Finally, the effectiveness of the barriers, evaluated in Step 3, and therefore the peculiarities of the barriers plant by plant, is taken into account, using the following formula:

$$PFD(corrected) = \left[(PFD(theoretical) - 1) * EFFECTIVENESS / 100 \right] + 1$$

In other words, this expression makes it possible to correct the EFP of barriers by modifying it within a range from PFD(theoretical), in the case of 100% effective barrier, to 1 in the case of 0% effective barrier. It has therefore been hypothesised that the variability within this range is

Table 9.12 Quality scores and judgements on the effectiveness of barriers.

	Range	
Effectiveness	Min	Max
Very poor	0%	29%
Poor	30%	59%
Good	60%	79%
Very good	80%	100%
Not applicable	NA	NA

Table 9.13 Standard performance scores (PS).

PS	Score	Description
++	1	Present and effective
+	0.66	Present and medium effective
−	0.33	Present and poorly effective
−	0	Absent
IM	0.33	Missing information

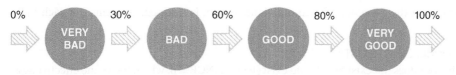

Figure 9.35 Scale of the effectiveness of a barrier and the relationship between effectiveness and PFD (correct).

directly proportional to the percentage value of effectiveness evaluated in the previous phase, as represented in Figures 9.35 and 9.36.

The two reasons why it is important to introduce the LOPA, as a method that can be combined with an event tree (Figure 9.37), in the context of this book are the following:

LOPA can be coupled with Bow-Tie risk analysis to perform a quantified assessment (as briefly shown previously in the book).

LOPA is crucial to evaluate those systems or installations with higher availability.

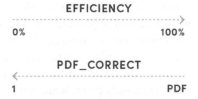

Figure 9.36 Relationship between effectiveness and PFD (correct).

They are system or installations with a higher level of availability than the minimum required by the reference standards of the system or installation. The definition of availability is provided in the UNI EN 13306 Standard. Definitions of availability, reliability, maintainability, maintenance support performance, degraded state, down state, failure and mean failure rate are provided in UNI EN 13306.

Higher availability for systems or installations may be achieved through

greater *reliability*, for example by using components with lower failure rates, redundancy of power supply sources, extinguishers, critical components, introducing measures to reduce human error, specific protection measures against the effects of fire, etc.;

greater *maintainability* and *maintenance support performance*, for example, by reducing fault recovery times, scheduling maintenance for sectors of the system, periodic checks and testing, etc. From this point of view, the NFPA 25 standards provide a useful reference for inspection, testing and maintenance of active protection systems.

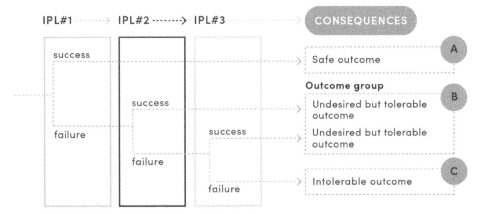

Figure 9.37 A comparison between ETA and LOPA's methodology.

In order to maintain the level of safety required for the activity, for systems or installations with higher availability, the management of degraded states or of the down states of the system must be anticipated, for example by limiting the severity of degraded states, compensatory management measures, operating conditions or limitations, etc.

9.3.4   FERA and Explosion Risk Assessment and Quantitative Risk Assessment

As anticipated in this book, analytical modelling and physical effects modelling play a fundamental role in the assessment of specific fire and explosion scenario. Given the results of this incidents in terms of fatalities, injuries, property damage, business continuity interruption or disruption and environmental impacts, it is fundamental to define the consequences levels associated with the identified scenarios.

As the same time, the magnitude results should be coupled with the estimated frequency. Frequency assessment and consequence assessment form the basis of quantitative risk assessment (QRA), often used to perform risk analysis and risk evaluation subsequent activities, including cost–benefit analysis.

In general terms, this workflow is known as QRA and, while limited to fire and explosion risks it is known as fire and explosion risk assessment (FERA), a subset of a full QRA aimed at the discussion of fire risk with some peculiar perspectives as discussed in the next sections. In both the cases, a fundamental element is connected with PEM activities to model scenarios outcomes and any eventual secondary or domino effects that may increase the overall risks.

A general workflow of QRA and FERA activities is given in Figure 9.38.

9.3.5   Quantitative Risk Assessment (QRA)

QRA is a systematic way to assess the significance of hazardous situations associated with the operation of a plant through the calculation of individual risk and social risk given specific identified scenarios.

The analysis is related to major hazards (as per the definition of major accidents given in ISO 17776 International Standard) leading to unplanned releases that can cause harm/damage to people, asset and environment.

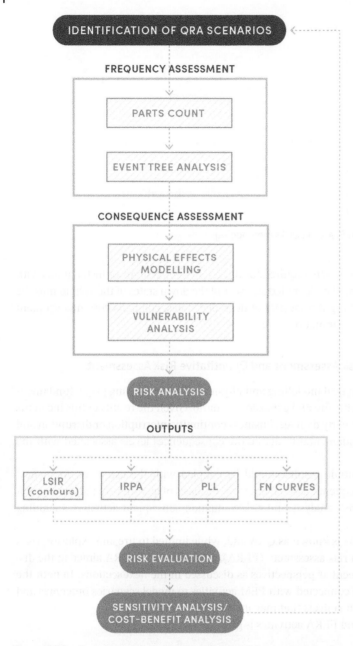

Figure 9.38 General workflow for a quantitative risk assessment.

The main steps through which the quantitative risk assessment should be carried out are as follows:

1) Identification of major credible accidental events (major accidents).
2) Calculation of occurring frequencies of main accidental scenarios.
3) Estimation of physical effects of main accidental scenarios.
4) Damage limits identification and mapping against pre-defined thresholds.

5) Individual risk quantification in terms of location-specific individual risk (LSIR) and individual risk per annum (IRPA).
6) Social risk quantification.

The previous steps are discussed in detail in the following sections.

Usually, QRA activities follow a structured hazard identification phase and they consider the main results of any eventual detailed studies available (Figure 9.39).

Identification of major credible accidental events (major accidents)

Quantitative risk assessment considers accidental events with release of hazardous materials that can cause damage to operators and population.

The QRA considers that release of hazardous materials is due to random rupture and not directly related to process deviations, covered by a specific HAZOP study.

The releases caused by random failures of the equipment are evaluated using hazard analysis.

In order to identify and record the leak sources associated with a system and to determine an overall release frequency, it is recommended to subdivide the plant into 'isolatable section' (also called 'parts count section').

It is specified that only sections considered potentially dangerous, based on considerations such as the flammability and explosivity of processed fluid, operating conditions and location, were counted in the analysis.

Each system is identified with the use of the equipment list and process flow diagrams and is subdivided based on the capability to isolate the process in event of an emergency.

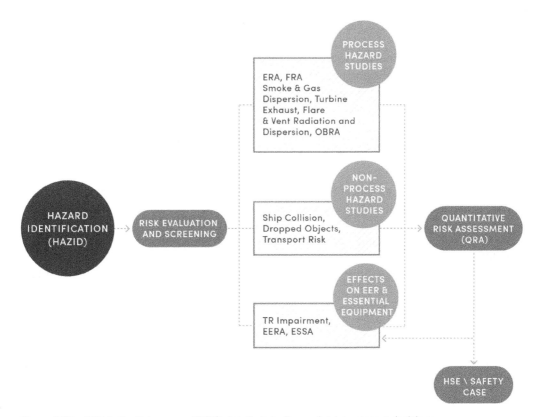

Figure 9.39 QRA is the link among HAZID, detailed studies and risk treatment decisions.

The limits of each Isolatable Section are defined by isolation valves, such as

- emergency shutdown (ESD) valves;
- PSV valves;
- BDV valves;
- locked close valves (LC);
- normally closed valves (NC);
- spectacle blind or spade in closed position;
- control valves limiting the flow or pressure of fluids, provided they can be driven to safe position (closed) by acting remotely (this case shall address a longer time to isolate due to human operation);
- chemical injections.

Check valves and rotating equipment while providing isolation in certain circumstances (they can limit the inventory) will not be given credit to provide positive isolation as this is not their designated function and they are not used to define the isolatable section itself.

Isolatable Sections are split for different fluids as separate sections, i.e. gas or liquid. In vessels that contain fluid in two phases, i.e. gas and liquid, only half is counted since it is assumed that only the top half is in contact with gas under normal operating conditions, while the lower half is included in a separate count for the liquid stream.

Calculation of occurring frequencies of main accidental scenarios

The information about the possible leak sources associated with a system will ultimately be used to determine the predicted leak frequency from each section considered.

In hazard analysis or random rupture analysis the whole process has been divided into sections, which are characterised by the same pressure and main involved substance.

Loss of containment (random releases) have been associated with each section. The number of basic components (pipes, flanges, pumps, valves, vessel, etc.) present in each section has been considered and a failure frequency has been associated with each of them.

Particularly, the 'parts count' methodology has two basic steps:

- Define parts count systems.
- For each defined parts count system, count the number of parts and record in a parts count table.

When Isolatable Sections are split at isolation valves, only half of these valves and one flange connection is counted to upstream iso-section, and the other half is counted to the downstream iso-section.

Isolatable Sections that include vessels with fluids in different phases are split on the basis of their different phases.

Moreover, if equipment with spare is present inside a single section, only one equipment has been considered in parts count.

The parts counts have been carried out using P&IDs, supplemented by process flow diagrams where necessary.

Equipment will be counted as follows:

- *Flanges*
  Flanges are classified according to the size.
  A basic flanged joint consists of two flanged faces, comprising a gasket (where fitted), and two welds to the pipe.
- *Valves*

Valves are differentiated within the count according to the following groups:

- Size.
- Operation mechanism (manual or actuated).

A valve consists of the body, stem and packer, but excludes flanges, controls and instrumentation.

Since the database leak frequency of the valve does not account for flanged connections, no. 2 flanges are counted on each valve except for valves that isolate a section: in fact, only half of them are counted, with only a flange connection being at contact with the hazardous fluid. The next scheme (Figure 9.40) is a summary of the parts count for valves and flanges.

- *Spectacle blinds and orifice plates*
Spectacle blinds are counted as shown in the following Figure 9.41.

Flanged Valves (normally open)	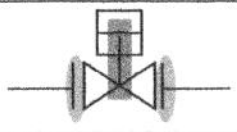	1 valve 2 flanges
Flanged Valves (normally open, e. g. SDV)	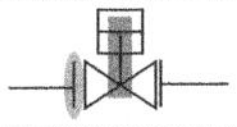	0,5 valve 1 flange
Flanged Valves (normally closed, e.g. BDV)	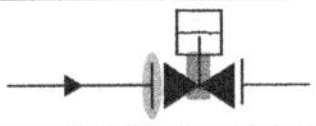	1 valves 1 flange
Welded valves (normally open)	a)	a) 2 valves
	b)	b) 1 valve
Welded valves (normallyclosed)	a)	a) 1 valve
	b)	b) 1 valve
Interlocked relief valves	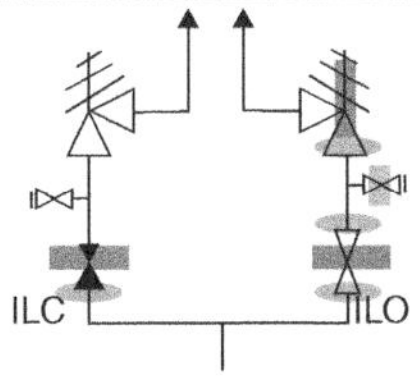 ILC ILO	3 valves 4 flanges 1 instrument (small bore connection)

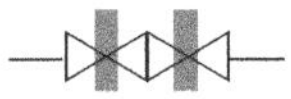

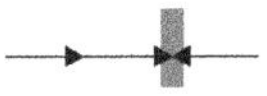

Figure 9.40   Summary of the parts count for valves and flanges.

Spectacle Blind (**open**) - 1	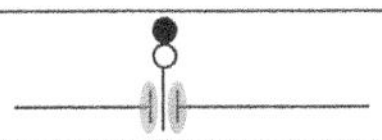	2 flanges
Spectacle Blind (**open**) - 2	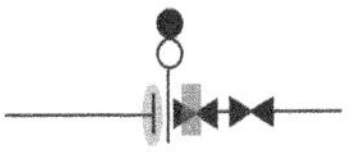	1 valve 1 flange
Spectacle Blind (**closed**)	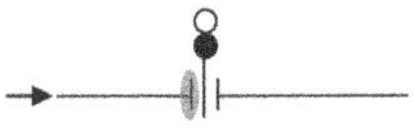	1 flange
Orifice plate / restriction orifice	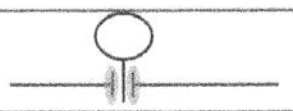	2 flanges

Figure 9.41   Spectacle blinds and orifice plates.

- *Instruments and small bore connections are counted as shown in* Figure 9.42
  One Instrument could comprise the instrument itself, plus up to two valves, up to four flanges and associated small bore connections (usually 1′ diameter or less).

  All non-intrusive instruments (such as some particular types of flow transmitters) are not counted.
- *Process vessels*
  Process vessels are counted as a generic item independently from their type (adsorber, KO drum, reboiler, separator, etc.) and independent from their orientation (vertical or horizontal).

  They are considered as 'independent of equipment size', and the largest pipe diameter that is connected to them is taken to be the size.

  A vessel comprises the vessel itself and any nozzles or inspection openings, but excludes all attached valves, piping, flanges, instruments and fittings beyond the first flange.

  If a vessel contains both liquid and vapour phases, then one-half of the vessel shall be counted on the gas section count and one-half on the liquid section count.
- *Atmospheric vessels*
  Atmospheric vessels category includes types of vessels at atmospheric pressure (i.e. oil storage vessels).

  They are considered as 'independent of equipment size' and the largest pipe diameter that is connected to them is taken to be the size.

  A vessel comprises the vessel itself and any nozzles or inspection openings, but excludes all attached valves, piping, flanges, instruments and fittings beyond the first flange.
- *Filters*
  Filters are counted in a single category, independently from their type or size.
  They are considered as 'independent of equipment size', and the largest pipe diameter that is connected to them is taken to be the size.
- *Shell and tube heat exchangers*
  Shell and tube heat exchangers are classified within the count as hydrocarbon in either shell or tube side.

  They are considered as 'independent of equipment size', and the largest pipe diameter that is connected to them is taken to be the size.

  A heat exchanger count is inclusive of the heat exchanger itself (in the case of shell and tube, the particular side under consideration) and excludes all attached valves, piping, flanges, instruments and fittings beyond the first flange.

Pressure measurement device / Thermowell	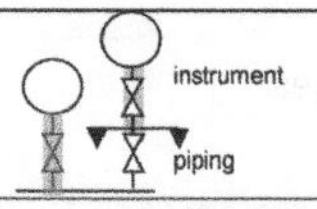	1 instrument connection
Differential measurement device (Level / Pressure) – 1 (welded)	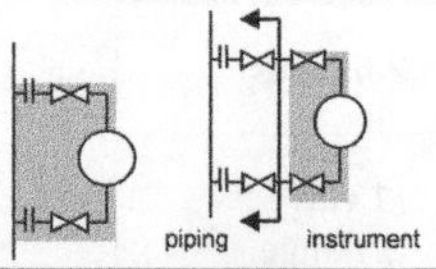	1 instrument connection
Differential measurement device (Level / Pressure) – 2 (flanged)	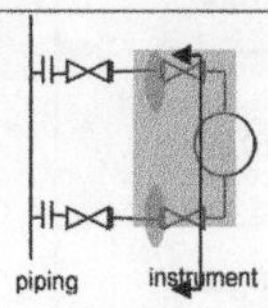	1 instrument connection 2 flanges

Figure 9.42 Instruments and small bore connections.

Where both sides of a shell and tube exchanger contain hazardous inventories, one tube side and one shell side heat exchanger are counted.

- *Plate heat exchangers*
  They are considered as 'independent of equipment size', and the largest pipe diameter that is connected to them is taken to be the size.

  A plate heat exchanger count is inclusive of the heat exchanger itself and excludes all attached valves, piping, flanges, instruments and fittings beyond the first flange.

- *Pumps*
  Pumps are classified within the count according to type (reciprocating or centrifugal).

  They are considered as 'independent of equipment size', and the largest pipe diameter that is connected to them is taken to be the size.

  The count includes single-seal and double-seal types and excludes all attached valves, piping, flanges, instruments and fittings beyond the first flange.

- *Compressors*
  Compressors are classified within the count according to type (reciprocating or centrifugal).

  They are considered as 'independent of equipment size', and the largest pipe diameter that is connected to them is taken to be the size.

  The count includes single-seal and double-seal types and excludes all attached valves, piping, flanges, instruments and fittings beyond the first flange, which is excluded itself.

- *Process pipes*
  Process pipes are classified according to diameter.

  The count excludes all valves, flanges and instruments.

  Considering the preliminary phase of the project, the count includes 3 m of process pipe upstream and 3 m downstream the relevant equipment.

  Release frequency is calculated by failure rate data of components using the statistical data taken from officially recognised references, such as

- DNV Failure Frequency Guidance 2012. 'Process equipment leak frequency data for use in QRA';
- TNO – Purple Book;
- OGP Report No. 434-1 'Process release frequencies';
- API 581, ed. 2016 'Risk-Based Inspection'.
- The frequency of releases and their consequences are calculated for
- two representative hole sizes, 10% diameter and full bore rupture, if TNO – Purple Book database is used;
- four representative hole sizes, small/medium/large/full bore rupture, if DNV, OGP or API databases are used.

In all cases, full bore rupture is modelled with a hole of 150 mm maximum diameter.

The release size taken as representative is a key factor in the release parameters and subsequent consequences in each case. However, the use of representative releases is inherent in FERA/QRA, and the frequencies are assigned according to each of the defined leak size ranges such that the overall risks should not be sensitive to the specific values selected.

Estimation of physical effects of main accidental scenarios

The accidental events can give rise to different effects, according to various parameters concurring in determining the consequences of the accidental event itself, as listed in the following:

- Hazardous substances released.
- Process conditions (pressure and temperature).

- Location where the hazardous substance is released and where it is likely to spread.
- Degree of confinement.
- Possible presence of ignition source.
- Meteorological conditions.

The possible accident sequences and the relationship between the initiating event and the subsequent events, which result in accidents, are represented by means of event trees.

An event tree is made up of nodes that correspond to different stages in an escalating incident sequence. The lines that lead out of node correspond to the paths of success or failure in the considered event. Each branch of the event tree represents a separate accident sequence, i.e. a defined set of functional relationships between the initiating event and the subsequent events.

The following Figure 9.43 illustrates the general approach used in this work for the identification of scenarios and relevant occurrence frequency for the release of flammable substances.

Ignition probability

Ignition is caused by external sources by supplying sufficient heat or energy to initiate combustion of a flammable mixture. In a QRA, it is necessary to estimate the probability of ignition if a leak of flammable substances occurs. Ignition of a leak may occur either at the point of leak or at some distance from it.

The cause of ignition may be the leak itself (e.g. a leak may generate static electricity) or an ignition source that can ignites the leak (hot surfaces, electrical motors, etc.).

The ignition probability depends on the properties of the material being released, location of the release, process conditions and leak size. In practice, ignition probability is related to the flow rate: the bigger is the flow rate, the more likely ignition occurs.

The immediate ignition probability has a direct influence on the risks associated with jet and pool fire to personnel. Delayed ignition is key influence in determining the likelihood of flash fire and explosion hazards and the extent of each (i.e. time of ignition relative to size of cloud).

An example of immediate ignition probability estimation, for process temperature lower than auto-ignition temperature, is shown in Table 9.14 (Reference 'TNO – Purple Book').

TNO 'Analysis of Vapour Cloud Accidents' (B.J. Wiekema) suggests a delayed ignition probability (PDI) equal to 0.1 for release higher than 1000 kg or 0.01 for release lower than 1000 kg.

Detection and isolation

Following an accidental release of flammable substance, the gas detection system can reduce the amount of released material providing alarm in the control room and intervention of the ESD system to isolate the section involved by the release.

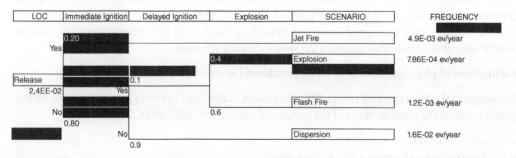

Figure 9.43   Event tree for the identification of scenarios.

Table 9.14 Example of immediate ignition probability estimation.

Source release		Substance	
Continuous	Instantaneous	Liquid	Gas, average/high reactive
<10 kg/s	<1000 kg	0.065	0.2
10–100 kg/s	1000–10 000 kg	0.065	0.5
>100 kg/s	>10 000 kg	0.065	0.7

The probability of detection and isolation depends on many factors such as field detection devices, proper response of operator in the control room after alarm and current functioning of actuators (isolation valves).

Considering the preliminary phase of the project, the probability of failure on demand of the detection system is assumed conservatively equal to 1 and the probability of failure on demand of the isolation system is considered conservatively equal to 1, by considering that the effect of detection and isolation is used for the definition of isolatable and the section estimation of release duration.

Physical effects of main accidental scenarios

The hazard potential posed by accidental releases (loss of containment) is principally governed by the material properties, process conditions, mass of material confined and possibility of section isolation. Consequences of hazardous material release are calculated for the main credible accidental events (with an occurrence frequency higher than a certain acceptability value).

Results will include consequences from

1) outflow;
2) dispersion;
3) jet/pool fire;
4) explosion (UVCE/VCE).

Damage limits identification and mapping against pre-defined thresholds

The following damage thresholds are identified for the most hazardous scenarios in the plant:

Flash fire scenarios

• Distance to lower flammability limit (LFL) concentration:	100% lethality
• Distance to 1/2 LFL concentration:	10% lethality – minor injury

Jet/pool fire scenarios

• Distance to 37.5 kW/m^2 radiation level:	Damage of assets and 100% lethality
• Distance to 12.5 kW/m^2 radiation level:	50% lethality
• Distance to 8 kW/m^2 radiation level:	6% lethality – minor injury
• Distance to 5 kW/m^2 radiation level:	0.1% lethality – negligible injury

Explosion scenarios

● Distance to 1.0 bar overpressure level:	100% lethality indoor and outdoor and damage of assets
● Distance to 0.83 bar overpressure level:	50% lethality outdoor – major injury and 80%/100% lethality indoor
● Distance to 0.3 bar overpressure level:	10% lethality outdoor – minor injury and 60% lethality indoor – major injury
● Distance to 0.1 bar overpressure level:	1% lethality – negligible injury
● Distance to 0.05 bar overpressure level:	0.1% lethality – negligible injury

Individual and social risk quantification

Individual risk is defined as the frequency at which an individual may be expected to sustain a pre-fixed level of harm from the deployment of specified hazards. The level of harm for risk analyses is conventionally assumed as the risk of death and usually expressed as the frequency associated with fatality events per year. The frequencies and the impacts of all the accident scenarios are to be taken into account in the calculation of the overall, aggregated, risk.

Location-specific individual risk (LSIR) is a measure of the risk calculated in a particular location, assuming the continuous presence (24 hours per day and 365 days per year) of a hypothetical individual at that location.

The LSIR is represented by iso-risk contour plots displayed on layouts of each deck.

LSIR takes into account the following parameters:

● Frequency of the release events.
● Likelihood of pair wind speed/stability class and their related wind sector direction.
● Likelihood of specific events leading to the hazardous outcomes (i.e. early/late ignition probability, probability of success of ESD/blowdown and probability of late ignition giving explosion).
● Human vulnerability, i.e. the fatality probability related to physical effects such as heat radiation, overpressure or toxic concentration.

The LSIR at a given location T is then calculated as follows:

$$LSIR(T) = \sum_l \lambda_l \cdot \sum_s P_{\text{scen}_{s,l}} \cdot \sum_w (P_{\text{wind}} \cdot V(T))_{l,s,w}$$

where
λ is the release frequency (summation over l release events);
P_{scen} is the probability of the scenarios, given the release (summation over s outcome scenarios);
P_{wind} is the wind direction probability (summation over w wind directions);
V is vulnerability.

A second measure of risk is the individual risk per annum (IRPA), defined as the risk of fatality to an exposed individual belonging to a working category. IRPA value is calculated from the LSIR value at the specified working location (outdoor and/or indoor) multiplied by the workers exposure, i.e. the average yearly fraction of time spent at a specific work area or rest area factor obtained from manning level data.

The societal risk represents the frequency of having an accident with N or more people being killed simultaneously. It is presented as an *FN* curve, where N is the number of deaths and F is the cumulative frequency of accidents with N or more deaths. An example of it is reported in Figure 9.44.

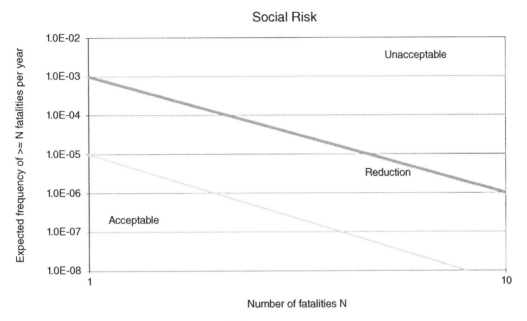

Figure 9.44 Example of risk domains in an *FN* chart.

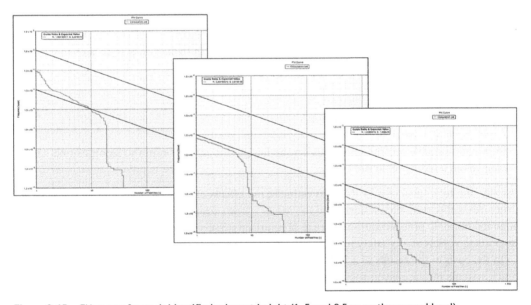

Figure 9.45 *FN* curves for each identified relevant height (1, 5 and 8.5 m on the ground level).

The highest area of the above figure represents the 'unacceptable region'. If this is the case, it is mandatory to implement actions that reduce the frequency of occurrence or the consequences of the scenarios that mainly contribute to the risk. The intermediate area represents the 'reduction desired' area.

No objection can be raised to the design if the *F/N* curve falls in this zone; however, scenarios contributing to risk in this area may be subjected to remedial actions.

The lower area of the risk matrix represents the acceptable risk region.

Figures 9.45 and 9.46 show examples of possible outcomes of a QRA.

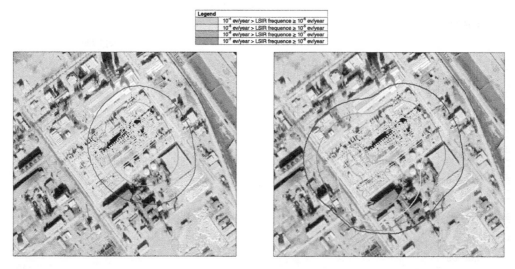

Figure 9.46 Comparison between LSIR curves referred to existing plant only (on the left) and to the complex of intervention (on the right), at 1 m on the ground level.

9.3.6 Fire and Explosion Risk Assessment (FERA)

The objectives of the fire and explosion risk assessment (FERA) study are as follows:

- To identify and quantitatively evaluate all credible fire and explosion events associated with flammable inventories that could have an impact on the facilities.
- To provide input to decisions relating to:
  - design of systems and equipment;
  - layout of main areas and equipment;
  - requirement to barriers.
- To provide suitable recommendations on the design and operation of the facilities that would bring about a reduction in fire risk.

Usually, the main goal of a FERA is to estimate the effects of fires and blast on equipment and structures, supporting the platform and the safety system design. With the term 'equipment', it is intended not only process equipment (e.g. pressurised vessels), but also life boats, living quarter, supports to vessels or any safety mean that need to be protected by the fire or explosion.

On the basis of the results of a FERA, it is possible to conduct a criticality assessment of emergency systems in place given their specific vulnerability to FERA scenarios considering the pair frequency–magnitude (Figure 9.47).

The need of possible changes in layout, safety systems and fire protection requirements may be addressed through the results from the fire and explosion analysis.

The escalation potential due to different incident scenario categories (i.e. fire and explosion) may be assessed separately for each category.

Escalation risk on each target, caused by a single incident scenario, may be assessed multiplying the incident scenario frequency by the damage probability, on the basis of the defined thresholds.

FERA is usually conducted for all hazardous equipment that can lead to fire and explosion, and the list of all such equipment is reported in the FERA report. Each apparatus is considered separately.

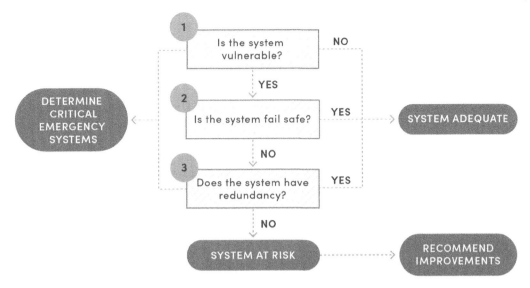

Figure 9.47 Emergency systems criticality and vulnerability definition.

At least three different types of leaks are evaluated.

The frequency of each single event is calculated based on worldwide recognised databases. For piping, equipment or unit components, recognised databases are used based on parts count methodology, as per the QRA.

Probability of ignition is calculated based on correlations, which usually depend on the released mass flow rate.

Same approach is usually taken for explosion probabilities (late ignition).

An event tree is used when propagating the probabilities that a sequence of events will occur ending to a specific outcome scenario.

Risk classification is performed for each event using the company criteria. Severity may account for the time needed to repair the damaged elements and their costs, including those for loss of production.

The criticality of the event in the company risk tolerability criteria highlights if the plant is sufficiently safe with respect to the type of event analysed. In case the risk is intolerable, risk reduction measures are taken, to shift the risk to a different region.

- *Risk reduction measure (medium–high risk region)*: The level of risk shall be mandatorily reduced applying suitable corrective measures.
- *Risk reduction measure (medium risk region)*: The level of risks requires generic control measures provided it is demonstrated that the implementation of such measures is not disproportionate to the benefits.

Measures to control and mitigate hazards are proposed by involving, if it is possible, slight modifications of the plant to conform to standard practice.

Otherwise, measures are evaluated carefully considering a wide range of possibilities as per the following hierarchy:

1) Prevention
2) Detection

3) Control
4) Mitigation
5) Emergency response

Results are supplied in terms of safety distances (maps), to be used for equipment and fire zone spacing as in the general PEM approach.

Besides, the FERA could also supply pressure set points for ESD valves as well as actuation time. In addition, these studies could assess the maximum acceptable delay in emergency system actuation such as equipment blowdown or fire protection systems.

FERA to be able to identify the design case scenario defined as a set of incident scenarios, both in terms of hazard likelihood and severity of consequences for assets, to be used as a reference for systems design (e.g. safety distances and spacing, active fire protection [AFP], passive fire protection [PFP], blast protection, etc.).

Usually, at least the following targets are investigated:

- Layout optimisation;
- Emergency shutdown (ESD) and emergency depressurisation (EDP) systems design;
- Active fire protection system design;
- Evacuation systems design;
- Passive fire protection system design;
- Blast loads design.

In order to optimise the design of the mentioned systems, all the identified targets are assessed analysing all potential jet fires, all potential pool fires and all potential explosions, separately, for all production and utility units, identifying potential escalation.

Various methods are available to present the asset risk, ranging from simple tables to graphs.

Graphs represent the iso-risk curves, and scaling and colours of tolerability criteria are used.

For each identified target under evaluation, impairment frequencies caused by different incident scenario categories separately are numerically presented in the form of tables. Furthermore, for each identified target under evaluation and for each incident scenario category, the contribution of different sources is numerically presented (both in the form of tables or bar charts).

In presenting results, the following sets of objectives, function of the risk reduction action the study aims to identify, are usually exhaustively addressed within the fire and explosion risk analysis.

Layout design

In order to assess the optimal equipment layout based on their vulnerability, all potential jet fire impairment frequencies, pool fire impairment frequencies and explosion impairment frequencies on the plant are graphically shown, conveniently separated, in terms of iso-risk curves.

ESD and EDP system design

In order to optimise the ESD and EDP system design, all potential jet fire impairment frequencies, all potential pool fire impairment frequencies and all potential explosion impairment frequencies on the plant are graphically shown, conveniently separated, in terms of iso-risk curves.

Firefighting system design

In order to assess firefighting system, all potential pool fire impairment frequencies and all potential jet fire impairment frequencies on the plant, caused by each fire risk area separately, are graphically shown in terms of iso-risk curves.

In order to assess the optimal position of the firefighting manual devices (i.e. initiation stations, monitors, etc.), all potential pool fire impairment frequencies and all potential jet fire impairment frequencies on the plant, caused by all fire risk areas, are graphically shown in terms of iso-risk curves.

The outputs of FERA confirm the need to simultaneously extinguish the areas originating a pool fire and cool the adjacent areas.

Evacuation systems

In order to assess the optimal position of the evacuation systems (i.e. muster area, temporary refuge, life boats, etc.), all potential pool fire impairment frequencies, all potential jet fire impairment frequencies and all potential explosion impairment frequencies are graphically shown, conveniently separated, in terms of iso-risk curves.

Passive fire protection system design

In order to define the PFP required insulation duration, for each identified target under evaluation, the contribution of different sources is numerically presented (both in the form of tables or bar charts) to assess the longest duration of the most probable jet fires reaching the target.

Blast loads design

In order to assess blast protection, all potential explosion impairment frequencies on the plant are graphically shown in terms of iso-risk curves.

The risk contribution, due to each PES, is usually represented using specific diagrams to highlight the global risk reducing factor associated to the selected pool of control measures.

Furthermore, if blast design is required, exceedance curves are plotted with overpressure on a linear scale on the horizontal axis and annual exceedance frequency on a log scale on the vertical axis (SLB).

The design is based on usual target frequency levels of 10^{-4} strength level blast (SLB) and 10^{-5} ductility level blast (DLB) events/year. While the DLB level blast is the design level overpressure used to represent the extreme design event, the SLB load case provides a degree of additional asset protection.

Figure 9.48 shows an example of frequency maps for fire scenarios.

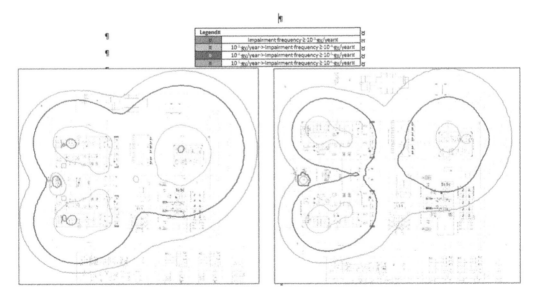

Figure 9.48 Impairment frequency maps for fire scenarios (for 8 kW/m² threshold on the left and 37.5 kW/m² threshold on the right).

9.4   Risk Evaluation

Risk assessment is the last of the three phases of the more general risk assessment process. Thanks to 'risk evaluation', the level of risk obtained through the previous analysis phase is translated, through the methods described in this section, into indices or values that can be compared with the thresholds of acceptability and tolerability defined at the preliminary stage together with the objectives of the analysis and the general objectives of the organisation, in order to determine whether the risk can be accepted (and therefore be found in the context of continuous improvement), tolerated (and therefore falls into the so-called ALARP region) or whether it should be treated further.

9.4.1   *FN* Curves

F-N curves, often used in industrial safety linked to the significant risks associated with fires, explosions and flashes, are a two-dimensional graphical representation of the cumulative frequency of the expected adverse consequences in relation to the extent of their magnitude (often expressed in terms of fatality numbers).

The construction of an *F–N* curve is known (note the different notation between the lowercase '*f*' and the uppercase '*F*'):

- *N* is the number of people estimated to be harmed (e.g. deceased) as a result of each event.
- *F* is the frequency attributed to each event.

Before building an *F–N* curve, it is good to make sure we have only one '*f*' value for each value of '*N*' since in the early stages of very complex analysis there could be multiple events with the same number of fatalities. So, if more than one event has the same number of fatalities, then we combine (sum) frequencies to have a single value. When we are finally ready to produce an *F–N* curve, we identify the *f–N* pairs.

However, diagramming the *f–N* relationship does not produce the desired result as the graph obtained is poorly informative. The result is not satisfactory even if we use a bar chart for groupings of *N* (e.g. Nos. 1–5, Nos. 6–10, etc.) because the result is highly dependent on arbitrary grouping performed.

It is therefore essential to work with the cumulative frequency *F*. The number '*F*', for a given '*N*', is the sum of the frequencies of all events whose fatality value is *N* or greater. In practice, the *F–N* curves report in orderly the expected frequency *F* that damage is given interests more than *N* people and in a sharp number of people, *N*. If *N* is the number of deaths, then the information provided by the *F–N* curve is related to the frequency of an incidental event capable of causing at least *N* deaths. Clearly, the estimate of the number of deaths is produced downstream of the studies on the analysis of the consequences, which are not the subject of this book. The analysis of the consequences should take into account the number of occupants, the extent of the damaged area, the time of the occupants' stay in the damaged area and so on.

Given the orders of magnitude in play, *F–N* curves are often represented on logarithmic axes. Within the Cartesian space, areas of tolerability, acceptability and unacceptable risk are identified. If an *F–N* curve intersects the region of unacceptableness (Figure 9.49), then the risk is considered unacceptable.

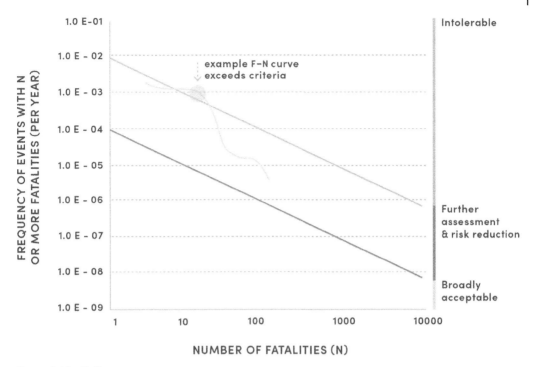

Figure 9.49 *F–N* curve.

9.4.2 Risk Indices

The risk index is probably one of the simplest and most intuitive risk assessment methods. The risk index 'R' is defined by the product between the frequency of occurrence 'f' and the magnitude of the 'N' consequences:

$$R = f * N$$

The risk index can then be defined from both quantitative assessments of the 'f' and 'N' parameters, as well as qualitative assessments.

For example, the frequency of an event could be defined using the following qualitative scale:

- $f = 1 \rightarrow$ very unlikely event;
- $f = 2 \rightarrow$ unlikely event;
- $f = 3 \rightarrow$ very likely event;
- $f = 4 \rightarrow$ extremely likely event.

Similarly, a qualitative scale such as the following could also be arbitrarily defined for magnitude:

- $N = 1 \rightarrow$ no impact;
- $N = 2 \rightarrow$ minor consequences;
- $N = 3 \rightarrow$ major consequences.

In this case, the R risk index is a value between 1 and 12. The higher the value, the higher the risk. The lower the value, the lower the risk is contained.

Of course, the risk index, regardless of its definition in a qualitative, semi-quantitative (ranges by order of magnitude) or quantity, then must be compared with the acceptable risk index and tolerable risk index, values that are defined from the beginning of the assessment risk.

9.4.3   Risk Matrices

Risk matrices are also a tool for risk assessment. For their treatment, please refer in full to Section 1.3 of this book. A risk matrix can be interpreted as the two-dimensional graphical representation of the risk index. Thresholds for acceptability and tolerability of risk are now translated into regions of the two-dimensional space, making the comparison between the level of risk reached and the aforementioned thresholds more immediate. An example of risk matrix is shown in Figure 9.50.

To create a proper risk matrix, some rules must be respected:

- It should include three different colours;
- No red cell can share an edge with a green cell;
- No red cell can occur in the left column or in the bottom row of the matrix;
- To pass from a red to a green cell, it is necessary to go through an intermediate one.

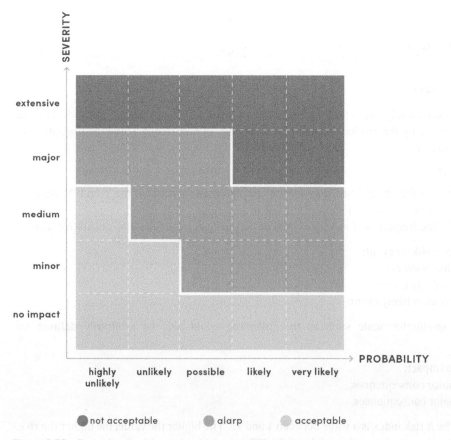

Figure 9.50   Example of a risk matrix with acceptability regions.

Table 9.15 provides guidance on practical frequency levels that can be used to develop a risk matrix in both qualitative and quantitative terms. Note that the rankings provided are examples, and these ranges may be revised for specific risk assessments.

Similar to frequencies, consequences can be categorised into different levels. Typically, these levels range from negligible to catastrophic. Given that consequences can be expressed in different terms (e.g. different units) depending on the application, it is practical to represent them with normalised values expanding various orders of magnitude. This has the following practical implications:

- Consequences levels are rarely linear. They often range from negligible, consisting of minimal impact, to catastrophic consisting of very large implications. A financial/monetary value can easily replace the normalised values, the number of injuries or deaths, etc.
- Normalised consequences allow for the interpretation of risk in frequency terms. This may ease the risk evaluation and communication process. For example, setting the catastrophic consequence to 1.0 allows for decision-making based on the likelihood of event occurrence as defined by the frequency levels, which have already been defined in terms of the expectancy of experiencing an event in an individual's lifetime.

Ultimately, each consequence level is mapped to a number between a value close to zero and one representing the damage caused by the fire scenario. An assessment should inform these numbers of the consequences relative to the potential number of injuries, fatalities, property damage and business interruption losses that may occur. As an example, the consequences in Table 9.16 are expressed as a normalised value. If the risk is represented as death or injury rate due to fire per year, the consequence term may be the number of death or injuries. Note that the rankings provided are examples, and these ranges may be revised for specific risk assessments.

Utilising the information in the previous two tables, two conceptual examples of a risk matrix are provided in Tables 9.17 and 9.18. In these examples, the header rows represent different consequence levels. The first two columns represent the frequency categories. The resulting risk level is then obtained from different combinations of frequency and consequences in the matrix. The matrix can be qualitative or quantitative. It is noted that quantitatively different frequency and consequence categories are populated with numerical values. Note that the rankings provided in

Table 9.15 Typical frequency levels in a risk matrix.

Ranking	Description	Frequency
Frequent	Likely to frequently occur during the lifetime of an individual item or very often in the operation of a large number of similar items	Greater than 1.0/yr
Probable	Will occur several times during system life or often in the operation of a large number of similar items	0.1/yr to 1.0/yr
Occasional	Likely to occur sometime in the lifetime of an item or will occur several times in the operation of a large number of similar items	0.01/yr to 0.1/yr
Remote	Unlikely, but possible to occur in the lifetime of an individual item. It can be reasonably expected to occur in the operation of a large number of similar items	1.0^{-4}/yr to 1.0^{-2}/yr
Improbable	Very unlikely to occur. It may be possible but unlikely to occur in the operation of a large number of similar items	1.0^{-6}/yr to 1.0^{-4}/yr
Incredible	Events that are not expected to occur	Less than 1.0^{-6}/yr

Table 9.16 Typical consequence levels in a risk matrix.

Ranking	Description	Normalised consequences
Negligible	The impact of loss is so minor that it would not have a discernible effect on the occupants, facility, operations or the environment. Negligible consequences may include the following examples: • No recordable/reportable event (i.e. event does not result in any work-related injury or illness requiring medical treatment beyond first aid). • Property losses consistent with failures address with routine budgeted maintenance activities.	Less than 1.0^{-5}
Marginal	The loss has a limited impact on the facility, which may have to suspend some ancillary operations briefly. Some monetary investments may be necessary to restore the facility to full operations. Minor personal injury may be involved. The fire could cause localised reversible environmental damage. Marginal consequences may include the following examples: • May be a recordable/reportable event (i.e. event included mitigation consistent with the state of industry practice such as sprinkler activation and no injuries beyond those requiring first aid). • Property losses consistent with those associated with damage limited by the effective operation of fire protection mitigation strategies.	1.0^{-5} to 1.0^{-3}
Major	The loss has a significant impact on the facility, which may have to suspend main operations for a limited time. Significant monetary investments may be necessary to restore to full operations. Multiple minor personal injuries and/or a single severe injury are involved. The fire could cause significant localised but reversible environmental damage.	1.0^{-3} to 1.0^{-1}
Critical	The loss has a critical impact on the facility, which may have to suspend operations for a prolonged period. Major monetary investments may be necessary to restore to full operations. Multiple severe personal injuries and/or a single fatality are involved. The fire could cause extensive but reversible environmental damage.	1.0^{-1} to 1.0
Catastrophic	The loss has a high impact on the facility, which may have to suspend operations permanently. Monetary investments reaching total facility cost may be necessary to restore to full operations. Multiple deaths may be involved. The fire could cause irreversible environmental damage.	1.0

Table 9.17 for reference purposes are examples, and these ranges may be revised for specific risk assessments.

Depending on the specific application, the matrix could be interpreted as follows:

• 'Negligible' or 'marginal' consequences can be accepted (or tolerated) as the 'cost of doing business' as it relates to routine equipment failures and personnel accidents not requiring medical treatment beyond first aid. However, repeated 'negligible' or 'marginal' consequences may indicate insufficient or deteriorating fire strategy and management and should be addressed.

• 'Improbable' and 'incredible' frequencies can also be accepted (or tolerated) as those events are highly unlikely to occur. The resulting risk values in this range consider the effects of fire protection features. The analyst should determine if there is a margin with and without these features. Fire protection features lowering risk values will need to be monitored routinely to ensure their effectiveness over the facility's operational life to minimise periods without protection that may be associated with higher risk levels.

Table 9.17 Qualitative risk matrix.

Frequency	Consequence				
	Negligible	**Marginal**	**Major**	**Critical**	**Catastrophic**
Frequent	Acceptable	Further evaluation	Not acceptable	Not acceptable	Not acceptable
Probable	Acceptable	Further evaluation	Not acceptable	Not acceptable	Not acceptable
Occasional	Acceptable	Acceptable	Further evaluation	Not acceptable	Not acceptable
Remote	Acceptable	Acceptable	Acceptable	Further evaluation	Further evaluation
Improbable	Acceptable	Acceptable	Acceptable	Acceptable	Further evaluation
Incredible	Acceptable	Acceptable	Acceptable	Acceptable	Acceptable

Table 9.18 Quantitative risk matrix.

		Consequence				
		Negligible	**Marginal**	**Major**	**Critical**	**Catastrophic**
Frequency		1.0E−06	1.0E−04	1.0E−02	1.0E−01	1.0E+00
Frequent	1.0E+00	1.0E−06	1.0E−04	1.0E−02	1.0E−01	1.0E+00
Probable	1.0E−01	1.0E−07	1.0E−05	1.0E−03	1.0E−02	1.0E−01
Occasional	1.0E−02	1.0E−08	1.0E−06	1.0E−04	1.0E−03	1.0E−02
Remote	1.0E−04	1.0E−10	1.0E−08	1.0E−06	1.0E−05	1.0E−04
Improbable	1.0E−06	1.0E−12	1.0E−10	1.0E−08	1.0E−07	1.0E−06
Incredible	1.0E−08	1.0E−14	1.0E−12	1.0E−10	1.0E−09	1.0E−08

- Risk values higher than 1.0E−3 may not be accepted or tolerated. Under this interpretation, which is based on normalised consequences, these values suggest frequencies of occasional events (or higher), which should address by improvements in the design or the fire protection strategy.
- Finally, fire scenarios associated with risk values in the range of 1.0E−5 to 1.0E−3 may require further evaluation. The scenarios that credit the fire protection systems to reduce the consequences may be in this regime. Therefore, the evaluation should ensure that the following conditions are fulfilled:
  - Risk insights are appropriately obtained. This specifically refers to the factors driving the risk numbers. These factors may point to the fire protection capabilities that may need to be improved.
  - There is a margin in the risk results. Sometimes, conservatism affects the input parameters' risk values or the models used to represent the scenarios. Identifying such conservatisms suggests a margin in the analysis that can be used to justify a final decision.

– An additional level of safety is provided if necessary. In situations where risk insights suggest low margins, additional fire protection features may be recommended.
– Appropriate fire protection strategies are monitored to maintain acceptable risk levels.
– If available, defence-in-depth measures address these fire scenarios (i.e. fire protection strategies beyond those explicitly included in the analysis).

For example, a facility has a plastics extruder with injection and blow moulding presses and a warehouse. The loss (i.e. consequence) associated with either the manufacturing or the warehouse may be classified as major for the facility operator. However, assuming a lack of clear space and fire-rated construction between manufacturing and the warehouse, the consequence of loss may be critical or catastrophic. Therefore, depending on the frequency level, reliance on fire protection strategies (e.g. fire-rated constructions, automatic sprinklers, etc.) is necessary for maintaining tolerable risk levels.

9.4.4 Index Methods

Fire and explosion risk evaluation in the industrial context is a fundamental tool for work owners and safety manager to individuate critical scenario and issues related to fire and explosion in industrial facilities and sites.

The primary objective of the risk evaluation is the definition of possible accident scenarios, their likelihood and consequences concerning damages to people and facilities, as to define an adequate fire strategy and preventive-protection measures.

The assessment of risk level in the industrial sector is an essential procedure that allows the employer to identify necessary interventions to be implemented to guarantee an adequate level of safety of the workplace.

Index methods for fire and explosion risk assessment are widely adopted in industrial plants and civil structures and may be defined as semi-quantitative since their outcome is a risk index (or an array of indices) that allows the definition of a risk class.

Among the most adopted worldwide methods, we can find the following:

• Fire and Explosion Index' (F&EI), by Dow Chemical Company.
• The Mond index, by Imperial Chemical Industries.
• The Safety Weighted Hazard Index (SWeHI) by Khan, Husain, and Abbasi (2001).

Indexing methods and logic procedure imply the definition of credits and penalties: while the first decrease the risk level, and the second increase it. Penalties are related to substances present in the site, to its structural characteristics and processes performed. Credits are obtained by implementing mitigation and protection measures and proper control of the processes.

A logic process leading to penalty evaluation is consisted of the following steps:

1) Identification of the units present in the site.
2) Individuation of the key substance (the most hazardous related to fire/explosion risk).
3) Evaluation of peculiar characteristics of the substances.
4) Evaluation of generic hazards related to the unit.
5) Evaluation of special hazards related to the unit.
6) Assessment of the structural characteristics of the site.

The analyst must consider 'logic units' according to the type of activities/processes performed, which of course is affecting the level of risk (storage, processing, loading, unloading, etc.), in parallel, materials or chemicals stored/processed in the unit have to be considered.

According to the SWeHI method, the units must be classified according to five types:

- Storage;
- Physical operations units (material handling, phase transitions, pumping, compressing, etc.);
- Chemical reactions;
- Transfer units;
- Other hazardous units like heating systems, heat exchangers with an open flame.

Classification of substances present in the unit could be performed according to EU Regulation CLP (1272/2008), which defines hazard statements (H followed by three numbers): H2xy stands for physical hazards, H3xy for health hazards and H4xy for environmental hazards. Another substance classification is proposed by NFPA 704, which identifies substances concerning health, flammability and instability issues.

NFPA 704 makes use of the so-called 'fire diamond', scoring hazards from 1 to 4 (major risk) according to the characteristics and properties of substances. The SWeHI and F&EI methods make use of NFPA 704 criteria to classify chemicals. Some peculiar characteristics of substances must be considered as they may introduce risk-enhancing factors: oxidising properties, a substance which reacts exothermically with water or highly sensitive to ignition, finely divided combustible materials prone to dust explosion, etc.

With regard to process conditions, considerable attention must be paid to temperature, pressure, hold-up, type of chemical reactions involved plus all 'special' elements that could characterise specific units: the possibility of unstable reaction, the presence of equipment in motion, the potential generation of flammable atmosphere and the closeness to other units.

Structural characterisation of the site deals with the size of the unit, its confinement and the density of equipment.

Other factors are represented by the density of workers within the surrounding area, the duration of their duties, the exposure to substances and external factors such as likelihood occurrence of environmental calamities in the geographic area of reference. Credits are attributed to all prevention/protection system in place that may reduce and mitigate fire and explosion risk, as shown in Table 9.19.

Table 9.19 Definition of credits and possible measures associated.

Credit	Measures
Planning of emergency procedures	Communication with surroundings, information to workers and medical structures
Planning of catastrophes management	Communication with authorities and healthcare structure management
Damage mitigation measures	Fire-safety devices (e.g. sprinklers, foam curtains, hydrants, ...)
Installation of control systems and detection devices	PLC operational parameter control and flammable and chemical detection devices
Emergency systems	Shutdown and safety relief devices
Worker formation and operational procedures	Responsibility and tasks of workers, safety awareness and the degree of automation of operations
Reliability of the plant	Maintenance and inspection of systems

These methods, extensively used worldwide for preliminary and initial risk screening (often before a full FERA or a QRA), may also be used for the risk assessment associated existing plants and even after a real accident to understand the risk level of the facility and fire risk reduction factor of additional measures that should have been implemented to prevent and/or to mitigate the accident.

9.4.4.1 An Example from a "Seveso" Plant

Fire and explosion risk screening is a specific requirement for a number of chemical plants facing major accidents event threat, such as those under the European Seveso III major accidents prevention directive.

The 'so-called' Seveso III directive (Directive 2012/18/EU) is imposed on plant managers to perform a detailed risk assessment and to adopt adequate protection measures in case their facility is included among those considered subjected to major accident, i.e. if the amount of hazardous substances stocked and handled within it is superior to defined threshold limits. Fire risk evaluation needs to consider each plant's complexity and different regulations and codes it is subjected to. Meanwhile, a thorough approach is required, which does not base itself uniquely on qualitative methods (such as checklists) or semi-quantitative (such as fire-load-based approach) but should consider these latter as starting processes to develop a more comprehensive evaluation. Besides this, accident scenarios associated with chemical plants may differ significantly, according to the substances handled and the activities and processes implemented: typically, they could range from small to medium scale in terms of consequences depending on the impact on human operators and structures. Several 'risk screening' methods exist, differing from their fields of applications and limitations.

The principal methodologies to be compliant with the Seveso III directive are 30 years old (the first dated back to 1980) and updated in the early 1990s. Their development derives from the need for insurance companies to evaluate adequate insurance fees quickly. The most used are the Dow F&EI (Dow 1987) and the Mond index methods (ICI 1985), both thought to be explicitly applied to the oil and gas industry. In the years, many researchers proposed methods to assess chemical plants' inherent safety through various ways of indexing the hazard potential and the related risk. Among these, the contribution by Khan and co-workers, who developed integrated inherently safety indexing methods together with Amyotte as summarised in Amyotte (2020) and Amyotte and Khan (2021), and by Heikkila, who introduced the use of inherent safety indices, is relevant. Generally speaking, these methods are not suited for all those facilities where a relatively low amount of substances is present; the threshold limits are defined, as for the Dow F&EI method (Dow Chemical Company 1987), as 5000 lb (or 600 gal) of flammable substances, while Seveso III directive adopts threshold values accordingly to the types of substances. All plants below the threshold are not considered 'subjected to major accidents', though fire and explosion risk could be high if work owners and regulations impose any prevention/protection strategies; besides this, the cited methods lack in some aspects of the risk classification strategy.

To this purpose, indexing methods could be employed: they are devoted to risk evaluation screening procedures as their application is quick and cost-saving, and potential critical issues could be quickly underlined.

Among the most recent, the hazard identification and risk assessment (HIRA) method, RRHI, and the SWeHI method, the latter was adopted in this work and applied to a fine chemical industry in Northern Italy as a test case.

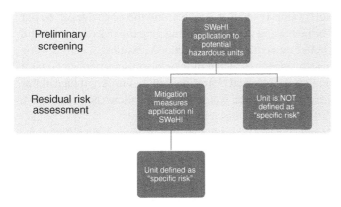

Figure 9.51 Approach to screen 'specific' risk process units.

The approach followed here is intended to incorporate an international validated method and the Italian regulations currently in law in the fire-safety field.

The process department of a chemical plant that is subjected to Seveso III could be evaluated according to this EU Directive regulations, while the SWeHI method could help to evaluate whether a processing unit does represent a 'specific risk' or not. In this approach, SWeHI is used as a preliminary screening tool: once identified the units at risk, these must be further evaluated with a more detailed approach (such as QRA) to assess the likelihood of accident scenarios and their magnitude consequences. Figure 9.51 reports the approach adopted in this work to assess the most hazardous units in the case study. A threshold value of the SWeHI index has been adopted: units with an index smaller than 5, which associate with a light to low risk in the method logic, are considered 'reasonably acceptable,' and no additional measures are requested. On the other hand, units with greater than 5 values are considered at 'specific risk,' and mitigation measures, in the form of additional credits according to the SWeHI methodology, are introduced to lower the risk to an acceptable level.

9.4.5 SWeHI Method

The SWeHI method has been developed by Khan, Husain, and Abbasi (2001) as a quick and user-friendly tool for identifying hazards and assessing fire and explosion risk in the process industry. It represents an evolution of a previous method developed by the same authors. The authors intended to create a more systematic tool that provides the user a more comprehensible and reproducible evaluation procedure than previous methodologies.

The SWeHI method allows defining a risk level index, called *SWEH index*, which represents, in quantitative terms, the radius of the area in which there are conditions of moderate danger or where there is a probability of fatality/damage equal to 50%. The index is assessed by considering all existing control and protection measures; the higher the index value, the more vulnerable the unit analysed is.

The index is evaluated as

$$\text{SWEHI} = B\big/A$$

where B represents the quantitative measure of the damage caused by the process unit on an area that considers 50% of the probability of damage (m^2) and A represents the sum of the credits attributed to the installed protection systems.

Each unit must tend towards an SWeHI value as small as possible, and this objective can be pursued either by reducing the value of *B* or increasing value *A*, i.e. mitigating the risks due to hazardous substances or processes adopting better prevention and protection measures.

Plant units are classified into five macro-categories, depending on operations performed (storage, chemical reactions, handling of materials, physical operations, etc.). A different logical flow for index evaluation is associated with different units: weighted factors allow the procedure to be as specific as possible to consider all unit peculiarities and critical points.

In this work, only term *B*1 (fire and explosion risk) is evaluated, which depends on different factors (process conditions and type of substances mainly), grouped within the definition of 'hazard potential'.

The *A* value includes factors defined as 'credits', depending on the company's safety management, the effective presence of control systems and failure prevention, protection devices, the characteristic of operators and operation, the reliability of equipment, etc.

9.4.6 Application

As a case study, the plant produces chemicals for polymers, cosmetics, and other manufacturing sectors, with different dedicated production lines.

Process equipment studied are essentially reactors and auxiliary equipment. The chemical processes include esterification, ethoxylation and mixing.

The units analysed are located in three departments, devoted to different production lines, identified as departments A, B, and C.

Main units represented in these areas are reactors, storage tanks, intermediary tanks, distillation columns, heat exchangers and other ancillary equipment serving different production lines; most of the processes are batch, reactants are charged with dedicated lines from supply tanks or trucks, after inserting with a nitrogen purge.

Table 9.20 reports the main hazardous substances present in the departments; in particular, ethylene oxide is stored in several tanks and used in ethoxylation reactors; methanol is mostly a by-product of process reactions; *n*-heptane and toluene are used as the solvent medium in different reactors; and hydrogen is fed to a hydrogenation reactor from a dedicated pipeline coming from outside the plant.

The most hazardous units that are identified in the safety report of the plant society are as follows:

- *n*-Heptane recovery units (C2, C5, C7 and C8).
- *n*-Heptane intermediate and process tanks (B15 and B36).
- *n*-Heptane charging line into reactors (B14, C1, C, C4 and C6).

The SWeHI index incorporates several factors, mostly related to the key substances identified in the unit. 'Energetic factors' are defined depending on the substance properties that could enhance or reduce fire hazards, such as flash point, vapour pressure and heat of combustion. These values are compared with the actual process conditions to identify whether a potentially flammable atmosphere could arise. The heat of combustion has a relevant impact on the SWeHI index. Figure 9.52 shows the SWeHI index calculated at the same units when different chemicals are handled (*n*-heptane and toluene).

The higher value is obtained at unit C5 when handling *n*-heptane due to its high combustion heat and greater inventory (almost double than the others, as C5 is a storage tank).

Table 9.20 Overview of hazardous substances handled in the plant.

Department	Key substance	Classification	The total amount in the plant (tonnes)	No. of units
A	Ethylene oxide	P2	25	10
B	Methanol	P5a	55	12
	n-Heptane	P5a	10	26
	Hydrogen	P2	0.01	2
C	Toluene	P5a	8.5	18
	n-Heptane	P5a	40	21
	Methanol	P5a	55	6

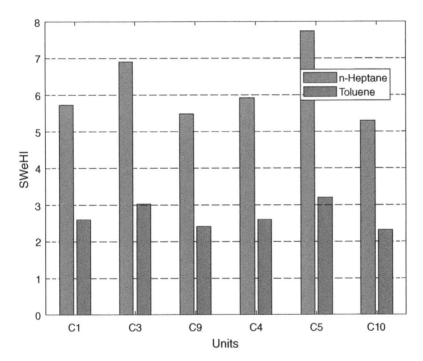

Figure 9.52 Effect of type of substance (heat of combustion) on the SWeHI index in different volume units.

An experience-based hypothesis sets the acceptable risk index threshold to 5. A detailed analysis must be undertaken for any units with a higher value to determine why the equipment safety is compromised and which protection/prevention measures are required. As they are identified and applied, the SWeHI index is re-calculated.

Figure 9.53 shows how the adopting of protection and mitigation measures influence the SWeHI index. In this case, additional improved flow and level control systems are implemented. In this way, the index C4 of the method increases. A high level of risk is still computed in the case of storage tank loading operations, while the risk of other process phases reduces below the acceptable threshold.

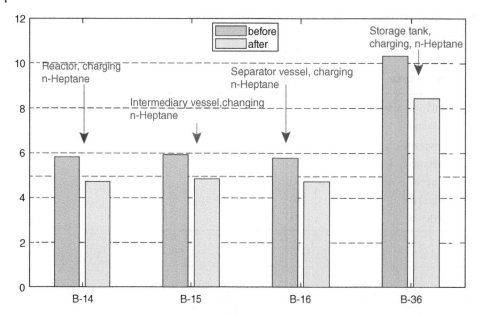

Figure 9.53   Hazardous units in department B, and the SWeHI index before and after control system improvement.

The unique solution to mitigate the latter is to reduce the hold-up of the tank or by implementing advanced improved system controls and fire-safety measures such as an improved control system (on other process variables), a higher degree of automatisation of operations (degree of human–machine interaction is lowered, e.g. with a dedicated feeding line from storage tanks to reactors), with additional fire protection devices (flame arrestors, water blankets and inert gas sprinklers).

Reducing the mass amount of *n*-heptane in the tank, without any other additional measures, would lower the SWeHI below 5, only if the tank is emptied for more than one-fourth of its hold-up, which is unpracticable for production yield reasons. According to the SWeHI results, further measures have to be taken into account to meet the acceptability criteria according to the SWeHI method (Figure 9.54).

Figure 9.55 reports the SWeHI index representation in department C of the plant, before and after, respectively, implementing the additional credit method due to improved control systems on the units.

As the SWeHI value is pictured as a damage radius, the additional measures' effect could be better appreciated. The radius is decreased in size with respect to 'no measure' configuration, and potential hazards from multiple item involved scenarios (domino effect) are almost prevented, since total equipment included in the damaged area is reduced, while also no domino effects are expected on other 'greater than 5 SWeHI' units in the after configuration. Units C1, C2 and C3 are downgraded to a low risk level (smaller than 5) following the inclusion of safety measures.

The SWeHI method, applied thoroughly on the totality of the units present in the chemical plant test case, helped in the preliminary identification and estimation of risks associated with the plant units.

The method also allowed to integrate additional measures and verify their effectiveness: in this case, implementing an improved control system can lower the risk level. The domino effect has been assessed with the adoption of the SWeHI graphic representation, and the damage radius could be compared to identify the involved units according to the scenario.

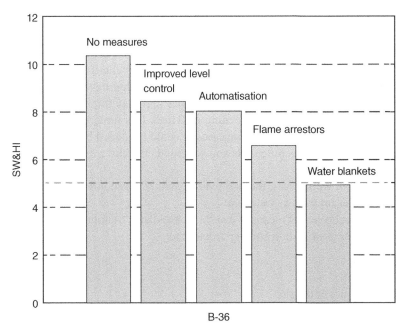

Figure 9.54 Effect of additional fire-safety measures on the *n*-heptane storage tank unit.

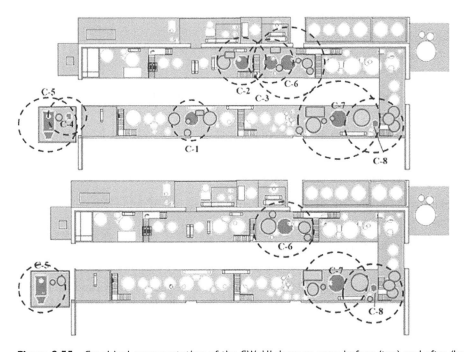

Figure 9.55 Graphical representation of the SWeHI damage area, before (top) and after (bottom) implementation of control measures, in department C (units containing flammables with greater than 5 SWeHI are depicted in red, and others in yellow).

9.5   Simplified Fire Risk Assessment Using a Weighted Checklist

In some cases, it is important to estimate the fire risk level associated with a number of fire compartments belonging to an activity. For existing buildings and industrial installations, it is important to estimate the actual fire risk level given the specific conditions, including the availability of the control measures in place and all the management factors that may impact on the fire safety. During audit, due to diligence and other screening activities, it is important to have a clear and fast overview of the actual situation. In some cases, it could be gained by the application of expert judgement supported by screening tools and weighted checklists, eventually supported by risk matrices.

In order to assess the fire risk of workplaces, the methodology characterised by the following steps, applied to each workplace examined, was adopted:

Analysis of the technical data collected by means of a special checklist is useful to comprehensively characterise the place under examination and in particular to

identify the site and workplace;
describe the layout of the site;
indicate the technical characteristics of the site;
assess the number of people present in the place under examination under conditions of maximum crowding;
identify the technical measures;
identify organisational and management measures;
identify heat sources;
identify escape routes;
enumerate extinguishing means and systems;
indicate fire detection and alarm systems;
assess the maintenance of safety measures.

During the survey, it is fundamental to gather all the evidences related to fire safety, including

occupancy type;
any eventual hazardous chemicals (storage);
fire load;
process and activities (including any supporting equipment and machinery);
tools;
technological plants and systems (e.g. heating, ventilation and air conditioning [HVAC] systems);
specific risk areas.

Specific risk areas may include the following:

- ATEX zones.
- Areas, belonging to the main activity, where the fire risk may be judged significantly different from the fire risk estimated for the rest of the activity. Some criteria are available to identify specific risk areas:
  - Areas in which hazardous substances or mixtures, combustible materials, are held or treated in significant quantities (these areas do not include the storage of limited quantities of flammable liquids in metal cabinets for functional uses in the main activity, which is generally not considered specific risk).
  - Areas where hazardous work is carried out for the purposes of fire.

- Areas where there is the presence of installations or their components relevant for the purposes of fire safety (e.g. specific fire detection and protection systems).
- Areas with a specific fire load of >1200 MJ/m^2, unoccupied or with the occasional and short-term presence of assigned personnel.
- Areas in which there is the presence of systems and equipment with process fluids under pressure or at high temperature.
- Areas where there is the presence of surfaces exposed to high temperatures or open flames.
- Areas where there are hazardous chemical reactions to fire.
- Areas of activity with significant risk posed to the environment in case of fire.

These areas should be considered by using the information that is also coming from

material safety data sheets (MSDS) for hazardous chemicals, including mixtures; specifications and producers/manufacturer manuals of machinery, tools and equipment.

Estimation of the fire risk index 'R_{FIN}' (high, medium and low) given the judgement about the status of the preventive and protection barriers in place.

R_1 represents the 'potential danger level of the workplace' that can take three values: $M1$, $M2$ or $M3$ (numerical values between 1 and 3);

S is a number from 1 to 90 expressing the degree of identified criticalities (this value expresses the compliance degree with existing regulation requirements and the actual degree of the fire-safety management in place).

Classification of the final risk value given the potential risk modified with the adequacy of the preventive and protective measures in place, as from the results of the surveys conducted.

9.5.1 Risk Levels

Potential risk (R_1) can be classified according to three levels as shown in Table 9.21:

Potential risk level (R_1) is estimated by expert judgement considering a number of attributes that include the presence of flammable chemicals, the number of people exposed, fire load, the probability of explosive atmospheres (ATEX), layout complexity, etc.

Calculation method for the evaluation of the final risk level:

Fire risk level $\mathbf{R_{FIN}} = \mathbf{R_1} \times \mathbf{S}$

where

R_1 is the potential workplace hazard (takes into account the tabular risk);

S is the compliance degree (takes into account the judgement about the adequacy of the controls in place to maintain the identified risk; this compliance should consider any requirements coming from the applicable regulations).

In the below example (Table 9.22), three different compartments have been classified in order to define the S value.

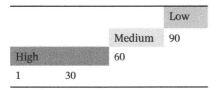

		Low	
	Medium	90	
High		60	
1	30		

Given the potential level of risk ($R1$), this could be modified considering the estimated compliance degree (S) according to the following table:

Compliance degree (S) →	High	Medium	Low
Potential risk level (R_1) ↓			(·) High/ Not acceptable
Low	Low (confirmed)	Medium	
Medium	Medium (··) High (confirmed)	High	Not acceptable
		Not acceptable	Not acceptable
High			

Table 9.21 R_1 level of potential workplace hazard.

'R_1' Level of potential workplace hazard

Level by danger	Workplace at risk of fire
Low	• Workplaces where low-flammability substances are present and local conditions present no risk of fire ignition and spread. • Workplaces where work processes involve the use of low-flammability substances and the local and operating conditions offer little likelihood of fire ignition and spread. • Workplaces where the crowdedness of rooms cannot jeopardise rapid evacuation in the event of a fire.
Medium	• Workplaces where flammable, combustible substances and/or local conditions are present that may allow fires to develop, but whose likelihood of propagation is limited. • Workplaces where work processes involve the use of flammable substances and/or local and/or operational conditions that may favour the development of fires, but where the likelihood of their spread is limited. • Workplaces where room crowding and the state of the premises may lead to reduced evacuation times in the event of a fire.
High	• Workplaces where work processes involve the use of highly flammable substances (e.g. painting equipment), or open flames, or the production of considerable heat in the presence of combustible materials. • Workplace where there is storage or handling of chemicals that may, under certain circumstances, produce exothermic reactions, give off flammable gases or vapours, or react with other combustible substances. • Workplaces where explosive or highly flammable substances are stored or handled. • Place of work where there is a considerable amount of combustible materials. • Workplaces where, irrespective of the presence of flammable substances and the ease with which flames can spread, the crowdedness of the rooms, the state of the premises or the mobility limitations of the people present make evacuation difficult in the event of a fire.

Table 9.22 Probability of accident related to the adapted safety measures in place of work.

Probability of accident related to the adapted safety measures in place of work

N°	Entry	Fire compartment 1			Fire compartment 2			Fire compartment 3		
		Suitable	Partially suitable	Not suitable	Suitable	Partially suitable	Not suitable	Suitable	Partially suitable	Not suitable
Technical measures										
1	Electrical installations	1					3	1		
2	Grounding	1					3	1		
3	Atmospheric discharges	1				2		–		
4	Ventilation	1			1			1		
	Organisational and management measures									
1	Order and cleanliness	1					3	1		
2	No-smoking rules	1				2		1		
3	Regulation on security measures to be observed	1			1			1		
4	Worker information and training			3			3			3
5	Fire training of workers	1			1			1		
6	Emergency plan			3			3			3
7	Emergency team	1			1			1		
	Heat sources									
1	Safety distance	–			–			–		
2	Insulation	–			–			–		
3	Shielding	–			–			–		

(Continued)

Table 9.22 (Continued)

Probability of accident related to the adapted safety measures in place of work

N°	Entry	Fire compartment 1			Fire compartment 2			Fire compartment 3		
		Suitable	Partially suitable	Not suitable	Suitable	Partially suitable	Not suitable	Suitable	Partially suitable	Not suitable
Exodus routes										
1	Length of exit routes	1			1			1		
2	Width of exit routes	1			1			1		
3	Signage		2			2		1		
4	Lighting			3			3		2	
5	Facilities for the disabled	–			–			–		
6	Crowding	1					3	1		
Extinguishing media and equipment										
1	Portable and/or wheeled fire extinguishers			3	1			1		
2	Mobile means	–			–			–		
3	Manual/automatic fixed installations	–			–			–		
4	Attacks VV.F.	1			1			1		
Fire detection and alarm										
1	Detection system	–			–					3
2	Alarm system	–			–			–		
3	Garrison staff	1			1			1		
Maintaining security measures										
1	Maintenance programme	1			1			1		
2	Functional control programme			3			3			3
3	Simulated emergency tests			3			3			3
	Partial total:	15	2	18	10	6	27	18	2	15
	Total General		35			43			35	

	Immediate intervention	(·) To be defined considering the specific situation
	Medium priority	
	Risk management, continuous improvement and actions to solve criticalities	(··) To be evaluated for possible reduction and to be put under strict control

This Table shows that in the case of low compliance the expected fire risk level should be increased up to one step/two steps depending on the reduction in compliance degree of the specific compartment.

Reduction degree may lead to non-acceptable situations (immediate intervention) or to increased level of fire risk (medium priority). Confirmed levels should lead to continuous fire risk management over time. These general indications should be coupled with the actions, in each single compartment, aimed to solve mayor identified criticalities.

Classifications of risk can be modified if a local applicable regulation defines, as per its requirements, a specific minimum level for defined activities/occupancies.

Following Tables present an example of calculation with the simplified method for a work place made up of three different areas whose compliance levels have been identified.

Areas should be identified in terms of fire compartments (physical portions of the workplace characterised by a specific fire resistance that can be identified in atomic-independent portions from the fire-safety point of view).

Area no.	Compartment	Risk level 'R1'	Level of criticalities	Conformity (<30)	Critical items Work placespecific	General	MODIFIED RISK (R_{FIN})
1	Compartment 1	1 (Low)	35	No (Medium)	2	4	Medium
2	Compartment 2	2 (Medium)	43	No (Medium)	5	4	High
3	Compartment 3	1 (Low)	35	No (Medium)	1	4	Medium

All the compartments do not comply with a sufficient conformity level, i.e. the fire risk level is higher than expected and not controlled by the organisation.

Organisation has some general criticalities extended to the entire site connected with

training activity;
emergency planning;
inspections;
emergency drills.

Each compartment also has specific deficiencies:

Compartment 1:
- High priority (Level 3):
  – lighting;
  – portable fire extinguishers.
- Medium priority (Level 2):
  – signage.

Compartment 2:
- High priority (Level 3):
  - electrical installations;
  - grounding;
  - housekeeping;
  - lighting;
  - crowding.
- Medium priority (Level 2):
  - atmospheric discharges;
  - no-smoking rules;
  - signage.

Compartment 3:
- High priority (Level 3):
  - fire detection system;
- Medium priority (Level 2):
  - facilities for disabled people.

As further considerations, it can be said that expert judgement can be coupled with matrix-based approaches and semi-quantitative estimations to define safety levels, priorities, compliance degree, etc. This kind of approach is very useful and common; in some cases, it can be applied using multi-attribute judgement in order to highlight the aspects having the greatest impact on the fire risk level.

One of the most used methods is the workflow proposed in Research Report 40 'Fire risk assessment for workplaces containing flammable substances' prepared in 2002 by W.S. Atkins Consultants Limited for the use Health and Safety Executive.

It could be a method to develop some insights for specific risk areas as these workplaces contain flammable substances.

The proposed risk-based workflow (Figure 9.56) considers the following elements of judgement:

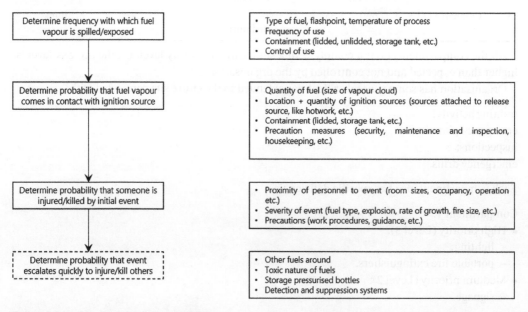

Figure 9.56 Proposed risk-based workflow.

Table 9.23 Fire risk governing attributes.

Attribute measured		Frequency	Ignition	Injury
A	Material classification (flash point)		▓	
B	Release quantity (container size and operation)		▓	
C	Frequency of use (constant, regular or occasional)	▓		
D	Containment (lidded, unlidded or fixed tank)	▓		
E	Spill control (bunding and ventilation)		▓	
F	Housekeeping (spill clean-up and removal waste)	▓		
G	Safety management (procedures and enforcement)	▓	▓	
H	Stagg awareness (training, familiarity and contractors)	▓		
I	Maintenance and inspection (containers)	▓		
J	Control fixed ignition sources (HAC and segregation)		▓	
K	Control static ignition (design and clothing control)		▓	
L	Control miscellaneous sources (lightning, friction, etc.)		▓	
M	Control deliberate ignition (arson and horseplay)		▓	
N	Proximity of staff (operation, room size and occupancy)			▓

Frequency of spillage and exposure due to a potential loss of containment.
Probability of contact among fuel vapours and effective ignition sources.
Probability of injury due to an initial incident.
Probability of injury due to a secondary event triggered by the first.

 The proposed method considers a range of different attributes (Table 9.23) that affect one or more aspects of the event (characterised in terms of loss of containment frequency, ignition probability and injury/fatality probabilities). These attributes may be linked to the fire risk management policy in place and individually estimated by the fire risk assessor using a quantitative scale (low, medium and high) or a quantitative scale (8 in the range of 0–1 for each attribute) as suggested by the original method, developed considering real statistics in the United Kingdom.

10

Risk Profiles

Fire-safety design for both new and existing civil and industrial premises is the ultimate result of a structured workflow, involving a number of professional figures and several stakeholders, aimed to the consideration of the outcomes of a fire (and explosion) risk assessment procedure, based on the use of methods and tools that have been selected among those that are pertinent with the complexity and detail of the premise characteristics and use cases.

Fire risk assessment should consider, especially in a performance-based approach, the relevant scenarios that can be very numerous, even if grouped in pools given their relevant similarities. This amplitude may result in several outcomes to be discussed and verified during the fire-safety strategy design at the light of the fire-safety principles.

In order to simplify the outcomes, it could be very convenient to define specific qualitative or quantitative values to describe the resulting fire risk profile impacting on the identified vulnerabilities (occupants, environment, assets, business continuity, reputation, etc.) according to the aim of the assessment and any eventual documentation requirements of the applicable regulations.

If these values have been used since many years in the industrial field where the use of quantitative risk assessment (QRA) and fire and explosion risk assessment (FERA) studies is common, for civil buildings the use of risk profiles as a summary of the fire risk identified is rather new.

Following sections describe some profiles commonly used for civil buildings to identify the classes of requirements to be adopted during the definition of the performances of the elements of the fire-safety strategy. Preventive and mitigating controls' global performance should be compatible with the risk profiles identified from the risk assessment, given the considered scenarios.

While generally the risk profiles for occupants and environment related vulnerabilities are mandatory to be satisfied by the selected fire strategy (among the various strategies deemed to satisfy), it is the responsibility of the owner and relevant stakeholders to define the follow-up of other risk profiles, such as those connected with the performances aimed to mitigating the risk for assets and for business continuity, in a resilience-based logic beyond traditional fire safety of people and natural resources.

Risk profiles should be seen as a summary of the fire risk assessment rather than a method to qualitatively and apodictically state a level of risk. Fire risk level, given fire scenarios, is the outcome of a complete fire risk assessment activity (identification, analysis, ...) and not the selection of a single profile.

In order to better assess (synthetically describe) the fire risk for an activity, the following types of *risk profiles* are set out:

Fire Risk Management: Principles and Strategies for Buildings and Industrial Assets, First Edition.
Luca Fiorentini and Fabio Dattilo.
© 2023 John Wiley & Sons, Inc. Published 2023 by John Wiley & Sons, Inc.
Companion Website: www.wiley.com/go/Fiorentini/FireRiskManagement

R_{life}: risk profile concerning *human life safety*;

R_{prop}: risk profile concerning the protection of *property*;

R_{env}: risk profile concerning the protection of the environment.

The R_{life} risk profile shall be assigned to *each compartment* and, where necessary, for each *open-air space* of the activity (for example, the assignment of the R_{life} risk profile in open-air spaces is required for the design of the evacuation for *outdoor activities*). The risk profile R_{prop} shall be assigned to the *entire activity* or to *settings (areas)* of it and the same shall be done with the risk profile R_{env}.

10.1   People

The R_{life} risk profile is assigned in relation to the following factors:

δocc: *Prevailing* characteristics of the occupants. (*'Prevailing'* refers to the characteristics of the occupants which, due to their number and type, are more representative of the activity carried out in the setting under consideration and under any operating condition. For example, an office in which there is only a small occasional and short-term presence of members of the public may be classified as δocc = A.)

$\delta\alpha$: *Prevailing* rate of the growth of fire characteristic, referring to the time $t\alpha$ in seconds, used by the thermal potential to reach 1000 kW (*'Prevailing'* refers to the characteristic representative of the fire risk under all operational conditions. For example, the presence in civil activities of limited quantities of flammable cleaning products properly stored is not considered significant and therefore not prevailing.)

Tables 10.1 and 10.2 show how the designer is guided in the selection of the δ_{occ} and δ_α factors. The designer may also select the t_α value by using one of the following options:

Table 10.1   Prevailing characteristics of the occupants.

Prevailing characteristics of the occupants δ_{occ}		Examples
A	Occupants are awake and familiar with the building	Offices with no public access, school, private garage, private sports centres, manufacturing activities in general, warehouses and industrial sheds
B	Occupants are awake and not familiar with the building	Stores, public parking garage, activity for exhibitions and public shows, convention centre, offices open to the public, bars, restaurants, medical offices, clinics and sports centres
C	Occupants may be asleep: [1]	
Ci	• in individual activity of long duration	Dwelling
Cii	• in managed activity of long duration	Dormitory, residence, student housing and residence for self-sufficient people
Ciii	• in managed activity of short duration	Hotel and alpine retreat
D	Occupants receive medical care	Hospital room, intensive care, operating rooms, residence for dependent persons and persons with healthcare
E	Occupants in transit	Train and metro stations, and airports

[1] When C is used in this document, those directions are valid for Ci, Cii, and Ciii.

Table 10.2 Prevailing characteristics of the speed of the propagation of the fire.

δ_α	t_α [1]	Criteria
1	600 s slow	Areas of activity with a specific fire load $q_f \leq 200$ MJ/m^2, or where there is a predominance of materials or other fuels that contribute negligibly to fire.
2	300 s medium	Areas of activity where there is a predominance of materials or other fuels that contribute moderately to the fire.
3	150 s fast (rapid)	Settings (areas) with significant amounts of stacked plastics, synthetic textiles, electric and electronic appliances and combustible materials not classified for reaction to fire (Chapter S.1).
		Settings (areas) where significant quantities of combustible materials are stacked vertically with 3.0 m $< h \leq 5.0$ m [2].
		HHS3-classified storage or HHP1-classified activities, in accordance with UNI EN 12845.
		Settings (areas) with technological or process systems that use significant amounts of combustible materials.
		Settings (areas) with the simultaneous presence of combustible materials and processes that may lead to fires.
4	75 s ultra-fast (ultra-rapid)	Settings (areas) where significant quantities of combustible materials are stacked vertically with $h > 5.0$ m [2].
		HHS4-classified storage or HHP2-, HHP3- or HHP4-classified activities, in accordance with UNI EN 12845.
		Settings (areas) where significant quantities of substances or mixtures that may lead to fires, or cellular plastics/expanded plastics or combustible foams not classified for reaction to fire, are present or are being processed.

Unless otherwise assessed by the designer (e.g. literature information, direct measurements, etc.), at least the quantities of materials in compartments with specific fire loads of $q_f \leq 200$ MJ/m^2 are considered *not significant* for the purposes of this classification.

[1] Prevailing characteristics of the speed of the growth of the fire.

[2] With stacking height of h.

data published by authoritative and shared sources,

direct determination of the rate of heat release (RHR) curve relative to the fuels actually present and in the configuration in which they can be found, according to the indications of eventually established by law or through measurements at the testing laboratory, according to consolidated experimental protocols (for example, the standards of the ISO 9705 series, the ISO 24473 standard and the ISO 16405 standard, are useful references for the experimental determination of the RHR curve, etc.).

The R_{life} value is determined as a combination of δ_{occ} and δ_α, as shown in Table 10.3.

Table 10.4 provides a non-exhaustive indication of the R_{life} risk profile for the most common types of *occupancy*. If the designer chooses values different from those proposed, the reasons for the choices must then be indicated in the design documentation.

10.2 Property

The assignment of the R_{prop} risk profile is performed according to the strategic nature of the entire activity or of the *settings* (areas) that constitute the activity, and of any historic, cultural, architectonic or artistic value it or its contents may have.

Table 10.3 Assignment of R_{life}.

Prevailing characteristics of the occupants δ_{occ}		Prevailing characteristics of the rate of the growth of the fire δ_{α}			
		1 Slow	2 Medium	3 Fast (rapid)	4 Ultra-fast (ultra- rapid)
A	Occupants are awake and familiar with the building	A1	A2	A3	A4
B	Occupants are awake and not familiar with the building	B1	B2	B3	Prohibited [1]
C	Occupants may be asleep: [2]	C1	C2	C3	Prohibited [1]
Ci	• in individual activity of long duration	Ci1	Ci2	Ci3	Prohibited [1]
Cii	• in managed activity of long duration	Cii1	Cii2	Cii3	Prohibited [1]
Ciii	• in managed activity of short duration	Ciii1	Ciii2	Ciii3	Prohibited [1]
D	Occupants receive medical care	D1	D2	Prohibited [1]	Prohibited
E	Occupants in transit	E1	E2	E3	Prohibited [1]

[1] To reach a permitted value, δ_{α} may be reduced by one level as specified in paragraph 3 of Section G.3.2.1.
[2] When in the text, if C1 is used, the relative indication shall be valid for Ci1, Cii1 and Ciii1. If C2 is used, the indication shall be valid for Ci2, Cii2 and Ciii2. If C3 is used, the indication shall be valid for Ci3, Cii3 and Ciii3.

Table 10.4 R_{life} risk profile for certain types of occupancy.

Occupancy types	R_{life}
School gymnasium	A1
Private garage	A2
Office not open to the public, lunchroom (canteen room), classroom, corporate conference room, file room, library and private sports centre	A2–A3
Commercial store not open to the public (e.g. wholesalers, etc.)	A2–A4
School laboratory and server room	A3
Manufacturing activity, light industry, process systems, research laboratory, warehouse and mechanics workshop	A1–A4
Stores of hazardous substances and mixtures	A4
Art galleries, waiting rooms, restaurants, medical offices and clinics	B1–B2
Public parking garage	B2
Offices open to the public, public sports centre, conference rooms open to the public, discotheques, museums, theatres, cinemas, detainment centres, library reading areas, exhibition activities and car showrooms	B2–B3
Commercial store open to the public (e.g. retailers, etc.)	B2–B4 [1]
Dwelling	Ci2–Ci3
Dormitory, residence, student housing and residence for self-sufficient people	Cii2–Cii3

Table 10.4 (Continued)

Occupancy types	R_{life}
Hotel room	Ciii2–Ciii3
Hospital room, intensive care, operating rooms, and residence for dependent persons and persons with healthcare	D2
Train and metro stations, airports	E2

[1] To reach a permitted value among those indicated in Table 10.4, $\delta\alpha$ may be reduced by one level as specified in paragraph 3 of Section G.3.2.1.

Table 10.5 Assignment of R_{prop}.

		Restricted activity or area	
		No	Yes
Strategic activity or area	No	$R_{prop} = 1$	$R_{prop} = 2$
	Yes	$R_{prop} = 3$	$R_{prop} = 4$

With regard to the application of this document:

an activity or setting (area) is considered restricted in its use because of art or history if it or its contents are considered such by law;

an activity or setting (area) is considered strategic if considered such by law or in consideration of public rescue and civil defence planning or upon indication of the activity manager. (At the request of the activity manager, in addition to the regulatory requirements, the designer may increase the value of the R_{prop} risk profile in order to ensure fire-safety objectives are achieved, such as business continuity after a fire.)

Table 10.5 shows how the designer is guided in the assignment of the R_{prop} risk profile.

10.3 Business Continuity

The organisations willing to succeed in the global market have to face the challenge of the business continuity. A fire scenario just like any other type of risk scenario could affect not only the tangible assets of the company but also its capability to be continuously present on the market. Supply chain, selling activities and customers acquisition are only three examples of targets that could suffer the negative consequences of a fire risk scenario: that is why business continuity shall be taken into account. A good reference to do it is the ISO 22301. ISO 22301 'Societal security – Business continuity management systems – Requirements' is an international standard related to business continuity management (BCM), which defines the requirements necessary to plan, establish, implement and operate a documented management system, and to monitor, maintain and continuously improve the management system to protect, reduce the possibility of occurrence, prepare, respond to and restore destabilising events for an organisation when they occur.

It specifies requirements to implement, maintain and improve a management system to protect against, reduce the likelihood of the occurrence of, prepare for, respond to and recover from disruptions when they arise.

The requirements specified in the standard are generic and intended to be applicable to all organisations, or parts thereof, regardless of type, size and nature of the organisation.

The extent of application of these requirements depends on the organisation's operating environment and complexity.

This document is applicable to all types and sizes of organisations that

- implement, maintain and improve a BCMs;
- seek to ensure conformity with stated business continuity policy;
- need to be able to continue to deliver products and services at an acceptable pre-defined capacity during a disruption;
- seek to enhance their resilience through the effective application of the BCMs.

Standard helps organisations to

- identify and manage current and future threats to business processes or to the entire organisation;
- adopt a proactive approach to minimise the impact of accidents;
- keep critical functions active during periods of crisis;
- minimise downtime during accidents and improve recovery time;
- demonstrate the company's resilience to customers, suppliers and requests for quotation.

Therefore, it is clear that there is a link among business continuity issues and risks and a strict correlation among the risk register and the business impact assessment (BIA).

The organisation is requested to define, implement and maintain a formal and documented process for business impact analysis 'AND' risk assessment that

- establishes the context of the assessment, defines criteria and evaluates the potential impact of a disruptive incident;
- takes into account legal and other requirements to which the organisation subscribes;
- includes systematic analysis, prioritisation of risk treatments and their related costs;
- defines the required output from the business impact analysis and risk assessment, and specifies the requirements for this information to be kept up-to-date and confidential.

The organisation, with reference to the risk management system supporting the BCMs, is requested to

- identify risks of disruption to the organisation's prioritised activities and the processes, systems, information, people, assets, outsource partners and other resources that support them;
- systematically analyse risk and evaluate which disruption related risks require treatment;
- identify treatments commensurate with business continuity objectives and in accordance with the organisation's risk appetite.

BCM is a vital process to ensure resilience during time: since disruptions could be determined by new and emerging risks it is fundamental to keep in place a dialogue with risk management process and also learn from experience, even considering soft event, and performance monitoring during real events and drills. A barrier-based approach can induce the assessment of common cause failures that could affect more organisation processes or determine the failure of several controls.

In this particular domain field, an easy notation, like the notation used by Bow-Tie/LOPA and by BFA, could be very effective in communicating relationships and describing business continuity procedures and in defining an incident response structure against identified disruptive events, with all the documents, roles and responsibilities associated (these elements guarantee the RRF associated with each control). Results of the assessment can be documented in business continuity plans and exercising and testing plans.

10.4 Environment

The designer assesses the R_{env} risk profile in the event of fire, distinguishing the settings (areas) of activity in which this risk profile is *significant*, from those where it is *not significant*.

The assessment of the R_{env} risk profile shall take into account the location of the activity, including the presence of sensitive receptors in outdoor areas, the type and quantities of combustible materials present and combustion products developed by them in the event of fire, and the fire prevention and protection measures adopted.

Unless otherwise determined as a result of a specific risk assessment, the R_{env} risk profile is considered *not significant*:

in settings (areas) protected by automatic total fire-extinguishing systems or installations with *higher availability*;
in civil activities (e.g. healthcare, scholastic and hospitality facilities, etc.).

11

Fire Strategies

11.1 Risk Mitigation

As illustrated in the previous chapters, fire risk assessment allows the identification of risk levels of an activity in its design, construction and operation concept with respect to certain profiles.

The task of the fire-safety engineer, in simple terms, is to identify the risk factors associated with the activity and then, in relation to these factors, to design the risk compensation strategy through the identification of a series of prevention and protection measures whose combination (Figure 11.1), in the overall management system of the activity, allows the residual risk to be considered 'acceptable'.

Prevention measures help prevent accidents and are largely attributable to organisational and procedural factors, as well as compliance with the relevant regulations in the design and construction of works and facilities.

Protection measures, on the other hand, mitigate the effects of a fire as a result of its development and are divided into two large macro-sets: 'passive' and 'active' protection measures; the former differ from the latter in that they are able to guarantee protection from the effects of a fire independently of the activation of a system or human action within the framework of the activity's management processes.

Preventive and protective measures are not alternatives to each other but compete with each other to achieve the safety objectives set by the designer, in the technical framework assumed as the basis for the design.

No matter how straightforward the process may appear, the definition of the fire strategy and its actual effectiveness pass through the evaluation of factors other than the mere technical consistency of the activity.

This, in fact, cannot be considered in isolation with respect to the operational and economic contexts, both internal and external to the organisation; as an 'assessment', the designer on behalf of the organisation must necessarily consider the factors in the domain of the commissioning organisation and those unavailable to the organisation itself but nevertheless interfering. In this regard, one thinks of the general safety provisions in Italian law, where the operator is required to ensure the continued absence of risks to the 'population'.

Consider also the technical and economic conditions to the realisation of a given plant solution, its integration in the broader complexity of the works, its interference with other plants and activities, the definition of its operating standards in this context and its technical and economic sustainability considered not only during its realisation but also in the perspective of its future

Fire Risk Management: Principles and Strategies for Buildings and Industrial Assets, First Edition.
Luca Fiorentini and Fabio Dattilo.
© 2023 John Wiley & Sons, Inc. Published 2023 by John Wiley & Sons, Inc.
Companion Website: www.wiley.com/go/Fiorentini/FireRiskManagement

FIRE SAFETY CASE

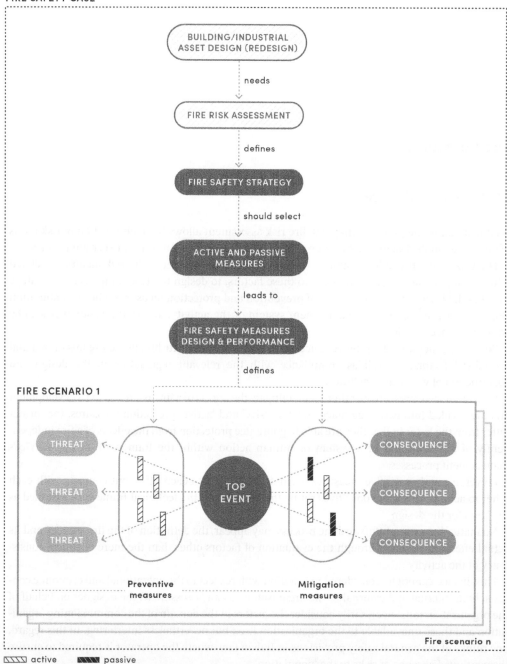

Figure 11.1 Risk preventive and mitigation measures.

management, also considering the planning of the possible partial release by phases of the work and its progressive commissioning.

In this regard, it should first of all be pointed out that there is no mathematical equation that makes it possible to univocally identify the strategy to be implemented, which remains the result of the experience and sensitivity of the designer; the same criterion of 'acceptability of residual risk' remains strongly conditioned both by the design input provided by the client and by the economic and social factors of the context in which we operate.

It is clear that the identification of measures, both preventive and mitigative, is the only one aspect of the fire risk reduction design. Fire risk reduction over time can be achieved if the strategy composed of the selected measures is accompanied by a fire-safety management system derived from the strategy, which supports the end user in ensuring that the performance of the identified measures is maintained. Given the expected performance of the barriers, all activities to maintain them become an integrated part of the set of actions described and regulated by the management system (Figure 11.2).

It is certainly possible to say that the goodness of design solutions is closely related to their relative simplicity and sustainability over time since particularly complex measures may be difficult to implement and, therefore, less effective or unavailable for longer or shorter periods, forcing in particular contexts, where, for example, it is necessary to guarantee continuity of operation, to identify alternative measures that may not, in any case, perfectly meet either the requirements of protection against fire risk or normal operation.

For example, let us think of the case of an activity where all the plant automation or protective measures are replaced by human intervention: a first intervention team of this type is certainly composed of a significant number of employees whose presence cannot necessarily be guaranteed at all times during the operation of an activity. Or let us imagine the case of a series of fixtures that can be opened in the event of an emergency, subservient to the fire detection system, the purpose of which is to guarantee the availability of an adequate smoke and heat evacuation surface, installed on the roof of a large room open to the public. Any failure of the fire detection system could compromise not only the protective measure itself but also the additional protective measure identified by the evacuation devices in question.

Now, if the environment being analysed is a mall in a shopping centre, it would also be possible to evaluate the possible suspension of the activity for the technical time strictly necessary to repair the fault, it being understood that, as far as feasible, the suspension of the activity would correspond to the cost of the company's loss of profit.

But if the environment is the hall of a hospital or the atrium of a large transport infrastructure, where for obvious reasons the continuity of operation is justified by the strategic nature of the activity, the owner will have to try to find alternative solutions that can guarantee an adequate level of safety in the time needed to repair the fault or carry out cyclical maintenance activities, regardless of the cost. For example, it would be possible to consider bringing the fixtures into the open position, effectively making the surfaces permanently available, albeit with the probable consequences of atmospheric events.

At the planning stage, it is necessary to consider, with adequate reasonableness, all profiles and therefore to identify solutions consistent with the needs of the activity and its operation over time.

Similarly, for modest activities equipped with a smoke extraction system, it is admissible to consider that the entry of counter air – necessary to prevent the environment affected by smoke extraction from becoming depressed – may take place through the entrance doors, which must be kept in an open position. Consider then the further complexity of implementing these solutions, where the activity involves further employers, as in the case of a shopping centre, where coordination between the various subjects would entail the adoption of articulated procedures spread over a large number of operators, which would be difficult to demand and implement in practice.

If we want to make a comparison, the designer struggling with his project is like a chef struggling with the preparation of a dish, who has to know how to skilfully mix the ingredients to meet the

FIRE SAFETY CASE

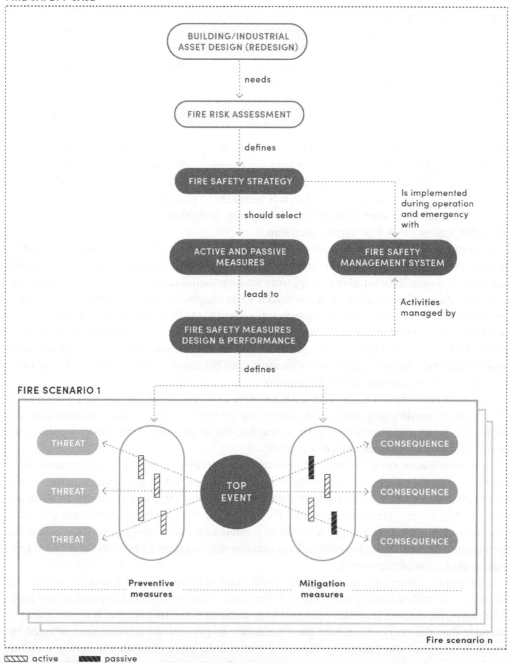

Figure 11.2 Role of the fire-safety management system.

customer's expectations, while also considering when and how the dish should be served so as not to debase its taste and quality, and the number of customers in the room.

Or rather, to return to the above equation and considering the profiles introduced on the subject of activity management, we can certainly state that the variables and their corresponding domains must be wisely identified by the designer as part of the broader process of activity design, which requires the fire protection designer's close collaboration with the client, the executive design and future management, in the knowledge that when a single variable is changed the entirety of the process must be reassessed.

The close cooperation between the planner and business owner is in effect the cornerstone of the process.

Without further dwelling on the complexity of the design process, it is evident how the designer tends to acquire design assumptions from the owner, in a 'closed' package on which to develop and complete his or her task on an assigned perimeter; this tendency is particularly noticeable in the case of entrusting fire design to professionals outside the organisation, given the specialised nature of the process.

The owner, however, as the one ultimately responsible for assessment and management, cannot afford to neglect the proper transfer of all information necessary for fire design to the appointed designer, especially that which finds its source within his own organisation.

Consider in this sense the variants of the work which are introduced, before or during its realisation, on the basis of commercial inputs to maximise the output associated with the investment. Similarly, consider constraints arising from financial availability or procedures specific to the sector in question (e.g. in the realities of transport infrastructure or hospital facilities).

It is undeniable that in complex organisations, but not only in them, dynamics of confrontation develop for the adoption of solutions and decisions, which are extremely articulated and time-consuming and which must be recovered in the fire design in order to guarantee the coherence of the result with respect to the sector regulations and the appropriateness with respect to the context of realisation.

Having said this, the owner of the activity, in his capacity as the commissioner of the fire prevention design and as the manager of the activity, must necessarily guarantee the punctual coordination of the design process by involving and giving integrated responsibility to all the stakeholders of his or her organisation; inadequate management of the process would necessarily entail the inadequacy of the assessment and design by the professional with the inevitable rise of responsibility to the owner.

The above process introduces a further and final general consideration with respect to design solutions, namely their appropriateness over time.

The process summarised above cannot but be projected over the life and operation of the activity. The dynamics of voluntary modification of the activity by the owner, already in itself complex and articulated in its organisational genesis, do not only concern the planning phase but also the operation over time of the activity, in which the modifications manifest themselves with less technical consistency but with greater frequency.

Also to be considered are contextual changes that are independent of the organisation's will, but which nonetheless fall under the onus of the owner's supervision, even if they are outside the operator's control. Figures 11.3 and 11.4 clearly show how the level of fire risk is highly dependent on different usage patterns than planned.

All of this inevitably leads to the extension over time of the fire assessment and design process, which cannot be considered limited to the initial fulfilment; on the contrary, and in line with the regulatory evolution in this sense, the assessment of fire risk and the adoption of active and passive protection remedies – in the responsibility of the employer or in broader regulatory and administrative frameworks provided for specific activities – must constitute an immanent process with respect to the life and operation of the activity, organically integrated in its broader management process.

Figure 11.3 Fire in waste collection area.

Figure 11.4 Area improperly used for temporary storage of waste.

11.2 Fire Reaction

Among the passive protection measures, particular attention must be paid to the choice of materials used for the construction of the building: the objective is to delay the propagation and development of the fire and the products of combustion, with immediate benefits on the time available for the evacuation of the rooms.

Reaction to fire, in fact, constitutes a kind of prerequisite that is inserted in a transversal manner throughout the design and management phase of a work, identifying inputs with respect to the choice of room finishes, including furnishings, and the definition of the materials with which to realise installations.

The materials fall into two broad categories:

- non-combustible materials that do not participate in combustion, i.e. they do not burn even when exposed to flame, such as concrete or marble (Figure 11.5);
- otherwise the materials are referred to as 'combustible'.

Not all combustible materials burn in the same way and therefore the reaction to fire of a material identifies its characteristics and performance with respect to the onset of fire and its development, extension and finally the production of smoke.

In summary, the reaction to fire defines the way in which a material burns in the initial phase of the fire, in its actual laying conditions. The clarification regarding the laying conditions is necessary because the same material burns differently in relation to the way in which it is installed; just think of a simple sheet of paper that burns 'differently' depending on whether it is laid horizontally or vertically.

On the international scene, there are various classifications that identify the performance of materials on the basis of certain parameters that can be measured in laboratory tests, such as ease of ignition, propagation of the flame front on the surface, and dripping and toxicity of fumes; however, the parameters observed and their weights are often not homogeneous, which does not make the various classification systems immediately comparable.

Figure 11.5 Non-combustible ceiling.

The identification of the reaction to fire performance of the materials to be used is correlated to the risk assessment; it is no coincidence that the various regulatory provisions generally link the reaction to fire characteristics of the materials to be used to the type of crowding expected, in terms of number and type of occupants. Within the same activity, moreover, it is possible to use materials with different reactions to fire performance depending on the precise point of installation, favouring the use of better performing materials along the escape routes.

As a result of the above, the need for the integration of the fire protection designer's role and his or her risk assessment in the process of analysing the technical solutions, materials and finishes in which the work is embodied, both in the initial phase and in the evaluation of its modifications during operation, is quite evident.

11.3 Fire Resistance

03 – Fire resistance

The fire resistance of an element is defined as its ability to guarantee a given performance under the action of fire, quantified in minutes.

The main performances considered are the load-bearing capacity, thermal insulation capacity and capacity to combat the propagation of fumes and hot gases of structural or non-structural construction elements; other performances are also the subject of attention, such as, for example, the capacity to limit thermal radiation or the propagation of fumes and cold gases, and the capacity to guarantee the transmission of electrical energy and data signals in the event of fire.

If we focus our attention on the concept of fire resistance – understood as the ability of the structural organism to maintain its load-bearing capacity under the effects of fire – we can see how this property contributes to the achievement of all the main objectives of the fire strategy, such as

- ensure the safety of occupants and rescuers by ascertaining that a building does not collapse under the influence of fire, allowing occupants to evacuate and rescuers to intervene;
- delay the spread of the fire to other parts of the structure and neighbouring areas, with obvious benefits in terms of protecting property and the environment.

Among structural elements, those in reinforced concrete have good fire resistance performance, the value of which is a function of the geometry of the element and the thickness of the cover, which represents the true fire 'protection' of the steel reinforcements.

Steel, on the other hand, shows a rapid decay of mechanical properties as temperatures increase, which makes its use difficult except in special contexts or in combination with protective elements (Figure 11.6).

Wood, finally, although unlike the first two is a combustible material, is a valid construction material because it possesses mechanical properties that are almost independent of temperature. In the event of fire, the carbonisation of the wooden structural elements proceeds slowly from the outside to the inside of the same, and the section of the structural element that has not yet been carbonised remains mechanically efficient; collapse only occurs when the dimensions of the 'resistant' section are reduced to the point where it is unable to withstand the stresses induced by the acting loads. The real weak point of timber structures is the joints, which, being made of metal elements, must be adequately protected in order to prevent their rapid collapse under the action of fire.

The choice of fire resistance value to be assigned as an activity requirement is derived from the risk assessment and is related to two mutually independent variables:

Figure 11.6   Protective treatment (fire-proofing) of a steel floor slab with intumescent plaster.

- the fire load, which is an objective fact;
- the performance to be targeted and identified by the designer.

Having in fact quantified the fire load, understood as the sum of the potential thermal energies of all the combustible elements present in a given room (including walls, floor and ceiling if made of combustible materials), fire resistance is defined either by minimum regulatory requirements or by more stringent objectives requested, for example, by the client.

In general, it is possible to state that the value of fire resistance to be ensured increases as the value of the fire load increases, the surface area of the environment considered and the 'residual' performance that the designer intends to ensure for the structural organism at the end of the fire. With reference to the latter, in fact, in some cases, it may be acceptable to ensure a resistance value such as to guarantee only the evacuation of the occupants while, in others, the performance to be achieved is to guarantee the complete functionality of the structures even after the development and relative extinction of a fire due to the nature of the activities installed in the structure.

On the other hand, the presence of active protection systems and firefighting personnel – being protective measures to mitigate the effects of a fire – are factors that allow the fire resistance value to be reduced.

The designer must therefore assign a fire resistance value to the structures by contextualising the expected fire in relation to the prospects of the work.

The fire resistance of an element can be determined according to three main approaches.

- The tabular method represents the most expeditious and conservative approach; it consists of assigning each element a fire resistance value on the basis of parameters such as geometric and construction characteristics and using specific reference tables.

The values they contain refer to the most commonly used construction types and materials and are the result of reliable experimental campaigns and numerical processing. This method is extremely simple to apply but at the same time extremely rigid since the tables only apply to the construction elements considered and do not allow for interpolation or extrapolation of data.

- The experimental method is the most onerous approach and consists of performing destructive tests in a furnace on real, loaded elements subjected to a conventional fire. Due to its nature, this approach is generally used for certifying the fire resistance of construction products or protective elements. The result of the test is reported in a certificate that also specifies the limits of the range of validity of the results obtained; for example, performance may only be guaranteed under certain conditions or may vary depending on, for example, the amount of material used.
- The analytical method, finally, allows the identification of the fire resistance of a structural body or part of it through the use of specific calculation models according to the type of construction being analysed.

Fire resistance, in fact, is not an intrinsic property of structural elements, being heavily conditioned by the context in which they are immersed and the type of fire to which they are exposed. The analytical approach (Figure 11.7), focusing on the specificity of the element under investigation, thus considering its punctual constraints and evaluating its performance within the overall framework of the structure and the owner's input, allows the fire resistance to be measured taking into account not only the intrinsic requirements of the element but also the relevant and competing stress factors.

Compared to the tabular method, therefore, the analytical method introduces a number of elements of flexibility by also analysing the installation and exposure context. To trivialise, a given value of fire resistance required for a structure, for example a beam, which would not be met through the application of the tabular method, could instead be met by examining the real contextual stresses, which could be such as not to compromise its resistance for as long as necessary and established as a requirement within the framework of the designer's overall assessment. In fact, it is intuitive to imagine that the same beam, subjected to the same fire, would be able to sustain its effects for longer as the distance from the fire increases and, therefore, for example, as the relative laying height increases with respect to the walking surface.

In order to do so, however, the analytical approach must be developed by the designer through a series of extremely rigorous assumptions and analyses; the designer must in fact

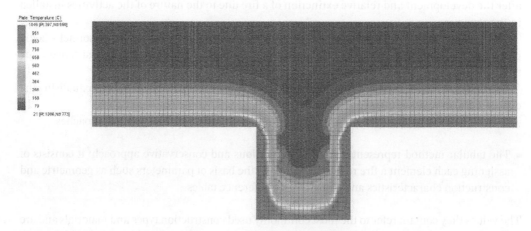

Figure 11.7 Analytical calculation of the fire resistance of a ceiling.

- verify – also by means of engineering simulations and specific fire scenarios – that all the risk factors foreseen in the design, and the corresponding protection measures foreseen in the overall design combination, develop in a manner and intensity that does not interfere with the stability of the structure for the necessary time/requirement;
- verify the same assumptions with respect to contextual risk factors outside the owner's domain;
- develop a specific technical summary, shared with the owner, in which the assumptions underlying the analysis are fully explained;
- represent the fire-safety management model that over time is able to ensure that the assumptions of the analysis and design are maintained.

It is quite clear that the analytical approach offers room for flexibility in the identification of design solutions, allowing, for example, risk factors and corresponding protective measures to be acted upon in a manner calibrated to the client's constraints and wishes.

It is equally clear, however, that the initial flexibility is countered by a strong element of subsequent management rigidity:

- to the constraint of maintaining all the assumptions underlying the analysis;
- to the possibility of its modification only through the complete revision of the process in the same manner.

Clearly, the design of the structural organism should take into account the fire resistance value to be targeted already in the early phases of the dimensioning of the structures in order to identify a structural organism that, on its own, is able to guarantee the achievement of the required performance. In cases where the fire resistance offered by the structural body is not sufficient in relation to the demand, it is possible to resort to the use of protective elements, such as calcium silicate slabs (Figure 11.8), plasters or protective paints, capable of improving the fire resistance performance of the elements to which they are applied.

The choice of protective elements depends on product specifications and their limitations of use, as well as the context of application in relation to which the installation and inspection criteria of protective coatings significantly contribute to the identification of the products to be used.

The above arguments make it easy to identify the importance of fire resistance in the development of the project design. The requirement in fact relates to central elements, structural in fact, of buildings, factories and infrastructure; if the assessment concerns newly built

Figure 11.8 Calcium silicate protection of an existing metal beam.

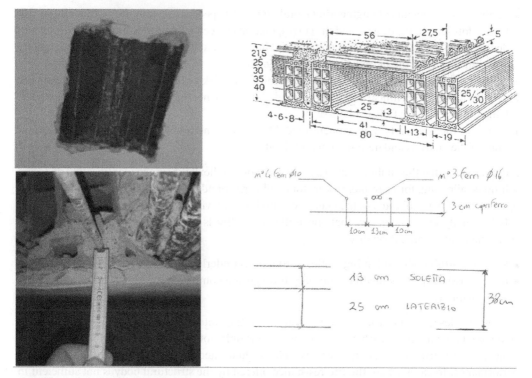

Figure 11.9 Investigations to assess the strength of an existing floor slab.

structures, the designer can act on the requirement that can easily be satisfied during construction. But if it is a question of assessing the risk on existing buildings or on works subject to monumental constraints, or if the elements to be protected are not immediately accessible (Figure 11.9), extremely complex problems emerge, in terms of both the verification of the requirement assumed in the project and the relative realisation if it is not verified on existing works.

Consider the impact of the profile in question with respect to building heritage restoration work, all the more so when subject to monumental constraints. It goes without saying that on the basis of the considerations formulated above, recourse to the analytical approach in such cases is almost obligatory, offering possibilities for immediate solutions, without however neglecting how, in fact, it introduces factors of rigidity and constraints, both technical and economic, that are extremely significant in subsequent management.

11.4 Fire Compartments

One of the fundamental principles on which fire design is based is the containment, within predetermined limits, of the effects of a fire through the construction of independent units, suitably spaced or constructed in such a way that each unit constitutes a fire compartment, i.e. a volume separated from all adjacent ones by elements with an assigned fire resistance value.

As anticipated, the objective is to counteract the spread of a fire to adjacent structural units or from one portion of a building to another.

As the term suggests, distancing is obtained by building structural organisms at a distance defined by local regulatory prescriptions or in any case quantifiable according to the performance to be guaranteed. The quantification can essentially be traced back to a value of thermal radiation induced on the structure not affected by the fire such that it does not trigger the fire. If the available distance is insufficient, it is possible to interpose elements that constitute real barriers against radiation such as curtains, water curtains and fire and smoke resistant baffles.

There are two factors that guide the practitioner in designing the compartmentalisation of a business:

- surfaces;
- the different use of the rooms.

With regard to the surface area, it should first be pointed out that there are no limits, at least from a technical point of view, on the maximum surface area that can be assigned to a compartment; on the contrary, specific limits on the surface area of a compartment are generally dictated by local regulations (there are various factors that can influence this aspect, just think of the most common local building types, average room heights, etc.).

With regard to the intended use, compartmentalisation arises either to limit the effects of a fire to a specific area (think, for example, of the compartmentalisation of an archive in relation to the workstations of an office) or to protect areas, such as escape routes or strategic areas, from the effects of a fire outside them.

Adjacent compartments, even of different fire resistance ratings, may be connected to each other through doors of homogeneous fire resistance to that of the compartment with the highest fire resistance rating.

To this end, a fundamental role is played by smoke-proof filters (Figure 11.10) that allow for the creation of connecting rooms between adjacent compartments, designed and built in such a way as to create a disconnection to the propagation of fumes; the objective is achieved either by means of chimneys that expel fumes thanks to natural draught (depression) or through the insertion of mechanical fans capable of generating an overpressure in the filter (Figure 11.11), thus preventing fumes from entering. It is understood that the overpressure value is generally defined by local regulations and in any case such that it does not compromise the simple push opening of the filter access doors, especially since they are generally installed along escape routes.

The former, as they do not require equipment that needs to be maintained over time, are certainly more economical and in some respects also more reliable as they operate 'naturally'; on the other hand, the design, in order to ensure their real effectiveness, must take due account of possible weather conditions that could affect their proper functioning.

The latter, on the other hand, are certainly to be preferred in existing rooms undergoing renovation where it is not always possible to have the necessary space for the passage of natural ventilation chimneys and whose paths are more rigid than those of the intake ducts upstream of the ventilators.

Due to the specific functionality, and irrespective of specific local regulations, it is advisable that the fire load in smoke-proof filters should be zero or completely negligible, and that fire hydrants covering adjacent compartments should not be installed in them, as unfortunately often happened in the past, since their use would presuppose the need to keep their doors open, which would in fact nullify their effectiveness.

While the representation provided so far does not introduce any particular criticalities, on closer inspection, however, the compartments introduce significant factors of technical complexity and even greater management difficulties.

The construction of stand-alone units must in fact be coordinated with the plant distribution serving the activity in general or the security installations.

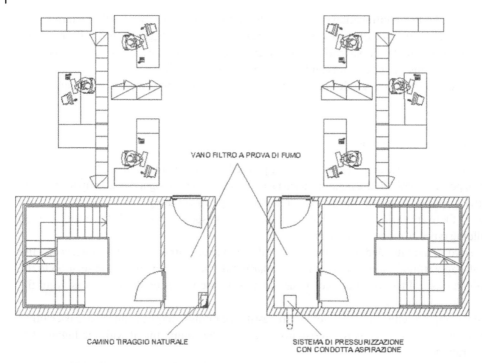

VANO FILTRO A PROVA DI FUMO

CAMINO TIRAGGIO NATURALE

SISTEMA DI PRESSURIZZAZIONE
CON CONDOTTA ASPIRAZIONE

Figure 11.10 Smoke-proof filter types.

Figure 11.11 Pressurisation system of a filter.

Figure 11.12 Restoration of plant crossings.

If, for the sake of immediacy and common experience, we think of the compartments in a shopping centre, then it becomes easier to glimpse the technical complexity involved in splitting a single business into several sections; indeed, try to consider

- the realisation of the crossings (Figure 11.12) determined, for example, by the distribution networks of the electrical installations, the cabling, the ventilation system and the smoke extraction system (as well as the dimensioning of the power of the latter systems in relation to the activation forecasts on the individual compartments);
- the introduction of mechanisms for the automatic actuation of the closing baffles of the distribution networks of ventilation and extraction systems (fire and smoke dampers);
- the use of self-closing fixtures on the boundaries of the compartment;
- the development and maintenance of the functional logics of overall and coordinated governance of the above-mentioned facilities.

The same complexity, if not more pronounced, occurs in the management phase where the functional requirements of the activity, especially if not adequately considered in the design, risk compromising the fire-safety level of the activity.

One thinks of the movement of people, goods and materials between different compartments and the impact on filters and fixtures; as mentioned, self-closing devices are often used, but it is not uncommon to observe the use of more empirical methods for the voluntary or involuntary locking of doors.

This includes the cyclic renovation of the commercial fittings of individual building units and the adaptation of centralised systems (ventilation, extraction and fire detection) and functional logics.

11.5 Evacuation and Escape Routes

The protection of the occupants' safety – a pivotal objective around which the entire fire design revolves – is pursued above all through the definition of an escape route system that, in the event of a fire, allows them to leave the building before the effects of the fire compromise their vital functions.

As is well known, the main cause of death in fires is not linked to the high temperatures reached or direct contact with the flames but rather to the time of exposure to the toxicity of the fumes; it follows that the sizing of escape routes cannot disregard the 'time' factor, especially since the evacuation process takes place at the same time as the fire develops and must be completed, except in special cases that are discussed later, within a few minutes after the fire is detected.

Fire prevention is, in the writer's opinion, the only discipline that allows the dimensioning of escape routes because fire represents the only quantifiable risk at the moment; in other words, if I can identify the instant in which the conditions of the environment are no longer compatible with human safety, then I know the time I have available to evacuate the activity and, therefore, I can correctly dimension the system of escape routes.

Let us therefore try to establish some essential concepts with respect to the topic.

- It seems first of all intuitive that the need to contain evacuation times can be associated with the speed of the fleeing users; but if speed and time are correlated with each other by space, and certainly not being able to rely on the design idea that all users are skilled runners, then the geometric characteristics of the escape routes assume a key role, not only in relation to lengths but also to relative widths.

Let us imagine, for example, that following the alarm all the users begin the evacuation process at exactly the same speed, leaving at exactly the same time; if this were the case, clearly the evacuation time would exclusively depend on space, which, as we said earlier, must be understood not only as the length of the evacuation route but also as the width available for routes and exits: the presence of bottlenecks may in fact determine the formation of queues that inevitably slow down the movement of users, thus increasing the overall evacuation time, apart from incredible configurations (Figure 11.13) we can find in our everyday life. The solution is to adopt routes as free of bottlenecks as possible, with widths commensurate with the levels of crowding expected to use that exit route. The possibility of reasoning in the opposite direction, i.e. extending the time available for safe exodus, remains understood: environments with high heights allow a greater accumulation of fumes, thus delaying contact with users and the negative effects in terms of visibility.

- Dimensional characteristics are, however, not the only properties that must be taken into account in the design of escape routes as other aspects such as relative visibility, signalling methods, number, characteristics and distribution of the final exits are equally determined: these are elements that are apparently distinct from each other but often interdependent, which must be skilfully interwoven by the designer in order to define exit routes that, as a whole, can effectively satisfy the demand for intuitiveness, simplicity and robustness that the final objective requires.

Having therefore set the objective of limiting the exposure time of users to harmful combustion products through the design of the exit routes, let us try to elaborate on the various elements briefly mentioned above.

- Clearly, shorter lengths correspond to shorter travel times. But different activities correspond to different characteristic fires, with different speeds of development and smoke production; therefore, there is no single threshold, since this threshold is the result of the specificity of the site, although the various local regulations may identify typical threshold values for certain types of activity. Paths protected against the effects of a fire, i.e. paths separated by fire-resistant structures from the compartment affected by the fire, such as protected or smoke-proof stairs, must be excluded from the calculation of lengths.

Figure 11.13 Emergency exit of a condominium garage.

- The widths of pathways and final exits to the safe space are essentially related to the design crowding, generally indicating the different regulations of simple relationships between the two elements; for the above mentioned, narrowing should be avoided as much as possible, favouring pathways with as constant widths as possible.

However, the design treatment of the subject cannot be as simplistic. Let us consider the case of a large environment, for example, a cinema, a mall of a shopping centre or a square destined to host a concert; the expected crowding will correspond, as above, to a minimum value of the overall width of the final exit to a safe place. If the lengths involved allow it, we could abstractly hypothesise the realisation of a single exit with a width at least equal to the minimum value imposed by the local regulations. A solution of this type is certainly to be avoided because having only one exit means having an extremely vulnerable system; the accidental event could in fact entirely or partially compromise the only available exit route and users would therefore not be guaranteed the possibility of evacuating safely. This is why regulations often prescribe the need to have at least two exits facing each other and to verify that the escape route system is able to guarantee the evacuation of users even in the event of the unavailability of a final exit. The same principle underlies the limitation generally imposed for the lengths of blind routes, i.e. the routes that evacuating users can take in one direction to gain the exit. The more robust and redundant the solution identified by the designer, the more flexible it will be. The designer can therefore already foresee the unavailability of one or more escape routes at the design stage, e.g. for maintenance work.

On the subject of the widths of the final exits, it should also be pointed out that, in reality, not all exits are used uniformly by users, who clearly tend to favour the closest or in any case the most immediately reachable exits, favouring the shortest and most direct routes to the detriment of diagonal ones, and above all the use of the exits already used by the users preceding them, which influence the choice of those following almost as if they were acknowledging the existence of a path to save themselves.

Several psychological factors come into play in the evacuation process, first and foremost fear and confusion, and therefore the instinctive choice of the easiest and most certain route appears to be the best choice. Similarly, the widths of routes and exits are not used to their full extent; in fact, users tend to keep a certain distance from walls and obstacles, effectively using only the central portion of the escape routes, thus making it inadvisable to design large exit gates.

The last, but by no means least important, characteristic of exit gates is their opening mode. As a general rule, the door leaves can open in the direction of the exit or in the opposite direction, or be of the 'sliding' type (Figure 11.14); the former are certainly to be preferred in the presence of large crowds, while the use of the latter should be limited to exits of areas for which informed personnel are expected or in any case scarcely frequented, such as technical areas or small workplaces.

The following is shown for the door opening mechanism.

- No special mechanisms can be envisaged, but only in small rooms with limited occupancy, for example, in the case of an office room.
- It is possible to use handles capable of directly unlocking the exit in the event of actuation, always in the presence of small crowds familiar with the environment (e.g. technical rooms for which segregation of the area from unauthorised access is in any case necessary).
- It is possible to use panic bars. These devices, which can only be applied to doors that open in the direction of escape, are devices that are operated by applying a slight push on a horizontal bar that allows the door to be unlocked; the advantage, in addition to the relative simplicity of operation, lies above all in its intuitiveness, simplicity and the possibility that it can also be operated by pressure exerted with the body by users, avoiding the dangers of crushing and crushing that

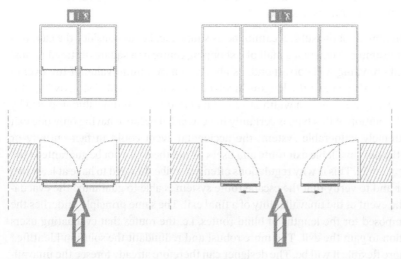

Figure 11.14   Doors with push-folding doors (left) and push-folding sliding doors (right).

could occur with less immediate opening devices. Because of the above-mentioned advantages, the use of panic bars is widely used in all those cases where large crowds are expected, or in any case where people are unfamiliar with the environments being designed.

- Finally, the further case of doors with sliding leaves located at escape routes must be considered. In this case, it is possible to use doors with sliding leaves in ordinary conditions that can also be pushed down, otherwise known as 'break down' doors, by simply pushing on the leaf or to provide special redundant safety systems that can always and in any case guarantee that the leaf is always open (for example, with leaves that move into the open position in the event of a power failure, photocells with a high capacity for detecting people approaching, local release buttons, etc.).
- A separate case, although it has several similarities with sliding doors, is represented by turnstiles for controlling access to or exits from certain rooms. Although in some cases it is possible to exclude turnstiles from the exit route system, providing normal exit doors in the immediate vicinity, in a sufficient number and width with respect to the expected crowding, for the considerations set forth above with respect to the intuitiveness of the exit routes and considering that users tend to privilege the use of the route followed for access to the room during exodus, in the writer's opinion, it is always preferable, where possible, to include them in the system of escape routes by providing special precautions similar to those indicated above for doors with sliding doors and also resorting to on-site personnel dedicated to unblocking them in environments characterised by high crowding and with little familiarity.

But for routes to be effectively identified and used safely by users, they must be effectively signposted and illuminated. Life-saving signage is a design aspect that is often not addressed with due attention by the fire prevention planner, but delegated to the professional figures in charge of designing the information and directional signs to be installed in the activity, also for obvious aesthetic opportunities, and who generally have little knowledge of the design criteria underlying the definition of signs that are visible and comprehensible to users; in many cases there is no real design of the signs, which are simply affixed in the rooms on the basis of unqualified choices. Life-saving signage, on the other hand, is the result of careful design choices designed to guarantee:

- the visibility of signs – signs must be immediately visible and to this end they must be given due prominence in relation to other signs that may be present, both in terms of size and relative location, preferring isolated installations in relation to signs of a different nature or, if they are integrated, relying on colour and size distinction;
- immediate communication of the route to be followed – pictograms must be intuitive and must effectively communicate to users the direction to follow or the location of the emergency exit, regardless of their cultural background;
- the simple comprehensibility of the indications – this is the case of all the information that must be provided to users in order to ensure their safe evacuation (Figure 11.15), such as, for example, instructions for unlocking the sliding doors of a door installed along an escape route or for the use of communication devices to signal one's presence in a safe place and to request the support of rescuers for evacuation.

Generally speaking, the pictograms of life-saving signs, codified by local and sector regulations, are white on a green background; the same regulations dictate relationships that allow the distance of visibility of the sign to be linked to its size and, in some cases, also to the mode of illumination. With reference to the last aspect, in fact, it is possible to resort to the use of both luminous signs

Figure 11.15 Backlit escape sign visible from several directions.

and signs illuminated by external sources, or a combination of both depending on the context of installation; a luminous sign is quicker to identify and better readable by users and is therefore visible from a greater distance than a sign of identical dimensions but not independently illuminated.

But what needs to be reported? Again, local regulations may dictate more or less stringent requirements although common sense often guides the designer. Routes and final exits must certainly be signposted, highlighting all changes of direction and any routes reserved for persons with reduced or impeded mobility and communicating all the information and instructions needed to ensure that the routes are used in accordance with the design idea. It is understood that in the case of environments frequented by users from different parts of the world, it would be necessary to use, in addition to the local language, English and the languages of the countries from which the visitors mainly come.

In addition to following the correct escape route, users must clearly walk along it in complete safety; this is ensured by providing them with adequate lighting and the absence, as far as possible, of elements that could constitute a tripping hazard.

- The first point is achieved by designing and implementing an emergency lighting system, either distinct from the ordinary one or derived from it by providing, for example, that a part of the lighting fixtures be equipped with buffer batteries or continuously powered by an independent power source, such as an uninterruptible power supply. The duration of the system's operation is the result of the risk assessment or, in any case, a minimum requirement indicated by the local regulations; in addition to guaranteeing the exodus of the occupants, which, as mentioned above, should be exhausted within a few minutes from the detection of the fire, emergency lighting is also instrumental in guaranteeing the operation of the rescue teams, both internal and external to the activity. In addition to guaranteeing the identification of routes and exits, the safety lighting system must allow exiting users to see any obstacles and narrowing of changes of direction and safety and rescue signs where they are not independently illuminated.

- The second point conditions both the initial design and choice of furniture elements as well as the maintenance of the identified design requirements in the event of temporary or permanent changes; consider:
  - seats used in entertainment events in permanent or occasional settings, such as squares and gymnasiums, which must be distributed in such a way as to ensure quickly accessible and sufficiently wide exit aisles, and bound together in groups to minimise the risk that in the event of an escape they can easily be moved, preventing the smooth flow of people;
  - the barriers or turnbuckles used, for example, to route users through security checkpoints at stadiums, concerts or airports and stations;

Any single element can nullify the goodness of the design solution identified, which cannot therefore disregard the analysis of the context in all aspects of the construction and operation of a work, even the apparently most insignificant ones. Not infrequently, in fact, local regulations prohibit the installation of mirrors along escape routes that could mislead fleeing users.

Having listed the main characteristics of the exit routes as above, let us analyse different design approaches to the issue of exodus which guide the designer in the technical choices to be implemented. We have in fact indicated so far that the exodus must be completed within a few minutes from the detection of the fire; the design approach is certainly correct although in some cases it is not actually possible or in any case difficult to implement in relation to the characteristics of the activity and the type of users present. But is it really necessary to immediately evacuate the entire building in the event of a fire according to the so-called 'simultaneous' exodus approach?

At least as a general rule, it is always necessary to ensure the almost immediate evacuation of the occupants of the compartment affected by the fire and then to proceed with the evacuation of the occupants in the other compartments of the activity according to a 'phased' evacuation approach.

An alternative is 'progressive' evacuation, which involves moving the occupants of the compartment affected by the fire to an adjacent compartment capable of containing and protecting them until extinguishment or subsequent evacuation.

The last approach, finally, does not involve the exodus of users but the protection of them in the compartment where they are until the emergency event is resolved.

The goodness of the escape route system can certainly be verified with the help of more or less complex calculation models. In order to address the problem fully, we have said that we cannot disregard the time factor. We therefore indicate

- with ASET (Available Safe Egress Time) the time available for users to evacuate safely;
- with RSET (Required Safe Egress Time) the time required for users to leave the building.

The designed system will be able to meet the expected performance if and only if RSET < ASET, i.e. if the time it takes for users to evacuate is less than the time available in relation to the design fire; therefore

- since the liveability conditions of an environment are attributable to several factors, the ASET is generally assessed using several performance parameters, calculating the time in which the threshold set for each parameter adopted is integrated and then cautiously choosing the lower limit;
- the time required for escape from the fire ignition, RSET, is made up of several contributions: to the time strictly necessary for the movement of people, the time between the ignition and the detection of the fire must be added and the time required to spread the alarm and the so-called 'pre-movement' time, i.e. the time taken by users to actually recognise the alarm and decide to escape.

While ASET, for an assigned domain, is a physical datum that the designer can quantify with more or less simplified calculation models (e.g. the 'zero exposure', 'toxic gas', 'irritant gas', 'visibility' and 'heat' models), RSET is instead the end result of many factors such as the alarm detection and alarm detection system, the number, distribution and type of users, the geometry of the environments, the fire-safety management system and is heavily conditioned by psychological factors; just imagine the differences that can occur between the evacuation of workers in a small office and the evacuation of large masses that, for example, crowd stadiums or large transport infrastructure nodes.

Over the years, various studies have been carried out that have enabled the characterisation of the properties of motion and the definition of calculation models that are today easily usable thanks to the aid of computers. Various software is therefore available to the designer, which can be of great help, but which certainly cannot replace experience: there are no universal models and each software has its limits that the designer must necessarily know in order to avoid arriving at erroneous results; if the designer knows the margins of error, then his experience can compensate for them.

Let us dwell on this aspect with a simple example: various calculation programmes make it possible to simulate the process of user exodus from a given environment by determining, the initial position being known and the exit fixed, for each user, n-paths, each of which is associated with a cost in terms of time: the user, in order to reach the destination, will always follow the path with the lowest cost associated with it. In the course of the simulation, each user always knows the cost assigned to each path, a cost that may vary over time due, for example, to queuing phenomena that may be generated at a door or stairway.

Let us try to explain this with an example.

In Figure 11.16, if P is the departure, and A the arrival, to go from P to A user U can follow path 1 and path 2, each of which is associated with an initial cost Ci1 and Ci2.

If there are more than one user, at the initial instant they will all take the same route Ci1 because it is more convenient; at the narrowing of door P1, however, a queue may be generated such that, at a given instant, it may be more convenient, in terms of time and only for some users, to cross door P2 to reach arrival A. Each user, therefore, recalibrates the choice of route according to what is to come. But in reality, not all users know the possible routes and their 'costs' – think, for example, of first-time users in a shopping centre or tourists in a railway or underground station. The only way to guarantee the adherence between reality and the calculation mode results (Figure 11.17), and thus to be able to consider the results of simulations reliable, since it is not possible to act on the model, is to set up clear, simple and intelligible signposting in reality.

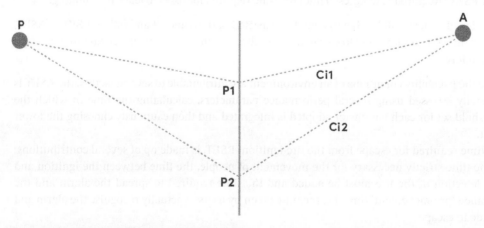

Figure 11.16 Choice of exit route with lower 'cost'.

Figure 11.17   Modelling escape routes through doors and turnstiles with software.

The intent is to highlight how the consistency of the model to the reality it is intended to simulate, and therefore the goodness of the method with respect to the aim pursued, is strongly conditioned by the knowledge of the calculation tool and the reality being investigated. The professional therefore acts as the creator of a simulated reality to explain and manage the material reality, with the claim of anticipating and replicating its dynamics and interpreting its causes and effects; depth and balance of the analysis, both in the preventive investigation and in the evaluation of the results, all the more so on the basis of the power of the instrument, remain an unavoidable factor for the validity of the method and the result. Therefore, the need emerges to adopt a process methodology that guarantees the critical review of assumptions and results; in this sense, once again, it appears necessary to involve the figures who will manage the work right from the design phase as this can ensure that the assumptions underlying the project will become a parameter of the future exercise, linking 'design' and 'management' in the construction of the model and in the analysis.

The concept briefly mentioned above becomes even more evident at this point: the design of the escape route system is a combination of many factors, geometric, architectural, psychological and organisational, and none of these can be overlooked by the designer.

We have so far dealt with the subject of exodus from a more strictly technical and basic point of view; let us now try to deal instead with some more conceptual and problematic profiles linked to the application experience of the discipline of exodus.

First of all, let us try to ask ourselves, at root, what the term exodus means: that is, if 'exodus' certainly means removing people from the physical place where they are directly or indirectly in a state of danger, when can the action be considered accomplished?

In the first instance, and as mentioned above with regard to the 'progressive' exodus, the exodus could be considered to have taken place when the persons have been removed from the dangerous condition also by means of their transfer to a compartment adjacent to the one affected by the emergency event. Similarly, in such a case, the exodus could be said to have been completed when the persons are led onto the roof of a building adjacent to the one affected by the fire, provided that this roof constitutes a safe place, therefore separate from the compartment in flames and free from sources of danger, regardless of whether or not it is served by stairs and paths that permit the final removal of the persons.

Although abstractly conceivable, such a solution certainly does not appear desirable and should certainly be avoided. It is undoubtedly correct to pursue the safety of persons in any way, but it must also be emphasised that the residual risk margins of an even adequate assessment and the opportunity to develop robust and redundant solutions at the planning stage require that the hypothesis put forward can only be considered acceptable to the extent that the use of another compartment is made to reach routes and stairs that permanently remove persons from the activity affected by the emergency event.

It is worth providing a final consideration on this subject; as mentioned at the beginning of this section, the definition of escape routes in relation to fire is the only area in which the measurement of fire risk allows the measurement of escape. Let us not forget, however, that the same escape routes are then used by the employer with respect to the entire emergency management system and therefore also with respect to hypotheses in which fire compartmentalisation is totally irrelevant. It therefore seems entirely appropriate to pursue the exit as the exit of persons from the activity, without considering the problem of rescuing any injured persons on the roof with no stairs.

Secondly, it is worth recalling a few considerations regarding the performance approach with respect to exodus.

Exodus regulations are normally rather limited; it is easy to understand the difficulty of formulating general and abstract forecasts in standards when, in reality, architectural trends and engineering solutions offer increasingly daring and complex geometric configurations.

In this sense, the engineering approach to exodus certainly appears to be a valid contribution due to the degree of flexibility offered and the breadth of factors considered in the analysis, in addition to the broader study of the levels of accessibility and ergonomics of flows, especially in large infrastructure with an indistinct public presence. However, an important point of attention must be introduced: the goodness of the result, as in any case where engineering analysis is used, is directly related to the inputs provided to the system. However extraordinary the technical capabilities of the software used may be, nevertheless the adequacy of the product and the coherence of the analysis to the intended purpose always depend on certain starting data such as the estimated crowding and its distribution. This information, however, remains the responsibility of the owner and planner, regardless of whether it belongs to their domain (think of events or infrastructure frequented by the general public). In other words, with this annotation and in confirming the usefulness of the performance approach, it is nevertheless intended to emphasise that it cannot be understood a priori as the solution to all evaluation problems.

11.6 Emergency Management

Emergency management planning is an essential organisational measure to contribute to the achievement of adequate safety standards; it is implemented by the person in charge of the activity to cope with accidental events, in addition to the technical solutions indicated by the designer to compensate for the fire risk and is closely related to them. It is a responsibility linked to management and cannot and must not be delegated to the intervention of external helpers.

The person in charge of the activity must therefore provide an adequate internal organisation, consisting of suitably trained personnel, capable of implementing all the actions necessary to contain an emergency event and to facilitate the evacuation of all occupants.

The extent of the resources and their training depend on two main factors:

- the degree of complexity of the activity and its context;
- the relevant plant equipment.

In fact, one can see how a complex reality – such as a railway station or a hospital – cannot be matched by the same internal organisation as a simpler reality, such as a medium-sized or small office: the technological solutions and the characteristics of the occupants have completely different specificities and therefore a different specialisation cannot correspond to them.

While in the latter case, a team of a few workers may be sufficient to verify the actual presence of a fire and, if necessary, attempt to extinguish it with the help of the extinguishers present, in the former it is necessary to identify a room that constitutes the actual emergency management centre. This room must be permanently manned by dedicated personnel, and the supervision and control of all the safety systems with which the activity is equipped, including for example the video surveillance system, as well as the means necessary to communicate with both internal and external rescue teams, must be available within it.

The type of team composition will also be different: it may be necessary to have personnel dedicated to the maintenance of the building and its installations in order to try to ensure the operational continuity of the activity as well as the safe intervention of rescuers (just think of the need to disconnect live power lines before using hydrants).

Finally, staff training must also be aligned not only to the specifics of the activity, as in the case of the turnstiles installed along the escape routes mentioned in the section on exodus, but also to the technical solutions adopted by the fire prevention planner: using a fire hydrant is certainly not the same thing as using a fire extinguisher.

But the organisation needs specific indications regarding the actions to be implemented in the event of a fire; hence the need to plan them in advance through the elaboration of an emergency plan that defines the structure and organisational logic of intervention, in order to allow, through timely action to be implemented according to the current scenario, the management of the various emergency situations, so as to protect people, property and the environment.

The purpose of the emergency plan is therefore to collect, disseminate and keep up to date the procedures adopted for the management of emergencies so that any events that may occur can be dealt with in an organised manner, putting in place measures to ensure the safety of workers and, more generally, of all persons present in the activity.

The plan must be structured in such a way that the main safety features of the activity, the measures and procedures in place to detect, report and counteract the outbreak of fire in a timely manner, and the rules of conduct to be observed in the event of an emergency, are known.

It is, therefore, an essential tool to control and circumscribe possible accidental events by identifying the response to be provided through the definition of the roles, responsibilities and tasks of each resource involved in the process. Its minimum contents concern:

- emergency activation and management;
- coordination procedures between the actors on the site;
- coordination procedures with the operators of adjacent sites, particularly in the case of activities that are communicating with each other or whose emergency management is closely interconnected, e.g. sharing of escape routes or protective measures;
- the arrangements for requesting the intervention of the competent public security services and the information and support to be provided upon their arrival;
- evacuation procedures from places affected by the emergency (Figure 11.18);
- the specific measures taken to assist persons with reduced mobility;
- the information to be provided, in the event of an emergency, to all persons in any capacity present within the activity, concerning the measures prepared and the behaviour to be adopted in the event of an emergency.

Figure 11.18 Orderly exodus down a staircase during an emergency evacuation simulation.

If the purpose of the emergency plan is to be certain that everyone knows what to do in the event of a fire, both to deal with the event and to request the intervention of external rescuers, as well as to abandon the activity, it is necessary that it is the subject of specific training and/or information activities in relation to the various roles, that it is always available and that it is compiled in such a way as to be easily and effectively understood by everyone. It is clearly understood that the articulation of the procedures mirrors the degree of complexity of the activity.

Let us think, for example, of the case of a railway station, adjacent to a metro station, with which it shares part of the escape route system and some active protection measures; let us think, for example, of smoke-proof filters with opposite directions of travel in relation to the location of the emergency event. Without entering into the merits of the methods for implementing the technical solutions and their activation, the site managers will have to discuss and agree on specific procedures to guarantee, in the first instance, the exchange of information already in the first moments following the detection of the emergency event and, as a result, the correct activation of the active protection measures, as well as the effective exodus of the occupants in relation to the scenario in progress. These procedures should be the subject of shared protocols and form an integral part of the respective emergency plans.

But within the railway station there may be operators of other activities, such as offices or commercial services, which in turn are obliged to draw up an emergency plan with reference to their own premises and internal resources, and which must necessarily indicate procedures harmonised with those of the broader site emergency plan (Figure 11.19) in a coordinated approach (Figure 11.20).

And if the manager of the railway station complex is a different legal entity from the manager of railway traffic, the issue becomes even more complicated. In all cases where more than one manager coexists on the same site, it is desirable that the emergency plan be a single one shared by all managers, albeit in awareness of the objective complexity of the undertaking.

Finally, in the most complex cases, a direct and preventive discussion is also desirable with the public services bodies responsible for security matters.

Figure 11.19 Large-scale simulation with involvement of relief agencies.

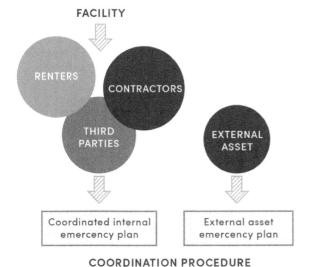

Figure 11.20 Coordination of emergency plans.

In fact, the definition of roles and the planning of actions to be implemented in the event of a fire will be all the more effective the more the personnel in charge are trained to cope with accidental incidents; hence the need to carry out regular tests, testing the organisation's degree of response at least once a year, if not every six months or quarterly in the case of particularly complex activities.

The simulations should be the subject of specific audits: the opportunity is also to identify in advance any criticalities in the system set-up, adopting the appropriate corrective actions and thus preventing them from occurring in the event of a real fire. In this sense, for example, it could be useful to simulate scenarios in which a protective measure is unavailable, such as a safety system or an escape route, and assess the readiness of rescue teams in response to the unforeseen condition. As an example, Table 11.1 shows the typical contents of an emergency plan for an offshore platform.

Table 11.1 Contents of an emergency plan of an offshore platform.

Industry	Offshore oil and gas production platform
Key information required to develop an ERP	**Details**
Offshore	
General information and layout	Production rates, reservoir conditions and fluids, layout of the levels, main production lines, phase of the facility (including those that may induce specific issues, such as pipelines) or interconnections (e.g. for large integrated offshore installations)
Hydrocarbons (and hazardous chemicals)	Type, hold-up, chemical and physical properties, and safety-related characteristics
Equipment/ machinery	Safety significant equipment/machinery and level of congestion
Emergency systems	Prevention and mitigation measures in place including the process safety systems (ISO 2019) and all the other relevant barriers for fires, explosions, spills such as, from ISO (2016b), emergency shutdown systems and blowdown, ignition control measures, spill control measures, emergency power systems, fire and gas detection systems, active and passive fire protection measures, explosions mitigation and protection measures, etc.
Critic areas, pedestrian routes, refugees and muster locations	Details about the accessibility of designated refuge and muster areas, means of evacuation and life-saving appliances, including any eventual estimation of travel times and maximum acceptable mustering time for personnel not involved in emergency team
Internal and external communications	Details about communication systems including resources, public address system, radio and satellite systems. Protocols and templates to be used (e.g. list of evacuated personnel) as well as the list of the milestones to be considered in the evolution of an emergency situation
Detection and alarm systems	Details about automatic and manual warning systems (lamps, sounders, optical signs, fire alarm and GMDSS) and their location. Details about continuous control and fire guard activities
Staffing level and occupancy	Crew and variation in the crew, location of accommodations (living quarters) and the number of third parties onboard

Table 11.1 (Continued)

Industry	Offshore oil and gas production platform
Key information required to develop an ERP	**Details**
Installation location (including meteo and marine conditions)	Distance to shore and environmental conditions (including probability of severe events and variations throughout the year)
Nearby installations (if any) including connections	Distance to the nearest installation
Operating experience	Experience from site near misses and incidents and reference incidents on comparable installations; experience from past ER drills and practices; operational experience that may suggest improvements to the ERP
Legal requirements	
Applicable codes	Details about specific requirements coming from legal requirements about the content, structure and elements of the ERP in case of specific regulation, such as the 'offshore directive' or voluntary frameworks that the organisation would adhere to, such as an SEMS programme

11.7 Active Fire Protection Measures

Extinguishing systems constitute active protection measures aimed at extinguishing or controlling a fire. These measures are divided into two main groups, manual and automatic.

Among the manual measures – leaving aside simple yet effective remedies such as buckets of sand, which can still be seen in some petrol stations today – are fire extinguishers (Figure 11.21), which are used to extinguish the initial fires.

Fire extinguishers differ according to the extinguishing substance and weight, parameters that guide the fire prevention planner in the choice of equipment to use. Mainly used are types with limited weights, not exceeding 6 kg, which allow them to be easily used even by non-specifically trained personnel, and which are used for extinguishing the principles of fire that generally occur in civil activities. Heavier weights, from 9 to 100 kg, are used in industrial contexts or technical environments, as well as in the transport sector; greater weights clearly correspond to greater extinguishing capacities but less practical application, so much so as to require the use of trolleys with wheels – it is no coincidence that they are referred to as wheeled extinguishers (Figure 11.22) – and at least two suitably trained personnel.

The distribution is strictly correlated to the risk assessment and the prescriptions of the local regulations; as a general rule, they must be positioned in such a way that the routes to reach them from any point of the activity are between 15 and 30 m, foreseeing the presence of at least one per room and a sufficient number also in the connecting spaces through which the various rooms of the activity are accessed in order to facilitate their immediate identification and easy use by anyone.

On the other hand, a firefighting water system, consisting of hose-reels or hydrants (Figure 11.23), is designed to extinguish any fire that may develop in the activity. The system can be schematised as a

Figure 11.21 Portable fire extinguisher.

Figure 11.22 Wheeled fire extinguisher.

network of pipes fed by a pump and to which reels or hydrants are connected. The two fire-fighting devices have the same purpose, although reels are easier to use even by a single operator (since it is not necessary for two people to be present to completely unroll the hose, as is the case with hydrants) and are character-ised by greater precision of the water jet, which is why the former are required to guarantee a much lower flow rate than the latter. The system is completed by one or more motor-pump connections (Figure 11.24), depending on the complexity of the site to be protected and, consequently, of the relevant system, which allow the network to be pressurised by external tankers.

The two types of protection are used for the internal protection of the activity; the installa-tion of hydrants outside is also envisaged, at a sufficient distance from the building struc-ture to be protected, to ensure their safe use by rescuers. External hydrants may be of the 'underground' type, in dedicated sumps, or above ground, of the 'column' type, and have significantly higher hydraulic performance than internal hydrants, both in terms of flow rate and pressure, and their use is exclusively by professional rescuers from the competent public services. Performance, distribution and type of water supply, both with respect to the water reserve and the characteristics of the pumping units (Figure 11.25), are detailed by local regulations; it is understood that activities characterised by a higher level of risk correspond to a system that must meet higher performance, with a type of water supply and pumping system characterised by greater reliability.

With regard to the design of the network, two final aspects must also be specified, one concerning the two types of installations that can be realised and one concerning the dis-tribution of the equipment.

In all those cases where it is foreseen that the system may be exposed to very cold tem-peratures or where there is a high proba-bility of accidental contact of the water jet

Figure 11.23 Hydrant UNI45.

Figure 11.24 Motor-pump connections serving a firefighting water system of a complex activity.

Figure 11.25   Fire pump room.

with live systems or equipment, with a consequent risk for rescuers (for example the electric trac-
tion lines of underground trains), it is possible to opt for the construction of a system with 'dry'
piping, i.e. without pressurised water. In the first case, the advantage is to avoid possible unavailability of the system due to the presence of ice, while in the second case, the risk of electrocution
is excluded by subordinating the pressurisation to the prior securing of the sites and systems,
according to specific procedures that must be indicated in the activity's emergency plan. On the
subject of the risk of electrocution, it must be pointed out that only in particular contexts similar
to the one referred to above, does the designer indicate the need for a dry water system? In the vast
majority of cases, the installation of permanently pressurised systems is used, which, for obvious
reasons, must coexist with the electrical systems serving the activity. The designer must therefore
ensure that there is a simple and effective system of disconnection buttons, which does not lead the
rescuer into false expectations and which can therefore intervene safely or propose its implementation. This is the case, for example, with electrical installations serving shops in a shopping centre
or those serving buildings constructed in past eras and subject to more recent renovation, where
the quantity and articulation of the electrical installations present, together with architectural constraints, often oblige the power supply cables of a sales unit or area to run in adjacent sales units or
compartments.

Finally, with regard to the distribution of hydrants or reels within the activity, it is necessary to
verify that the entire area is covered by at least one hydrant, net of any obstacles that must be easily
circumvented in compliance with the pre-set objective, otherwise it will be necessary to position
additional hydrants to protect any uncovered areas. Finally, in the case of more than one room

with access from a common corridor, it is a good rule that the hydrants are positioned in the common corridor in order to exclude the impossibility of using the hydrant in the event of access to the room where it is located; this is the case, for example, of technical and service areas or shops in a shopping centre, where it is necessary to find a balance between safety and security requirements and the prohibition of certain areas to non-authorised personnel.

On the other hand, there are different types of automatic systems that mainly differ in the type of extinguishing substance used.

The most common systems are water-fed sprinkler systems (Figure 11.26); these are pressurised networks on which there are heads fitted with a heat-sensitive bulb that, by breaking when a preset temperature is reached, allows water to escape. Since the spillage is punctual, and not diffuse, sprinkler systems 'follow' the fire, allowing it to be controlled and not extinguished, thus facilitating the intervention of rescuers who can then extinguish the fire thanks, for example, to the hydrants present in the activity. For such installations, one or more motor-pump connections must be provided outside the protected activity, for the relative pressurisation by the means of the intervening external rescuers. Depending on the installation context, it is also possible to provide dry or wet systems, for the same reasons as outlined above for hydrant systems. Local regulations govern the technical and functional specifications.

Deluge' systems also use water to control a fire, but unlike sprinkler systems, the water supply is not localised, but diffused over areas predetermined in the design. In some cases they may be used to create a water curtain (Figure 11.27). Activation can be delegated to the fire detection system or to some closed sprinkler heads that act as pilot heads whose rupture commands the opening of the

Figure 11.26 Incorrectly positioned sprinkler head.

Figure 11.27 Water blade.

deluge valve upstream of the section. This type of system is generally used for the cooling and protection of tanks or process plants, and for the cooling of metal structures, which have poor fire resistance if unprotected.

As in the previous cases, water mist systems use water as an extinguishing substance, delivering it in the form of mist formed by droplets in the order of tens or hundreds of micro-metres. Such systems are made with either permanently open heads ('deluge' type) or closed heads ('sprinkler' type), and, thanks to the tiny size of the droplets, they control or extinguish the fire because they wet the fuel and evaporate, thus removing heat from the system and saturating the space around the flame with water vapour (mist), reducing the concentration of oxygen. A further advantage is that compared to traditional sprinkler or deluge systems, working with much lower flow rates, they require extremely low water reserves. Usually these systems are controlled by a specific control unit (Figure 11.28), installed at safe position and eventually remotely connected to control room and/or mobile devices.

There are also deluge or sprinkler systems that use mixtures of water and foam or only foam as extinguishing substances, automatic CO_2 or powder systems: these are systems whose regulatory framework is being progressively developed and that tend to saturate the environments where they are installed; due to these characteristics, they are suitable for use above all in all those rooms where the presence, if not occasional, of people is not foreseen.

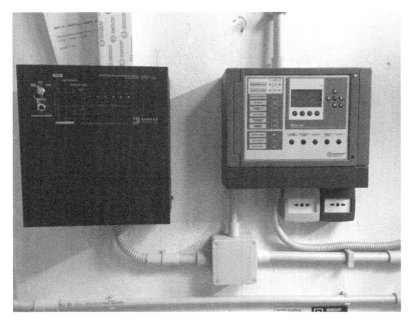

Figure 11.28   Controlling an automatic active protection system.

However, the undeniable advantages of using water as an extinguishing substance are often matched by extensive damage to goods and equipment in the environments served by such systems. Over the years, a great deal of research has been carried out to identify equally effective substances that can not only control but also extinguish a fire, while minimising damage to the equipment in the business and protecting the environment. This is the case with 'Clean Agents', a term used to refer to several gaseous extinguishers on the market today.

Lastly, oxygen reduction plants are particularly interesting. Installed in contexts of particular interest, such as museums, libraries or strategic archives and laboratories, they make it possible to maintain the concentration of oxygen in the environment at values compatible with human life, but at the same time contained to such an extent that any combustion is prevented.

From reading the above, it is easy to understand that the presence of an automatic fire control system – obviously with operation compatible with human presence, such as a sprinkler system – by delaying its development – allows an extension of the time available for the evacuation of ASET users with obvious benefits in terms of verification of the system of exit routes.

11.8   Fire Detection

Early detection of a fire is undoubtedly the best weapon of defence against its effects. Depending on the assessment of the risk – and in particular in consideration of the strategic nature of the activity with respect to the type of users or the protection of the asset – it can be delegated to manned detection or to a system dedicated to the purpose, periodically tested to verify its efficiency (Figure 11.29). It should be pointed out from the outset that these systems do not always make it possible to identify the actual extent of the scenario in progress, which often, within the scope of the procedures envisaged by the emergency plan, remains subordinate to the prior verification of

Figure 11.29 Testing a smoke sensor.

the rescue teams or, in any case, to the combination of one or more consents in order to exclude false alarms or breakdowns, which not infrequently occur for this type of system; we shall return to this subject, however, later on when we discuss its operating methods.

Generally, we speak of fire detection and fire alarm systems as being able to detect the presence of an incident and reveal its location; in relation to the precision with which the detection is carried out, these systems are distinguished into 'zone' systems if the information is referred to a more or less extensive area and into 'addressing' systems if the information is returned in a punctual manner and therefore the event is univocally determinable in relation to its location.

A typical system can be broken down into four sections that communicate with each other: the one that receives the inputs, which constitutes the detection part proper considering the relevant effects of the fire such as temperature and or smoke (Figure 11.30), the one that sends the alerts to the personnel in charge of their reception, the one that manages the alarm outputs to the users and finally the one that controls other systems for the activation of specific automatisms instrumental to the management of the emergency in progress.

The system consists of a series of sensors capable of detecting a fire in progress; the type of sensor varies according to the type of environment to be protected – in particular with respect to its geometric characteristics – and according to the type of fire that the designer expects to develop. The most widespread are smoke detectors, which consist of an analysis chamber capable of verifying the opacity of the air and thus highlighting the presence of a possible fire; they are indicated in contexts that do not present high heights and preponderant planimetric development in one direction, and in any case when the type of fire expected is characterised by the production of smoke, such as fires of solid materials.

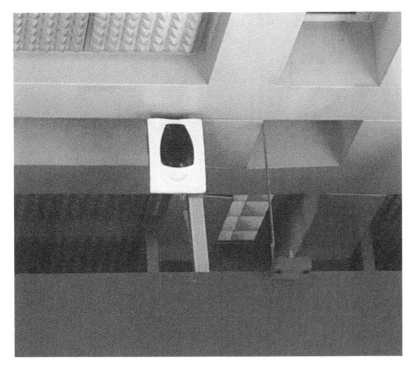

Figure 11.30 Linear sensor.

In environments characterised by elongated plan forms, it may instead be technically convenient to install linear detectors that differ from those indicated above in that they are able to detect the presence of a fire even tens of metres away from the point of installation due to the interruption of a laser beam by smoke. Alternatively, in the same environments, it is possible to use sampling detectors (Figure 11.31), consisting of an analysis unit and an aspiration tube that draws air from the environment for subsequent verification. The types of detectors just mentioned are often subject to false alarms due to dust generated, for example, by work in progress, or due to the interruption of the laser beam by the transit of a bird, alarms that have been drastically reduced over the years thanks to the many technological advances made.

In other contexts, with the presence of flammable materials, it is preferable to use different types of sensors, such as thermal, thermo-velocimetric and the more sophisticated flame detectors; the former monitor the temperature at the sensor and the latter the speed at which the temperature varies, while the latter are able to detect the presence of a flame in the monitored area. Depending on the environment in which they are to be used – as with linear smoke detectors as opposed to point detectors – the alternative is temperature-sensitive cables, which are suitable for detecting temperature trends in environments characterised by a predominantly one-direction layout or for monitoring concealed environments with particular geometric configurations. There are also types of combined sensors, such as smoke detectors and thermo-velocimetric sensors, which are particularly suitable for use in technical environments, generally located in remote and sparsely frequented locations, where the availability of several parameters can allow the monitoring operator an initial assessment of the extent of the signal received.

The last type of sensor is represented by traditional manual alarm push buttons (Figure 11.32), through the pressure of which people present in the activity can indicate the presence of a fire in

Figure 11.31 Analysis chamber of a sampling detector.

Figure 11.32 Manual push-button and optical-acoustic alarms.

progress. Depending on the risk assessment, the designer may deem it acceptable to have only alarm buttons or a specific combination of buttons and detectors.

All detector status signals pass through a control and command unit, which is the real 'head' of the system and which cyclically interrogates the sensors to check their status. Alarm monitoring can also be carried out through dedicated stations equipped with software capable of returning the precise position, on a graphic map, of all the detectors and their status; from the same it is also generally possible to disable the operation of one or more sensors and test the operation of any devices in the system or connected to it.

In addition to the personnel in charge of monitoring the status of the system, the presence of a fire is communicated, according to the times and modes defined by the designer, to the users present in the activity; the alarm modes may vary according to the installation context. It is necessary to associate an audible signal with a luminous signal, marked by an explicit indication of the type of alarm (such as 'fire alarm'); the presence of the visual component also allows people with hearing disorders to be alerted.

In this regard, some reflection is in order. The clearer the signal, the less time it will take users to understand what is happening and therefore begin to leave the premises; also considering that the effectiveness of the system in identifying the fire in progress contributes to reducing the time it takes to alert people, it is easy to deduce that fire detection systems play a fundamental role in reducing the RSET, significantly influencing all the contributions except for the time of movement. It is clear, therefore, that in particular contexts characterised by users unfamiliar with the environment and complex geometries, such as underground and railway stations or hospitals, special care must be taken in the design of the system, favouring in particular clear and direct acoustic signals, not simple sirens, through the use of vocal alarm systems, known as *EVAC systems* (Figure 11.33). Quite often these systems are coupled with other control measures (Figure 11.34).

These systems are characterised by a series of loudspeakers capable of broadcasting pre-recorded messages or messages from a microphone station, for the use of the internal rescue team or the intervening external rescuers. It is understood that if acoustic signalling is provided by an EVAC system, it is still necessary to set up components for visual signalling, obviously coordinating their

Figure 11.33 Microphone station EVAC system of a control room.

Figure 11.34 Magnet for automatic release of an EI door.

operation. It is not infrequent, however, that we see the presence of optical-acoustic signs, which are readily available on the market, and loudspeakers of the EVAC system; the two systems, in fact, although distinct, are closely interconnected and contribute to a single purpose. In these cases, therefore, in order to exclude a reduction in the intelligibility of the audio messages of the EVAC system, it will be necessary to verify the possibility of disabling the acoustic function of the number plates, where technically compatible with the product specifications, or coordinating the operation of the two types of devices, foreseeing, for example, an alternating function and maintaining the visual component always active.

The last section of the system is the one capable of governing the automatisms through the command of the activation of the active protection systems or the preventive shutdown of those normally active, depending on the scenario in progress; this is the case, for example, of the release of the EI doors kept in the open position by means of magnets or the start-up of the smoke extraction fans or non-automatic extinguishing systems. This is made possible thanks to 'output' modules inserted on the loops of the fire detection system to which specific activation rules are associated; the logic programmed in the fire detection system control unit will correspond, for example, to the activation of module X as a consequence of pressing a button or the activation of two or more sensors.

Unfortunately, the logics to be implemented in the control units of fire detection systems are in many cases developed independently by the installer of the system itself, who has little knowledge of the fire prevention design, the active protection measures, the specifications considered in the simulated scenarios for projects developed with a performance approach, the performance to be achieved by the systems and, above all, the management implications.

On the other hand, the logics must always be developed in agreement with the fire prevention designer, if not directly by him or her, since he or she is the only one who is fully familiar with the project strategy and all its application declinations. They must also be agreed in advance with the client for the necessary attention with respect to the management implications; a typical example is the definition of the alarm areas, i.e. the areas of an activity that must be evacuated at the same time in the event of a fire, particularly in sites of considerable extension in plan or in cases where, for example, the exodus has been designed according to a 'phased' approach or must be coordinated with the exodus of adjacent sites.

The same considerations apply to the management of the consents required to start smoke extraction, whose systems are often designed to be activated only after the sprinkler system has been triggered; or even in the case of a programmed delay between the first detection and the evacuation alarm to allow for the prior verification of the scenario in progress and to avoid a false alarm, a delay that is certainly a function of the complexity of the site and the readiness and composition of the first intervention team.

The various local regulations define the specific aspects of the system, such as the distribution and type of detectors to be adopted, the services to be guaranteed for alerting users and the areas to be monitored. With respect to the latter, all the areas of an activity are generally to be monitored, with the exclusion of those in which the fire hazard is practically negligible, such as external or internal stairwells that do not have a fire load and in which there are only systems serving the staircase itself (e.g. the lighting system); this means that in such areas it is possible to exclude detectors, but not that it is possible to avoid installing alarm devices, since any users present there must also be informed of a fire in progress. However, full compliance with the sector regulations cannot guarantee the goodness of the final solution; the system designer in any case must necessarily consider the fire prevention project in all its aspects in order to exclude, for example, misalignments between compartments and alarm areas.

Like all systems, the efficiency of fire detection over time is the result of correct management, both from the maintenance point of view and in terms of managing the changes that may occur over time with respect to the installation environments; for example, a change in layout due to the separation of a large room into two or more rooms must necessarily be matched by a prior verification of the compatibility of the distribution of the detectors, providing for modifications and additions to the existing system if necessary.

A final consideration is related to the reliability of such installations; if all components and connections must be certified in accordance with local regulations, it must be verified that the entire process, from detection to alarm and control of automatisms, is certified. This implies that, for example, the supervision stations and the way in which the control units are connected to them must also be properly certified.

In some contexts, although supervision software or devices for relaying detector statuses through telephone lines may exist on the market, their use would not be fully compliant with the above-mentioned requirement as they do not present the same performance as the 'physical' connection lines; in such cases, it is absolutely necessary to investigate the overall articulation of the system in order to guarantee its technical alignment with the reference standard.

A case in point are the wireless devices recently introduced on the market and adequately certified in terms of reliability; their use can be fundamental, especially in the case of retrofitting existing buildings characterised by particular architectural value, where it is often not possible to realise all the cable routes necessary for the capillary distribution of the devices.

11.9   Smoke Control

The evacuation of combustion products represents a strategic node in firefighting design, which unfortunately in many cases still receives little attention. The possibility of disposing of the smoke and heat of the fire slows down its development and mitigates its effects, favouring in particular the evacuation of users – improving living conditions increases the time available for ASET evacuation – and the intervention of rescuers, as well as contributing to the containment of damage to the building.

The design solution that the professional intends to apply derives from the results of the risk assessment or from any regulatory prescriptions, with performance gradually increasing as the risk as identified above increases. While, in fact, for environments frequented only occasionally, in the event of a fire, it may be sufficient to provide minimum conditions to guarantee the intervention of rescuers, for other more complex environments, such as an airport, it may be necessary to ensure the persistence of a smoke-free layer throughout the development of the fire in order to facilitate the evacuation of users and the intervention of rescuers, as well as clearly containing, as far as possible, the damage to an infrastructure of a strategic nature.

Once the objective has been defined, it can be pursued through the availability of surfaces that allow for the 'natural' evacuation of combustion products (Figure 11.35) or through a system that guarantees their extraction or at least their 'forced' movement (Figure 11.36). All solutions are equivalent with respect to the objective while presenting technical specifications that make them extremely different from each other.

As the name suggests, natural evacuation exploits the lower density of fumes that tend to stratify upwards; the presence of openings at the top of the rooms allows them to escape outside the activity. A direct consequence is the longer time it takes for the fumes to stratify at such a height that they are incompatible with the permanence of the users or with the intervention of rescuers, thus disposing of part of the heat produced and delaying the achievement of flashover. It is good to keep in mind three essential aspects in the design of surfaces, which are generally defined by local regulations, namely extension, distribution and availability. Although we will not go into detail, it

Figure 11.35   Natural smoke evacuator connected to the fire detection system.

Figure 11.36 Multi-compartment smoke extraction duct.

is clear that larger surfaces correspond to greater disposal capacity; at the same time, it is necessary to provide for the presence of make-up surfaces at the base of the compartment for the entry of clean air and to allow for the formation of an overpressure zone, below the neutral plane, to facilitate the expulsion of fumes upwards.

The distribution of the openings, both vertical and horizontal, must be as uniform as possible and distributed over the entire surface of the compartment, considering a radius of influence for each opening that could vary depending on local requirements and in any case in the order of 20 m. Finally, openings may be available at all times or fitted with fixtures, the latter of which can be opened manually from a position indicated and reserved for rescue teams or, better still, connected to the fire detection system. For the opening, it is desirable that both evacuation and reintegration surfaces are available at the same time to avoid turbulent smoke movements that hinder the escape of persons in particular. It is understood that if the surfaces are fitted with frames or grating elements, it must be ensured that the minimum surface area is actually available and that the operation of the closing elements is also ensured with respect to weather conditions, such as wind or snow.

The alternative to natural evacuation is represented by the forced evacuation of smoke and heat thanks to a system dedicated to this purpose and which can be schematised with a network of ducts connected to an extraction fan, all realised with fire-resistant elements in accordance with the highest resistance class of the compartments served. These types of systems can in fact be designed to serve a single compartment (single-compartment systems) or several compartments

Figure 11.37 Mechanical smoke extraction plant fans.

(multiple compartments); in the second case, it will be necessary to provide ducts that are suitably certified for this purpose, equipped with smoke dampers and possible smoke extraction grilles that, appropriately controlled by the fire detection system, will be able to concentrate the extraction flow exactly in the compartment affected by the fire.

The extraction fan(s) may also be equipped with inverters to adjust the extraction flow rate depending on the location of the fire – in some cases, critical issues may arise with respect to the tightness of the ducts due to the significant depressions generated within them. As with natural extraction, forced extraction also requires the reintegration of the extracted air flow rate by means of sufficient surfaces on the lower part of the compartment or a suitably designed system (Figure 11.37), also made of fire-resistant ducts in order to guarantee its operation in the event of an emergency.

The last method is represented by the mechanical movement of smoke; thanks to suitably sized fans, it is possible to move smoke, mitigating, for example, critical situations at the access points of rescue teams. This is the case, for example, in road tunnels where, appropriately controlled by a fire detection system, the fans are activated or not in accordance with the planned schedule to facilitate the intervention of rescue teams.

It is understood that the systems, in their entirety, must also be powered by a security energy source and must be connected to the fire detection system for all necessary automation.

11.10 Firefighting and Rescue Operations

The design of a safe structure cannot disregard the way in which external rescuers intervene; the safer the structure, the more the designer has taken measures to facilitate their intervention (Figure 11.38).

The main aspects on which attention should be focused concern

- the access of rescuers to the activity and the area on which it stands;
- the modalities of intervention.

With respect to the first point, the location of the activity and the precise position in the context of the settlement play a fundamental role; for these aspects too, it is the risk assessment that guides the planner in the choice of the most suitable solutions for the case under consideration. For example, for an office with a few dozen employees it may not be necessary to verify the criteria for accessibility of emergency vehicles to the area, a requirement that for obvious reasons is instead essential for activities with a higher risk, such as offices with hundreds of presences, shopping centres or theatres. Although it may seem a trivial concept – since 'ordinary' accessibility to an activity is a requirement that derives from quite different needs – in reality it is not always the case that the verification is fulfilled due to the stringent verification criteria, geometry and load, prescribed by local regulations. As a general rule, it is necessary to ensure that rescue vehicles can get as close as possible to the building, installing specific and explicit signs to indicate to rescuers the perimeter of possible collapse of the building if it has been designed exclusively to guarantee the evacuation of the occupants.

Figure 11.38 Emergency vehicle approach.

It is also necessary to verify the possibility that a rescue vehicle, and more precisely a ladder truck, can approach the building allowing the rescuers access to every floor of the activity; the verification, essentially due to a simple geometric problem, is linked to the characteristics of the vehicles used by the rescue teams. It is understood that for high buildings, where the vertical development of the ladder truck would in any case not allow all floors to be reached, the designer may resort to the construction of at least one rescue lift, i.e. a lift that is adequately designed and built (for example, equipped with safety power supply, built in a smoke-proof compartment with assigned fire resistance …) in order to allow rescuers to quickly reach the floor affected by the fire, while at the same time transporting the equipment necessary for the intervention more easily.

With regard to the modalities of intervention, the planner must pay particular attention to the availability of extinguishing substances and the communication methods of the members of the rescue teams.

In the event of the unavailability of a hydrant network, both external and internal, it will be possible, respectively, to verify the presence of a public hydrant in the vicinity of the activity and to envisage the realisation of a 'dry' column, i.e. an empty pipe, which reaches all the floors of the building, and which in the event of an intervention can be used by the fire brigade much more quickly and easily than the laying of several hoses.

The need to evaluate the communication methods of rescuers during the intervention phase also appears to be fundamental, especially in those contexts in which they may be compromised or significantly limited due to the geometry and location of the intervention areas (e.g. underground technical tunnels of a hospital complex or a metro station); in such environments, the reliability of communications plays a key role not only in firefighting operations but also for the safety of the rescuers themselves.

11.11   Technological Systems

The technological and plant engineering component, both in civil construction and production sites, has progressively grown over the years to the point of absorbing a significant technical and construction effort as well as a conspicuous part of the costs of realisation of the works; suffice it to think, for example, of all the measures that facilitate the liveability and sustainability of environments, such as escalators, lifts, air conditioning systems and photovoltaic panels, which increasingly characterise new and old constructions when they are subject to renovation. Or the traditional energetic systems, such as gas distribution systems supporting buildings utilities (Figure 11.39). A direct consequence of this is the need to take these components into account when defining the fire-safety measures of the activity in order to limit ignition phenomena or in any case contrast the propagation of fires or fumes within the various rooms of the same. The systems must therefore be designed, built, operated and maintained in accordance with the provisions of local regulations and the indications of the sector's good technical standards.

Figure 11.39   Isolation devices of a thermal power plant.

It is worth clarifying two aspects: the first concerns the performance to be targeted, which, inevitably, can only be the same as that defined in the firefighting strategy, irrespective of the risk assessment of the components introduced. A system must be 'safe' and certainly cannot be so at different levels. The second concerns the specificity of certain technological installations for which, precisely because of their relative specificity, safety measures cannot be identified a priori but must necessarily be the subject of careful identification and design by the designer; we are thinking, for example, of stop systems for assembly lines or 'unique' machines intended for the production of 'typical' products.

Let us first put the magnifying glass on the fire objectives to be achieved and then try to dwell on some of the main aspects of the most common installations.

Technological and service installations:

- They must limit the likelihood of fire and explosion and of being ignited by a fire due to causes external to them.
- They must counteract the spread of fire and smoke.
- They must not constitute an obstacle to the full functionality of safety installations or a danger to the evacuation of occupants or the intervention of rescue teams.
- They must be deactivated or manageable at any time, should the need arise.

It is understood that the possibility of steering the facilities must be made explicit in the activity's emergency plan (Figure 11.40) and made possible by protected, well-marked and easily accessible positions for both the activity's internal rescue teams and external rescuers.

In order to guarantee the above-mentioned objectives, the electrical systems must be designed in accordance with the rule of art, with cables whose reaction to fire must be consistent with the characteristics of the occupants of the activity, favouring the use of cables with higher performance in environments susceptible to large crowds and open to the public. Release buttons, effectively signalled and distributed in protected positions, must guarantee the certain disconnection of all circuits passing through a given environment – the reference circuits must be specified on the buttons – thus avoiding that circuits belonging to different disconnection buttons pass through the same spaces,

Figure 11.40   Electric release buttons.

precisely to guarantee the certainty of the absence of live lines in the event of intervention with water extinguishing systems.

In principle, and in any case in compliance with the minimum widths to be guaranteed, it is possible to provide for the installation of switchboards along escape routes, provided that there is at least the possibility, especially in areas open to the public, of preventing their operation by means of locked doors. Installations intended for emergency operation must have a safety power supply, which can be switched off if necessary. This also includes photovoltaic systems – which should be installed on fire-resistant or in any case non-combustible roofs – and electric vehicle recharging stations; the design, construction, operation and maintenance of which must be in line with local regulations or in any case with the manufacturer's instructions.

Lifts, goods lifts, escalators and treadmills, if not adequately designed for the purpose in accordance with the firefighting strategy, must be deactivated in the event of an emergency and the holder must take all measures, including management measures such as information signs (Figure 11.41), to prevent their use in the event of an emergency.

Storage and distribution facilities for combustible substances (e.g. methane, medical gases, fuels, etc.) should, as far as possible, be located outside the activities, adopting specific measures to

- limit the dispersion of these substances;
- exclude the possibility of accidental impacts and triggers;
- ensure their safety by means of shut-off devices (Figures 11.42 and 11.43) and stopping any feed pumps.

Lastly, air conditioning systems deserve specific treatment, not only because local regulations may dictate specific requirements regarding the toxicity and flammability of the fluids used, but above all because these systems may constitute one of the main paths of smoke propagation; it is no coincidence that these systems are generally closely interconnected with the fire detection and alarm system, both for the detection of events and for the compartmentalisation of environments. In ducts, and in particular near air treatment units, specific detectors must be installed, in addition

Figure 11.41 Sign prohibiting lift use in the event of fire.

Figure 11.42 Channel detector.

Figure 11.43 Fire damper.

to the 'room' detectors, dedicated to analysing the treated air in order to exclude the presence of smoke; in the event of a fire, these detectors cannot indicate the exact location of the fire – and, therefore, cannot ensure the correct activation of safety measures – but they can certainly provide an initial indication of the presence of the event in progress and ensure the immediate activation of the first countermeasures, such as turning off the fans and closing the fire dampers placed along the ducts in correspondence with the penetrations of fire-resistant elements such as floors and walls, while waiting for the 'room' detectors to identify the exact position of the fire and ensure the activation of the automatisms in accordance with the operating logics designed based on the firefighting strategy identified by the designer.

12

Fire-Safety Management and Performance

12.1 Preliminary Remarks

In this section, we deal with the management of fire safety, framing its prerequisites and application, at all stages of development and operation of the asset under consideration, in order for it to be considered 'adequate' with respect to the purposes pursued over 'time'.

As a preliminary framework, we consider it useful to briefly dwell on the correlation between the concepts of management and performance on the one hand and time on the other hand.

- In the first place, we observe that management is the voluntary process that presides over the state of a given reality by directing its change over time; indeed, it belongs to common experience that any change in a state requires the passage of a given period of time and that the magnitude of the change is also commensurate with the time required for it to take place.
- Secondly, we note how performance is embodied in the service levels of our given reality, the indices of which measure its variation over time, taking into account the design and construction characteristics; in other words, we could say that if management orients the state of the work, the performance indices measure its magnitude and variation over time.

As indicated, both concepts imply the temporal dimension, as if time were the domain of a function, in which the management variable is determined through the deviation of service indices from expected performance values.

Beware however, looking at the process from the outside – at least for the uninitiated – one might be falsely misled into believing that, after a given time, if no change in the performance indices occurred, there was no need to proceed with management actions.

Quite the contrary in fact, the maintenance over time of the performance does not at all exclude the need for management since it is in any case necessary even if only to maintain unchanged the performance of a given work whose change may be caused by factors extraneous and independent of the operator's will.

Let us banally think, for example, of the variation of a car's tyre pressure from the winter to the summer season and the relative impact on performance in terms of speed and driving stability. It is quite evident that even the mere maintenance of the initial tyre pressure condition required a management activity, the extent of which would depend on the time of exposure to the changed environmental conditions and the characteristics of the structure as determined during the design/construction phase.

Fire Risk Management: Principles and Strategies for Buildings and Industrial Assets, First Edition.
Luca Fiorentini and Fabio Dattilo.
© 2023 John Wiley & Sons, Inc. Published 2023 by John Wiley & Sons, Inc.
Companion Website: www.wiley.com/go/Fiorentini/FireRiskManagement

This last reference to the design and construction characteristics of the artefact allows us to complete our preliminary remarks by stating that if the passage of time determines the need for management, its nature and entity are functional to the duration of exposure to risk and to the technical construction characteristics of the artefact and therefore, in the final instance, to the risk assessment initially carried out at the design stage.

To return to our car, therefore, if the risk of gear instability due to variations in ambient temperature has been taken into account in the design, the designer will have formulated specific operating prescriptions for maintenance work or will have adopted technical measures to exclude and mitigate the risk in order to maintain the safety performance of the vehicle.

It is a fact that this assessment, together with time, determines the characteristics of the management process, accentuating it in the first case (by prescribing tyre intervention) and minimising it in the second (by mitigating it upstream through technological solutions).

If what has been set out above fits almost intuitively with the process of operational safety management – in other words, when the work is in operation – the reference to risk assessment in the design phase allows us to understand the need for safety management to preside over, evidently in more organisational and less operational terms, the conception, decision-making and design processes of the work.

In this phase, at first glance, it would seem that one cannot really speak of fire-safety management and performance, which are normally associated with the operational phase of the work. In this first phase, fire-safety management and the corresponding level of performance are themselves a part of the technical and conceptual investigation that leads to the genesis of the work.

Indeed, the work does not exist in empirical reality but only in the dialectic of the organisation in which it progressively takes shape; but the rivulets and articulation of the organisation in which it is generated are no less 'dangerous' than material reality.

Progressively, in fact, the design of the work will be transferred from the technical tables to the administrative procedures to which its implementation is subject, and thus to the planning of investments and the articulation of contracts for its realisation.

As a result of the aforementioned planning, the work will enter into the commercial production plans of the relevant structures based on a series of time assumptions that should align with the planned construction and commissioning schedule.

But it is clear that the activities summarised above together constitute an articulated and complex process with serial and parallel phases and with the interaction of multiple roles inside and outside the organisation, functionally bearers of particular and not necessarily converging interests. The complexity of the process is then obviously destined to be further accentuated in proportion to the technical articulation of the planned work, the context of insertion and the technical regulatory framework of reference.

In such a scenario, the danger of incomplete, inexact or inadequate implementation of the process is absolutely real with consequent risks with respect to the degree of fire safety of the work and therefore, as a derivative, of its management and the level of performance during operation.

In light of the above considerations, it is understood that:

- The management and performance of fire safety must necessarily be framed within all the phases of a project, starting from the conceptual and technical development of the work, through its realisation, to its commissioning and operation, and on to the modifications that inevitably characterise it;
- It is therefore necessary to identify, in relation to all the above-mentioned phases, the types of change – voluntary and incidental – that may affect the work, the regulatory context in which it

is designed and exercised, and the organisational articulation of the owner that together condition the definition of the management model, its implementation and the analysis of performance.

12.2 Safety Management in the Design Phase

As mentioned earlier, in the design phase, safety management takes on markedly organisational and procedural connotations in order to ensure the completeness and correctness of all activities in the design and decision-making process of the work.

In this regard, it has already been mentioned how, in this phase, safety management also takes the form of fire prevention planning, highlighting the complexity and breadth of the 'spectrum' of risk assessment analysis in this phase; it must now be emphasised how the appointed professional plays an essential role in the safety management of this phase, for the proper implementation of which close coordination with the overall organisation of the operator is necessary.

To come to the point, it must in fact be made immediately clear that security management will be concerned not so much with change hypotheses (as is the case in the operational phase) but rather with possible omissions or incorrect/inadequate assumptions underlying the assessment of risk profiles and the corresponding mitigation measures as a result of flaws and distortions in the organisation's information and decision-making process.

Let us therefore try to identify some of the general cases that may undermine the adequacy of risk assessment.

a) *Absence or erroneous regulatory assessment*: Consider the hypothesis in which an activity is erroneously considered not subject to an authorisation procedure, or becomes so on the basis of new legislation, without the corresponding administrative procedure being carried out; consider the hypothesis in which a technical rule is applied that is not in line with the activity then actually affected in the building. It is evident that in the hypotheses considered, the safety management must guarantee the regulatory supervision of the design phase and promote the acquisition of all the inputs necessary for the correct analysis of the scenario, promoting the contribution of all the corporate skills involved; suffice it to think in this regard how normally the functional destination of the spaces of a tertiary building is negotiated with the market by the commercial structures, with timeframes that normally go beyond the closing of the project.

b) *Omission or incorrect assessment of the context and interferences*: Consider the hypothesis in which the firefighting water system of the work is designed by deriving its extension from the accumulation and pressurisation units of another work of the same organisation without considering that the pre-existing activity envisaged the operation of the dry system; again, consider the hypothesis in which the design envisages the use of escape routes common to other works without considering and regulating the promiscuity of the escape. In the hypotheses considered, safety management must guarantee the interfacing of all the competences inside and outside the organisation to allow the acquisition of all the information necessary for the preliminary investigation and its final approval; suffice it to think of the functional interaction between different activities that takes place in large transport infrastructure nodes and the need for the regulation of these interferences to be crystallised in specific processes, formally shared, so that the relative discipline can be borrowed in the design forecasts of the work being developed.

c) *Erroneous formulation of fire scenarios*: Consider the hypothesis in which the design is carried out using the performance methodology, and the incomplete assumption of data in the technical summary leads to the formulation of scenarios that are not entirely realistic; also in this case – especially considering the constraint of maintaining the assumptions of the technical summary during operation – fire-safety management must preside over the construction of the scenarios, actively supporting the designer in charge for the correct formulation of the performance to be guaranteed during operation. As more extensively exposed with regard to the application of the performance methodology, the flexibility offered by the application of the analytical method with respect to prescriptive rigidity tends to favour its use; however, one cannot underestimate the constraint that this methodology determines on management, and in this sense safety management must ensure the consistency of the technical summary with the operational reality in which the work will be exercised.

d) *Incomplete or incorrect formulation of project data*: Consider the hypothesis of infrastructures with the presence of an indistinct public and uncontrolled access; in this case, security management must promote the adoption of solutions aimed at guaranteeing the assumption of evidently realistic and precautionary data, for the benefit of security, given the condition of indeterminateness associated with the activity.

e) *Incomplete management of variant processes*: Think of the hypothesis in which the work is designed with methodologies to which the regulations associate management implications (such as the fire-safety management system in application of the performance methodology or in the case of alternative solutions) without these being adequately developed and evaluated in the corresponding technical–economic implications; should any of the above cases emerge during the commissioning phase, in addition to the very practicability of the work, the adequacy of the safety management model would be called into question, if for no other reason than the number and degree of the prescriptions considered, or worse, the model's lack of a systemic approach.

It is not at all inconceivable that in the context of business dialectics, the design of a building for tertiary use may see the proposition of commercial demands aimed at maximising its utilisation by increasing surface areas or crowding levels, with the consequent need for technical and administrative review of the project. Often, moreover, such requests emerge as a result of discussions between the organisation's business structures and external stakeholders that take place after the project has been closed and the administrative approval procedures have been activated. Just as often, the weight of these instances actually asserts itself over the technical instances, resulting in a technical elaboration burden on the project that is claimed in the same timeframe as initially envisaged without the variants.

The variants themselves therefore entail the need to reconsider the prospects for operating the work in overall terms, those purely operational and those concerning safety management. It is a fact that the complex dialectic between the technical design instances, the managerial ones and the commercial ones, and the difficult balancing act between the internal tensions of the different competences involved constitute an organisational risk factor immanent in the design phase.

f) *Administrative misalignment*: Consider the hypothesis in which the activity is subject to administrative authorisation and the adoption of variants in the design phase takes place after the administrative procedure has been activated without the same being consequently updated; this is a particular declination of the previous risk of '*incomplete management of variant processes*' as a further representation of the investigative complexity of the design process.

The cases depicted above can certainly not be considered exhaustive of the many facets that the project investigation of a work can take on in the articulation of an organisation's technical and decision-making processes. Nevertheless, the cases cited offer sufficient representation of the 'organisational risks' that must be controlled in terms of fire-safety management.

It is also no coincidence that up to now fire-safety management has been referred to in deliberately impersonal terms, preaching the activity but not identifying the subject.

The rationale lies in the fact that in the face of such extensive organisational risks in the company articulation, the idea of attributing responsibility for managing safety and maintaining performance to a single figure appears unrealistic, at best, if not downright evasive of the risk in actual fact. In other words, in order to effectively preside over fire-safety management in the scenarios described, a single figure would have to perform an apex function hierarchically superordinate with respect to all other internal players and be the direct guarantor of all external stakeholders: but on closer inspection, such a role cannot be found in modern complex organisations where governance models are developed in line with the need for distribution and segregation of responsibilities and balancing of powers through collegial bodies; such a role can perhaps only be found in individual companies and provided that their size remains sufficiently contained to allow it.

The direct consequence of the above is that the only way to effectively oversee fire-safety management in the face of the organisational risks in question is through the adoption of a corporate system that explicitly distributes safety responsibilities and integrates them into the operations of all roles and processes so that they are an integral part of the company's business and not a mere fulfilment of the duty of the safety manager on duty. Consider in this regard how these instances are already explicitly provided for in the ISO 45001 standard.

Of course, the adoption of such a management model is directly dependent on the safety culture that characterises the organisation and, for all intents and purposes, reflects a distinctly high level of sensitivity; but we will return to this subject later in the discussion, following the analysis of the fire-safety management of the operational phase.

Finally, in conclusion to this section, we would like to point out that inadequate safety supervision in the design phase is not only related to organisational risks but also directly reflected in technical risks.

The design of the work is based on the analysis and assessment of the fire risk that informs the fire-safety strategy and the adoption of active and passive protection measures. The incomplete and/or inadequate development of the project, according to the risk profiles listed above, will condition the subsequent operational phase of the work and with it the management of fire safety and the expected performance. The operator will have to assume, borrowing from the previous phase, the design and risk assessment and update it with respect to the actual operating conditions: but if the incomplete profiles in question emerge, what will the operator have to do? Even worse, if these profiles have prejudiced the expected performance, which is lower than that actually required, what must the operator do in order to maintain operations?

There is no doubt that the operator will have to promote the correction of the analysis and assessment, will have to intervene to correct the design and implement updated forecasts, while considering the adoption of mitigation solutions that allow the operation of the activity through the progressive realignment of the expected performance.

But what must be emphasised here is the continuity of the process between design and operation, and consequently the need for fire-safety management to preside over all these phases with the same continuity. Splitting the process into different areas of activity is certainly necessary (think, for example, of the different engineering competences during the design phase); it certainly helps to make it intelligible to the actors involved, to promote their conscious participation and to measure their progress. However, the risk of 'fragmentation' should not be underestimated and must necessarily be presided over in terms of safety management to ensure coordination with respect to the overall goals.

12.3   Safety Management in the Implementation and Commissioning Phase

The considerations set out in the previous section are also largely pertinent to the construction phase although they manifest themselves with specific profiles that we will briefly try to identify.

a) *Variations promoted by the works management*: Think of changes in the laying conditions of a certain material or the variation of certain finishes (floors, walls, etc.) during execution. Although they may appear to be minor occurrences in the economy of the works, nevertheless, as better analysed in the risk assessment, they can significantly alter the risk assessment scenario and change the effectiveness of the associated protective measures.

b) *Construction interfering with activities already in operation*: Think of the hypothesis of the development of an existing work such as an increase in the surface area or the modernisation and replacement of installations. These are absolutely frequent cases where the autonomy of the works management in implementing the new project risks interfering heavily with the work being developed and modernised but which nevertheless remains in operation and whose established fire-safety performance must be safeguarded. In such a scenario, fire-safety management will have to guarantee the verification of compatibility of the individual worksite phases with respect to the activity in operation, paying particular attention to profiles such as the maintenance of escape routes, the compartmentalisation of work/worksite areas with respect to adjacent ones and the maintenance of the residual functionality of the systems serving the areas in operation. It certainly seems desirable that such a scenario is developed with the analysis of interferences already in the design phase; but it must also be realistically considered that the interventions in question often take place in a diffuse manner (think of the replacement of smoke detectors of an IRAI system) and are obviously not compatible with closed construction sites, whose most efficient realisation passes through a flexible comparison between the execution of works and the manager of the spaces in terms of safety management for the maintenance of performance.

c) *Partial release and gradual commissioning of the work*: These are not infrequent hypotheses where, for works of considerable size and long construction times, a phased release and gradual commissioning are envisaged. Also in this case, if said forecasts have been carefully assessed in the design phase, provided for in the tender specifications and aligned in the commercial activation plans, the operator should see the delivery of areas that are functionally and plant-engineering-wise perfectly in line with the project requirements and therefore with the expected safety performance, in which to proceed with the settlement of any sub-activities and authorise the relative operation.

The brief description above actually shows us the complexity of the process, already net of the risks discussed for the design phase; it goes without saying that in such a scenario the safety management will have to proceed to the analysis of the individual release phases and verify the actual safety performance on the basis of the actual consistencies made available, not only considering the completeness of the individual system but also its operational functionality; think for instance of a fire detection system that, although perfectly realised, was nevertheless not connected to the supervision system as it was planned in the subsequent release phase.

To complete this section, let us briefly mention the *commissioning* phase.

By this term, and within the limits of what is of interest here, we mean the process that, in a manner innervated in the construction and commissioning phases, is aimed at promoting, verifying and acquiring the performance of a work, in all its components and with the involvement of all the actors involved.

Without further ado, it seems rather evident that this process – rather than a separate phase in the genesis of the work – constitutes a methodology that incorporates the instances of involvement and balancing of the organisation's roles and organicity and integration in the approach to fire-safety management referred to above.

Moreover, it is no coincidence that *commissioning* is more widespread in Anglo-Saxon countries, which are more inclined to manage liability in terms of risk assessment than countries with a prescriptive regulatory approach where adherence to the norm is often conceived as mere compliance in the interpretive spaces of the precept in force.

12.4 Safety Management in the Operation Phase

In short, the above considerations have provided us with elements to reflect on the continuity of the macro-process from design to commissioning, on the organisational safety risks that affect these phases and on the need for safety management, in the face of the complexity of the scenario, to be developed through a systemic approach of explicit and conscious distribution of responsibilities with a view to guaranteeing performance in a shared and participatory manner.

Let us now turn to the phase most traditionally associated with safety management, that of the commissioning of the work, on the assumption – which we reiterate – of the continuity and dependence of this phase with respect to the risk assessment carried out at the design stage.

Unlike in previous phases, where risks to performance are a consequence of incomplete or inadequate management of the process, in this phase, with the work coming into material existence, the issue of managing safety and maintaining performance as a consequence of changes emerges.

Let us therefore try to classify the main cases.

a) Distinct process changes in
   1) *Organisational change*: Change in the number of staff to oversee processes, the articulation of company structures, and the associated tasks and responsibilities (all the more so with reference to the roles and responsibilities defined in the SGSA).
   2) *Procedural change*: Change affecting only the organisation of business processes.
b) *Operational Change*: Change that affects the organisation and procedures for the operational operation of the work (e.g. tertiary building);
c) *Technical modification*: A modification, plant, technological and/or infrastructural, of the asset or part of it, also in terms of its implementation, that changes the configuration of the asset in such a way that it can no longer be considered consistent with the original design stage or in any case with the existing condition (revamping and upgrading are included in technical modifications).
d) *Technological change*: Any change in the operating conditions resulting from the technological upgrade of the asset or part of it, whether or not associated with technical changes, which varies the pre-existing design conditions (generally associated with plant modifications).

Depending on their duration, the above-mentioned interventions may constitute

e) *Permanent changes*: These include changes that result in a permanent change in the plant configurations or conditions of use of the assets or part of them.
f) *Temporary changes*: These include changes that are made for a limited period of time, after which the asset returns to its original condition.

The scenario envisaged opens up to very different cases that may have extremely diverse impacts on fire-safety performance in terms of technical intensity or duration over time; suffice it to say that the case of asset modifications could in itself already configure a project development phase with all the risks exposed for the project design and development phase (in fact grafting a process into the process).

In any case, safety management cannot disregard the identification of associated hazards and their evaluation as part of the initial risk assessment. The operator must proceed with its actualisation based on the identified hazards and then measure its impact on the required safety performance (as designed).

In this regard, in a very general way with respect to risk, the changes can be categorised as follows:

- Fire-relevant modifications, which are in turn divided, also for the purposes of administrative regulation, into
  - changes that aggravate pre-existing risk conditions;
  - modifications that do not lead to an aggravation of pre-existing risk conditions.
- Non-substantial modifications.

The relevant character of the change is qualitatively attributable to changes concerning

- the presence of hazardous substances;
- the change in fire resistance class;
- modifications to installations;
- functional modifications;
- changes in protective measures for persons.

It is hardly appropriate to return to the subject of the continuity of the operational phase with respect to the development phase of the work in order to better define the terms of the process of identifying hazards and updating the risk assessment based on the change; it is no coincidence that the classification reported introduces the comparison with pre-existing risk conditions.

If the safety performance expected and necessary for the operation of the work is the result of the initial risk assessment, the assessment of the hazards introduced by the modification can only be carried out with the same methodology in order to make the path taken consistent and intelligible.

This methodological consistency is necessary in order to be able to correctly compare the variation of the initial safety performance to that resulting from the risks introduced with the change: if this were not the case, it would necessarily be flawed and no judgement of the 'acceptability' of the performance variation could be correctly formulated.

Or rather, in that case there would be a marked risk of misalignment, if not outright arbitrariness, of the risk assessment and its consequences on the fire-safety level of the work.

For the sake of completeness, it is just the case to introduce a few more profiles that characterise fire-safety management in the operational phase. Here too, after all, safety management cannot be relegated to the exclusive and vain domain of a 'stand-alone' safety manager.

Managing the fire safety of a modification to a work in operation involves additional activities, which are briefly listed below:

- the regulatory oversight for the correct framing of changes and the identification of changes required as a result of new legislation;
- the assumption of changes in the processes of operation and maintenance of the work;
- information and training activities with respect to personnel directly involved in the implementation and exercise of the amendment;
- the supervision of administrative processes possibly associated with the changes;
- reviewing and updating the emergency management process;
- the review of critical operational parameters.

13

Learning from Real Fires (Forensic Highlights)

Fire events occur every day and each one of them is different for a number of reasons (ignition, environmental conditions, immediate/latent and root causes, dynamics, resulting severity, etc.), while at the same time similarities can be identified. This is utmost true considering emerging fire and explosion threats across multiple sectors, including those connected with specific aspects of our life such as the issues posed by complex architectures, very tall buildings, waste treatment plants, renewable energies, new energy solutions from the de-carbonising goal including energy storage, etc.

In full compliance with risk-management approaches, lessons learnt from real events (both accidents and incidents, as well as near misses and anomalies) are a fundamental way to increase fire safety and mitigate the reoccurrence. This activity is not only supported by the information and knowledge sharing activity but also by the methods available to conduct the analysis of real events. Some of them are very structured and standardised.

In the next paragraph, some of the anticipated cases, already presented in this volume, are assessed to identify the main failures that led or escalated to a fire event in different sectors ranging from the maritime industry to heritage buildings.

The use of the fire safety concepts tree is also shown as a method to allocate the observed failures to the main elements of a fire-safety strategy with an obvious advantage in the identification of corrective actions for new projects.

13.1 Torre dei Moro

13.1.1 Why It Happened

'Torre dei Moro' is a tall building located in Milan (Figure 13.1), it has been affected by a severe fire on 29 August 2021. While the fire investigations and forensic activities are still ongoing by the different involved parties, it appears that the ignition of the fire in the starting balcony has been promoted by cigarette butt thrown from one of the upper floors, which led to the combustion of a certain amount of combustibles deposited on the terrace, the source of the subsequent spread. An analysis of the records shows that ignitions had occurred in the past, fortunately without evolution, for precisely the same problem.

Fire Risk Management: Principles and Strategies for Buildings and Industrial Assets, First Edition.
Luca Fiorentini and Fabio Dattilo.
© 2023 John Wiley & Sons, Inc. Published 2023 by John Wiley & Sons, Inc.
Companion Website: www.wiley.com/go/Fiorentini/FireRiskManagement

Figure 13.1 Torre dei Moro before the fire.

13.1.2 Findings

Based on an initial examination of the event, it is possible to identify at least three phases of the fire, which, in any case, was extremely fast. The three phases can be traced back:

- to the ignition of combustibles on the terrace at a height of a residential building unit and the subsequent development of the same with the involvement of a very large portion of the façade, within a confined space or more properly a balcony cavity;
- to the involvement of the entire external façade exposed to the severe initial fire given the high fire load on the balcony, combustible and subsequently also of the building's insulation with the consequent spread of the fire to the entire height of the building in a few moments, also in relation to the wind, the temperature of the façade and the ambient temperature. This development has been promoted by two different fire propagation effects: the 'chimney effect' and the 'corner effect';
- the involvement of the building's residential units as a result of the penetration of the effects of the fire (given the wind, the geometry of the tower and the layout of the facade) and the involvement of neighbouring buildings as a result of the melting and falling of pieces of the facade panels and insulation.

13.1.3 Lessons Learned and Recommendations

The fire event, resulted in severe external effects (Figure 13.2) and internal effects in several compartment, of the high-rise building known as the 'Torre dei Moro' highlighted a number of peculiar aspects of fire risk management for such types of settlements that are unfortunately absolutely common to numerous fires in entirely similar areas that have been observed in recent years and that, in some cases, have unfortunately led not only to consequences for the building but also for its occupants and the rescuers who intervened.

Figure 13.2 Torre dei Moro after the fire.

It is possible to state that

- the fire risk must necessarily be managed during the use phase of the building (with correct housekeeping and appropriate management of fuels) since the initial fire of the balcony has been determined by a huge uncontrolled fire load resulting in a fire directly impinging the facade with great thermal effects and direct flame;
- during the design phase it is essential to critically evaluate the design choices, verifying their impact on fire safety, both in terms of geometry and the materials used in the construction also considering that fire tests used to certify products often consider the reaction to small fires not the reaction from severe fires;
- components that are critical to fire safety (e.g. mobile firefighting equipment) must be known, properly maintained and regularly inspected;
- the fire load is only one of the parameters used to determine the fire risk, and this can only be defined after a detailed analysis of a large number of aspects, including the vulnerabilities of the occupants;
- this punctual analysis must involve all the stakeholders and different expertise, each expert for the aspects for which he or she is competent in order to guarantee a multidisciplinary approach but recomposed in a judgement and overall assessment of the degree of fire safety;
- the requirements of the relevant legal regulations must be fully known and complied with;
- the documentation pertaining to the work must be kept up to date, congruent and consistent throughout the building's life cycle and shared with all stakeholders without delay in the utmost transparency;
- the experience derived from the historical analysis of similar events calls for reflection and requires the implementation of more stringent requirements that may exceed the general legal standard in force;
- approval and certification activities for critical materials, components and technical systems must be timely and appropriate to the use case;
- the technical standards selected must be appropriate to the use case;

- changes in design, construction or use must be thoroughly reviewed for their impact on the degree of fire safety;
- works of a complex nature require timely verification, possibly conducted by an independent third party (peer review) at the various stages of the design and construction life cycle;
- the periodic renewal of fire-safety compliance is a key moment for the review of the degree of fire safety, and not a mere bureaucratic fulfilment;
- activities of a purely residential nature carried out, however, in particular areas (such as high-rise buildings) require specific and periodic information, education and training of the occupants;
- the particular areas referred to in the previous point impose a series of attentions and precautions for rescuers in order to guarantee, on the one hand, firefighting operations and, on the other, the protection of the rescuers' lives;
- the materials and components selected must also be paid attention to aspects related to the type and quality of installation, which may, in some cases, affect their degree of participation in the fire event.

13.2 Norman Atlantic

13.2.1 Why It Happened

The reconstruction of the facts that led to the Norman Atlantic Fire, including its dynamics and the research of the root causes, has been mainly based on the data from the Voyage Data Recorder (VDR), the testimonies from the interrogatories, the documentation taken on board and from the ship owner, the transcription of the audio communications, the census operations about the vehicles on garage decks and the collected evidence.

The investigation team performed a root cause analysis to investigate the Norman Atlantic Fire. The recursive questioning of 'why', starting from the 'main event' (i.e. the Norman Atlantic Fire), has brought the team in driving the investigation, including the collection and the analysis of the evidence, to find the immediate and the root causes of the incident. It is clear that the main event has been the consequence of the failures of different sets of safeguards, which are now discussed.

The fire dampers for the garage ventilation were found opened, so favouring the fire propagation to the other decks different from the 4th, where the fire started (Figure 13.3).

One contributory cause is the positioning of the local commands for closing the dampers: indeed the majority of them is placed on deck 4 (Figure 13.3), and only a limited number of them can be controlled remotely, in a safer position. The possibility of continuing to feed the drencher system (manual deluge) after the occurrence of the blackout is guaranteed by the emergency pump, which

Figure 13.3 Left: Open fire damper of the garage ventilation. Right: Local command at deck 4 for closing the fire dampers.

Figure 13.4 Closed intercept valve between the emergency pump and the drencher collector.

never started. This was because of the emergency generator that, even if its engine started, was not capable of supplying the energy to the final utilities, because of an electrical fault due to the propagation of the flames in other spaces of the ship that damaged the electrical cables. Moreover, it should be noted that even a correct supply of energy during the blackout at the emergency pump could not pump the water inside the garage deck because the intercept valve between the emergency pump and the drencher collector was found closed (Figure 13.4).

However, even if the intercept valve would be found open, the zones activated by the operator in the drencher room (i.e. the valve house, where the distribution of the drencher water is set) were wrong. Indeed, the four zones activated were on deck 3 (Figure 13.5) while it is clear that the fire should be faced on deck 4 (as correctly ordered by the Master).

A possible contributory cause is the drencher plan (Figure 13.6), provided as documentation inside the drencher room to the operator that intends to activate the system. In this scheme, the decks of the ship are named with their English names (e.g. deck 4 was named 'Weather Deck'), while the order given by the Master was to 'activate the drencher at deck number 4'. Therefore, there is not a full alignment between the order of the Master, that needs to be elaborated, and the documentation available in drencher room.

However, even if the drencher would be activated at deck 4, the operator opened four zones versus the maximum allowable of two, according to the drencher manual, to ensure its extinguishing performances: this incorrect operation could be addressed to an ineffective training. However, the timeline reconstruction of the event and the advanced simulations in computational fluid dynamics (CFD) revealed that even a correct activation of the drencher system would not have extinguished the fire, but only controlled it, because of its belated activation. The reasons for such late intervention are mainly attributable to a self-evident underestimation of the problem by the crew.

ZONA 4	n.42 ugelli B15SSP	Deck3 Garage P.te Principale	Fr.96 - 122	Locale Drencher
ZONA 5	n.42 ugelli B15SSP	Deck3 Garage P.te Principale	Fr.122 - 148	Locale Drencher
ZONA 6	n.44 ugelli B15SSP	Deck3 Garage P.te Principale	Fr.148 - 174	Locale Drencher
ZONA 7	n.36 ugelli B15SSP	Deck3 Garage P.te Principale	Fr.174 - 207	Locale Drencher

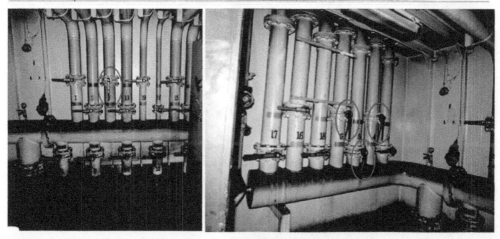

Figure 13.5 The valves opened in the valve house are those activating the drencher at deck 3 (not deck 4).

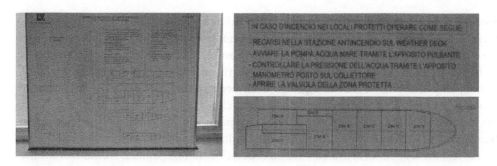

Figure 13.6 Left: The drencher plan located in the drencher room. Right: Details of the instruction on the plan.

The root cause analysis revealed that the crew members agreed in having reefer trucks not connected to the electrical supply of the ship (Figure 13.7) because they embarked a higher number of reefers respect to the number of available reefer sockets, violating the prescription of a correct embarkation. The malfunction of an auxiliary diesel engine of one of the loaded vehicles can be considered the most likely cause of the ignition. This probability must be regarded as higher for those vehicles equipped with an auxiliary diesel generator at the service of the refrigerator system or at the service of the oxygen pumping in the water tanks for the transportation of alive fishes.

The usage of these diesel engines is forbidden inside the garage of the ship because they should be used only when the vehicle is in motion, being cooled by air. The trucks, the oil in their tanks

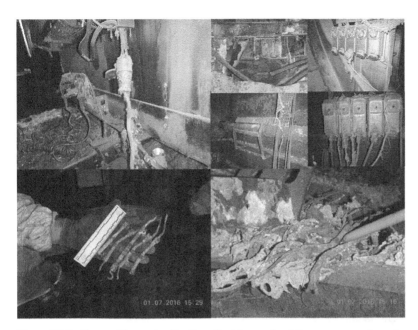

Figure 13.7 Recognition and collection of evidence about the power supply on board.

Figure 13.8 Localised bending of transversal beams and V-shaped traces of smoke on the bulkhead. The majority of the fire load is attributable to the olive oil tanks.

and the olive oil (including pomace) transported by some of them represented the combustible materials (Figure 13.8). The openings of deck 4 (Figure 13.9) continuously provided the oxygen, arising serious questions about its design. The fire triangle was satisfied.

The limited arc of time between the first alarm at deck 4 and the other decks is very short (e.g. three minutes between decks 4 and 5) but not incompatible with the literature. Moreover, this rapidity also emerged from the numerical simulations that have been carried out to validate the hypothesis advanced during the first stages of the investigation.

Some of the outcomes are shown in Figure 13.10.

13.2.2 Findings

A detailed RCA tree is shown in Figure 13.11, where the attention has been focused on the incapability to fight the fire and the ineffectiveness of the drencher system.

Moreover, the timeline of the event has been reconstructed taking into account several sources of evidence, such as the Voyage Data Recorder, the Fire Detection System, the interviews and so on. A part of the timeline, for readability reason, is shown in Figure 13.12.

Figure 13.9 Lateral openings on deck 4. *Source*: Agence France-Presse.

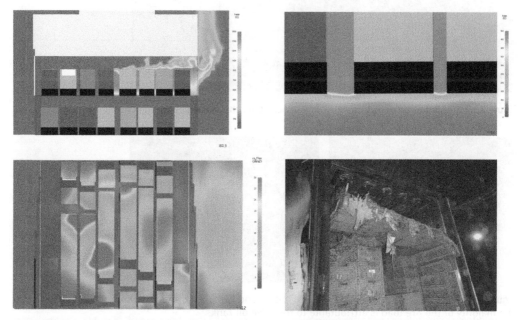

Figure 13.10 CFD simulation about the heat transfer by radiation through the metal plate between decks 3 and 4. Conditions of the plastic boxes inside a truck on deck 3.

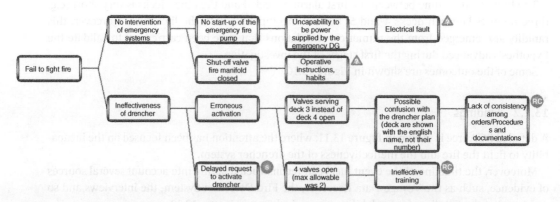

Figure 13.11 Detailed RCA logic tree.

	27 / 12 / 2014 15:39:48	27 / 12 / 2014 23:27:25	28 / 12 / 2014 03:23:05	28 / 12 / 2014 03:27:00	28 / 12 / 2014 03:38:00	28 / 12 / 2014 03:40 - 03:50	28 / 12 / 2014 04:39:01
witness						Steam coming from hydrants, not water	
Voyage Data Recorder					Engines stop		
Fire Detection System			Fire Alarm at Deck 4, frame 156				
Master				Flames are seen from openings of Deck 4 / Order to activate drencher			
Ferry boat	Departure from Patras	Departure from Igoumenitsa				Black-out	Arrival of the first rescuer

Figure 13.12 Part of the timeline of the incident.

Figure 13.13 Photos taken inside the ferryboat from Villa to Messina, 2016.

13.2.3 Lessons Learned and Recommendations

Sometimes, the lessons learned from an incident run so fast to be implemented before the final legal judgement. This case study is an example. During a travel to Sicily, one of the Authors' collaborator sent us the photos shown in Figure 13.13. They clearly depict two positive reinforcement from the management that were never noted before. The photo on the left shows the prescription 'shut down engine', which is written in big capital letters on the lateral wall of the main deck, where cars are parked. The photo on the right shows an information sign about the drencher system, where the distinction between 'Section #5' and 'Section #6' is clearly understandable: in this way, walking on that area, everyone in the crew is automatically refreshed about the correspondence between drencher section number and the covered area.

13.3 Storage Building on Fire

13.3.1 Why It Happened

One of the first conclusions of the investigation was that the fire had started on the building roof and then spread on the roof and later to the building inside (floor below the roof) through the skylights. Another version was that the fire started inside the building (last floor, the one under the roof) and later it was spreading from inside to outside (roof) through the skylights. The compatibility between PV system and roof (according to the Italian regulations) was OK with some 'distinguo' about the interaction between skylights and roof (PV system and other layers on the roof).

Fire spread rate (horizontal, on the roof) was about 2 m/min (value calculated/evaluated from real event observation).

After the event some experiments were carried out, also with the collaboration of firefighter national corps, and the fire spread rate (horizontal) on a 'sandwich' of layers with the same features of the one on the real event roof was about of 5÷6 cm/min. Obviously, the heat release rate (HRR) was much lesser than the real case value, but the difference between the fire spread values (experiments vs. real case) was really big and this feature may have occurred due to a strong heating on the bottom roof covering, especially in its initial steps. Later the great value of HRR could justify the great value of the fire spread rate.

Other possible causes of that rate were analysed, in particular, effects of age of XPS (old polystyrene) layer and effects of air in the roof layers (and effects of wind). Because of chemical features, the first one feature was considered not able to improve the rate to that value and the second one even. But a combination of wind (the fire occurred in a windy day) and of roof bottom heating was, on many opinions, able to improve so much the fire spread rate.

In particular, a first analytical modelling based on chemical kinetics and a more 'coarse' simulation carried out by the (thermal) network theory, both approaches have led to conclusions according to which that great rate (as shown in Figure 13.14) was possible due to a strong heating under the roof.

So, it is really possible that in the first step the fire raised up in the building last floor (the one under the roof) imposing a strong heating on the lower surface of the building roof and later the fire was spreading to the roof through the skylights.

13.3.2 Findings

With regard to the fire behaviour of the roof covering materials, from the experimentation carried out after the event it was noted that the PV thin film (Figure 13.15) was able, given the post-fire examination of the burned portions of the roof (Figure 13.16) to slow down the fire spread rate with respect to the covering package without such a film as the last/upper layer.

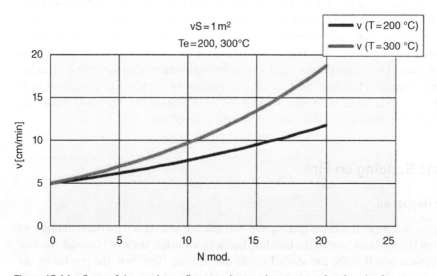

Figure 13.14 Curve of the maximum fire spread rate values *v* on roof surface (surface composed of modules of area equal to 1 m² placed continuously one to another one). Cases with bottom surface temperature – Te – equal to 200 and 300 °C. The case with more heating (300 °C) is clearly with a bigger rate.

Figure 13.15 The PV thin film.

Figure 13.16 The burned layers of the roof.

13.3.3 Lessons Learned and Recommendations

From the considered event, we can also draw some lessons for the future such as

- it would have been better if the activities had the formal authorisation of Ministry of Home Affairs to be operating;
- a fire detection system more effective (e.g. focused on the roof) could have signalled the fire earlier;
- the skylights are privileged ways to fire spread, so keeping proper distances between them and combustible layers and PV systems and limiting the fire load on the roof would be a good idea;
- in general, the assembly of combustible layers and skylight systems is a factor that can amplify the risk of fire in coverage in a non-negligible way.

13.4 ThyssenKrupp Fire

13.4.1 Why It Happened

A specific analysis has been conducted in order to understand the consequences of the accident and the level of risk for the operators. The fire scenario was modelled by means of a specific CFD calculation tool (Fire Dynamics Simulator, generally known as 'FDS', which was developed by the Building and Fire Research Laboratory e BFRL e of the US National Institute for Standards and Technology) on the basis of the evidence and information collected during the investigation. The numerical simulation of the consequences of an accident is a useful and recognised methodology to estimate the consequences of accidental releases of hazardous chemicals in industrial premises in terms of thermal radiation, temperature rise, presence and extension of flames, smoke production, the dispersion of combustion products and the movement of those species in the compartment/s under examination in order to verify what happened with a certain degree of certainty and to verify the modification of the consequences connected to the modification of the parameters that govern the accidental release. Simulation results can help technical consultants in the reconstruction of the accidental event. The well-known NFPA n. 921 standard recognises that fire behaviour numeric codes play a fundamental role in in-depth analysis in the forensic framework: both simplified routines and zone and field models are explicitly quoted. The 'FDS' chosen by the authors is, currently, one of the most specialised and frequently used codes to assess the consequences of a fire inside a compartment, even in industrial premises, and also for forensics purposes. An extensive amount of technical literature has been published by the authors of the code and the same technical reference guide of version 5.0 of the code presents a specific section (n. 2.3) on the reconstruction of real fires. A number of forensic activities are listed in this section (mainly concerning the reconstruction of the consequences and dynamics of real fires). One of the fires dealt with the fire occurred in the World Trade Center on 11 September 2011 ('The collapse of the Twin Towers'). In this technical consultancy, the NIST, on behalf of the FEMA (Federal Emergency Management Agency), investigated the danger of the release of flammable liquids in the form of sprays and this danger was also assessed by conducting a number of real tests. Those tests were in fact similar to the activity that was later conducted by the US Navy to test the consequences of the accidental release of hydraulic flammable oil at high pressure at a real scale and which was described in a specific report that qualifies the four main objectives of the tests: to investigate the consequences of fires from hydraulic flammable fluids in submarines; to investigate the potential for hydraulic fluid explosions; to estimate the event timeline; to acquire experimental data in order to allow a proper fire modelling to be used in engineering practice. In that report as well as in a subsequent paper, the danger of hydraulic oil, even for releases of limited quantities of fuel, is clearly described, along with the description of the facilities used to simulate the release in the experiments. The real, full-scale experiments conducted by the US Navy are comparable with the data used by the authors for the simulation of the ThyssenKrupp accident, e.g. a pressure range from 69 to 103.4 bar, a released fluid with a combustion heat of 42.7 MJ/kg and a similar viscosity. With this data, the authors found it very useful to validate the use of FDS against the results of the US Navy experiments and completed the dissertation with a specific example that showed full agreement of the simulation with the results of a series of experiments conducted by the US Navy, characterised by a release pressure close to 70 bar. The HRR and thermal conditions in the compartments can reasonably be compared. Based on this, FDS has been employed to reconstruct the accidental release and fire that actually occurred at the ThyssenKrupp plant in Turin. The details of the conducted simulations are not provided in this section; a short description of the adopted workflow is given in the following sections in order to provide the readers with a clear picture of the procedure that has been employed by the authors as Technical Consultants of the Public Prosecutor's Office to determine the hazard associated with such an accidental event for the workers that died during the activities adopted to govern the emergency. The simulation activities helped the authors to describe the consequences of the accident in order to define the real risk for the operators and, subsequently, to verify whether the calculated risk level corresponded to the level formally declared by the owner (in the risk assessments required by law) to the Authorities Having Jurisdiction (AHJ) and to verify whether different scenarios could have exposed the operators to similar risks (e.g. with limited releases, with retarded ignition, etc.). In particular, the authors quantified the consequences according to the law requirements pertaining to industrial risks and identified the fire risk with respect to national law (see the threshold limits given by the national Decree dated 9 May 2001 in Table 13.1). Several analyses were conducted for both 'jet fire' and 'flash fire' cases, as defined from the extensive literature available, with the aim of comparing the results with the threshold values established by to Italian law (Table 13.1). The simulation of the real case (i.e. a 'jet fire') is described hereafter.

The simulation involved a preliminary reconstruction of the analysis domain. Several surveys were conducted to obtain a precise description of the tri-dimensional layout of the portion of the compartment that had to be investigated (dimensions: 12 m × 10.8 m × 11.2 m). The domain is presented in Figure 13.17. The release point and the main dimensions are indicated in the plot plan in Figure 13.17. The release point was located at a height of 0.5 m and identified with a circular orifice (diameter equal to 1 cm) directed towards the front wall. The analysis domain was divided into a cubic cell mesh with 1 cm side.

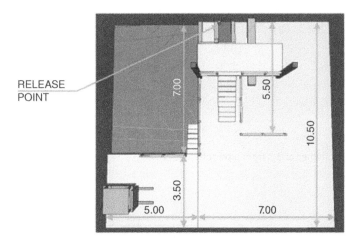

Figure 13.17 The domain used in the FDS fire simulations.

Table 13.1 Threshold values according to Italian regulations.

Accident	High fatalities	Beginning fatalities	Irreversible injuries	Reversible injuries	Domino effect
Fire (stationary thermal radiation)	12.5 kW/m^2	7 kW/m^2	5 kW/m^2	3 kW/m^2	12.5 kW/m^2
Flash fire	LFL	½ LFL			

The simulation of the 'jet fire' considered an initial pressure of the hydraulic circuit of 70 bar, although a number of different simulations were run in order to verify the sensitivity of the consequences in the area where the workers were believed to be at the release time, with variations in the pressure range (up to 140 bar which is the design pressure of the involved circuit) and a number of other parameters (e.g. direction of the release, total oil hold-up released, physical properties of the oil, etc.). This activity allowed the hazard level to be verified under different conditions and in particular to verify whether a release of a small amount of oil could have exposed the operators to danger (since a manual push button was present to limit the release via the isolation of the actuators of some hydraulic circuits).

An example of the results obtained through the use of FDS is given (the case considering a release pressure of 70 bar in the first instants from the release) in Figures 13.18 and 13.19.

P&A lines are commonly considered as plants at high fire risk. Nevertheless, the area at major fire risk is usually considered to be the annealing furnace where huge amounts of fuel gas (mainly natural gas) are used. Another huge fire that occurred in a P&A line in a plant located in Krefeld, Germany, in 2006, has shown that the fire risk can arise from the use of annealing basins and their covers made of plastic material.

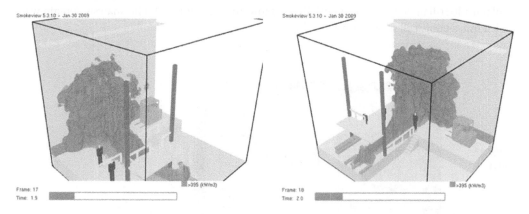

Figure 13.18   Jet fire simulation results: flames at 3 s from pipe collapse.

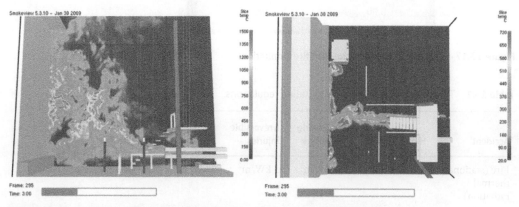

Figure 13.19   Jet fire simulation results: temperature at 3 s from pipe collapse.

Instead, the fire risk due to hydraulic circuits, in particular in the inlet zone of the line, where hydraulic circuits are present in huge numbers, seems to have been underestimated to a great extent in the present case but also in general in the steel industry. The inlet section of the line is a complex part of the plant as it is composed of many devices that are activated by hydraulic circuits. Each of these circuits is generally composed of a couple of pipes, a hydraulic piston and a valve that is activated electrically. The lengths of the pipes are mostly made of steel, but each single pipe is connected to a moving component (the hydraulic piston) and hence at least the last part of the pipe must be flexible. To accomplish this requirement, the terminal is usually made of a flexible, composed pipe, which is made of rubber and has a metal mesh that guarantees the mechanical performances. These terminals are connected to pistons and steel pipes via special fittings. Clearly this is, from the fire risk point of view, the weakest part of the plant. First, the hydraulic oil is combustible. Second, hydraulic circuits quite frequently suffer from leaks. There are generally two sources: the piston seals are subject to wear, which can provoke local spills, while flexible pipes are subject to fatigue which can cause cracks that can lead to sudden leaks under the form of sprays or liquid jets. Due to the frequency of these leaks, and to the huge number of hydraulic circuits, which can reach as many as several tenths in the entry section, the environment can easily become 'dirty' and prone to fire if a rigorous cleaning policy is not enforced.

Other sources of combustible material can also be present in the plant. The coils can, as in the present case, come from a cold rolling line. After cold rolling, coils are re-rolled with a paper strip between the metal coils to prevent surface damage. The paper should be recovered in the entry section of the P&A line to prevent its loss along the line. Sometimes the paper sticks to the metal and its recovery is almost impossible. In this situation, the paper is spread along the line or it enters the oven. The paper is usually impregnated by the oil used in the rolling unit. As a consequence, huge amounts of combustible materials can spread along the P&A line.

Ignition causes are also quite frequent. These are mainly due to mechanical or electrical faults or rubbing of the coil against some structural component. Scratching is far more probable in those plants, or in parts of them, where automatic coil position control devices are not present. The arc welding that is made to join each coil to the others is also a possible cause of ignition. However, this process is easy to control because welding is usually done by an operator and an eventual fire can easily be detected. Mechanical faults can occur in elements such as ball bearings, which are present in large quantities in such plants.

13.4.2 Findings

An accident can be the consequence of a series of undesired events, with consequences on people, objects and/or environment. The first element of the series is the primary event. There are usually many intermediate events between the primary event and the accident, which are determined by the reaction of the system and the personnel. The dynamics of an accident generally starts from a process failure, which is followed by the failure of automatic or manual protective devices. The common representation of this process, with logical trees, involves the use of logical gates (generally AND, OR).

Intermediate events are, in many cases, the condition in which the chain of events interacts with the action of protective devices. When these devices are successful, the chain of events is blocked, therefore the intermediate events correspond to conditions that contribute to decrease the likelihood of the top event. Protective devices can of course act either automatically or manually, which implies the implementation of procedures whose success depends on the level of training of the personnel. As a consequence, the expected frequency of a top event can be reduced by first and foremost adding protection devices, and then acting on the failure rate of the involved components or improving the training of the personnel, thus reducing the probability of human failure.

Hereafter, the representation of the dynamics of the ThyssenKrupp fire is proposed using a fault tree analysis (Figure 13.20), in which the INHIBIT gate is also used to represent the failure of the protective devices. This is substantially a variation of the AND door, which is used in the case of protective means. The INHIBIT gate can distinguish between the entering events since the event entering from the bottom can propagate to the event at the top outlet if the side event, which is represented by the unavailability of a protective device, has already occurred. The dynamics of the accident is represented considering the failure of the existing protective devices and also of those that the plant was not equipped with, which are indicated by the grey boxes.

These are

- automatic shutdown of the hydraulic circuits;
- automatic fire extinguishing plant;
- automatic coil control position in the inlet section;
- fire detection systems.

13.4.3 Lessons Learned and Recommendations

The ThyssenKrupp accident that occurred in Turin in December 2007 has offered some very important lessons about the fire risk for A&P lines. The dispute about the risk level of these plants was ongoing, with some technical experts (the minority) affirming that prevention and protection tools, such as automatic extinguishing plants, were necessary, while the majority of technicians declared that in most cases they were not necessary. It seemed that, at that time, there was a general conviction that only some parts of the line needed specific fire protection equipment. The pump station (which is often located in a separate compartment together with the oil reservoir) was of course considered to be a high-risk zone. The oven zone was also considered to be at a high risk, considering the elevated temperature reached there and the presence of large amounts of natural gas. In some cases, the pickling section was considered to be at a high fire risk, when the pickling pools and/or covers were made of plastic, as in the case of the fire in Krefeld.

However, the risk associated with high-pressure hydraulic circuits and with the potential release of huge amount of oil because of pipe failure had been largely underestimated.

The present case is a clear example of how simple it would have been to adopt measures to reduce risks which would have prevented the accident from occurring, also on the basis of what has already been stated in the extensive technical literature available (e.g. the NFPA considerations in the well-known 'Fire Protection Handbook' 1997).

In order to better understand the lessons that can be learned from this case, the dynamics of the accident was represented in the fault tree shown in Figure 13.20.

The sequence of the events has been reconstructed with a video, which was used to show the incident dynamics during the trial. The video environment has been built from the photos of the real incidental scenario to highlight the site conditions. It reproduces the sounds during the work activities, and it uses 3D images to reconstruct the movement of the victims, on the basis of the collected evidence (witnesses). Figure 13.21 collects some frames of the video.

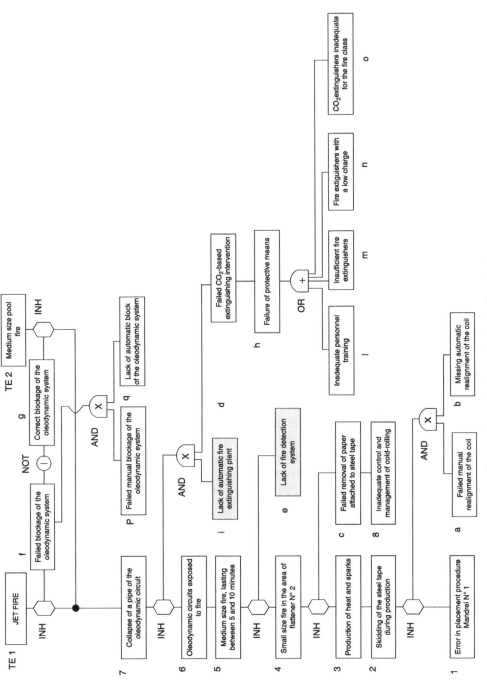

Figure 13.20 Event tree of the accident. The grey boxes indicate a lack of safety devices.

Figure 13.21 Frames from the 3D video, reconstructing the incident dynamics.

13.5 Refinery's Pipeway Fire

13.5.1 Why It Happened

The hydrocarbon leak, which originated in the above-mentioned pipeline, formed and extended pool with a length of approx. 60 m with respect to the leak point, also due to the slight slope in that section. A total amount of 830 tonnes of hazardous substances was involved in the accident. The amount of hazardous substances potentially involved is derived from the calculation of the quantities of products contained in the isolated segments of the pipe bundle, and corresponds to the hold-up of the pipe bundle, which is approx. 2200 m^3. The estimation in tonnes is complicated by the different densities of the products.

The causes of the accident are of technical nature: concerning the causes of the leak, it has been determined that the pipe was perforated due to corrosion processes which occurred externally on the pipe surface.

In particular, the incident report states that it is likely that the localisation of the fissure, with respect to the point where it formed, is linked to one or more of the following factors:

- localised damage in the original pipe coating;
- material defect in the original pipe coating;
- critical operative conditions (of the pipe section in which the fissure occurred) linked to the placement of the pipe near the ground and its exposition to atmospheric events (sea air).

The company declares that the pipe was periodically inspected; the last inspection of the pipeline had been performed in February 2005.

It is not possible to affirm that the maintenance of the pipe in question was insufficient, but it is pointed out that the pipeline examined had been built more than 40 years ago, and was bought from another company in 2002, that did not provide the technical documentation on maintenance operations on the piping bundle prior to the sale. This circumstance has not allowed, according to the technical consultancy report, to verify the compliance with the technical norms on visual tests, inspections and maintenance of the piping system concerned.

Concerning the cause of the fire, the company made the supposition that the contact between the vapours formed from the spilled hydrocarbons with hot spots of high-pressure steam pipes, reaching up to 280 °C steam temperature, in correspondence of the subway, may have contributed to the expansion of a vapour cloud; the vapour cloud ignited on an ignition source downhill from the subway where the first fire was detected.

Concerning the successive BLEVEs, which caused major damage to the persons involved in the emergency response operations, these are directly related to the permanent heat irradiation of the fire (which lasted approx. 48 hours), which overheated other pipelines containing hydrocarbons also in the gas form in the pipe string.

Concerning the direct involvement of the company, the same company states that the event may have had less serious consequences if the onsite fire brigade would have been alerted immediately at leak detection and not three hours later, as effectively happened. This assumption is confirmed by the technical consultancy report, which affirms that if the onsite fire brigade would have been alerted more rapidly the damages caused by the fire could have been contained.

The human error indicated relates to the faulty application of the emergency procedures foreseen in the onsite emergency plan. The fault during the repair operation (dismantling of the pipe insulation layer) relates to the failure to adopt appropriate safety measures during the operation. The design fault relates to the inadequate design and rationalisation of the pipe bundle.

The analysis carried out by the national fire brigade, who followed a structured investigative approach, as claimed by international guidelines for process industries, such as CCPS and NFPA 921, found the shortcomings shown in Figure 13.22.

13.5.2 Findings

For a better understanding of the global strategy provided to increase the fire-safety level, an extremely intuitive qualitative representation is shown in Figure 13.22, based on the conceptual tree provided by the standard NFPA 550, already discussed in the previous chapters. Starting from an incidental scenario, this representation allows identifying those critical aspects particularly significant for a fire and the compensative areas that can be hypothesised consequentially in order to reduce the magnitude or the probability.

13.5.3 Lessons Learned and Recommendations

The comparison of the tree before (Figure 13.22) and after (Figure 13.23) the incident allows to immediately visualise the application of the method to the peculiar situation. In particular, Figure 13.22 shows the found shortcomings, whereas Figure 13.23 shows the corrective measures developed after the incident analysis. From the graphical representation, it emerges that the identified strategy against fire promotes a global action of mitigation and that the different actions are effective for more than a single critical aspect. Indeed, it is highlighted how the incident has originated from a series of shortcomings (both in the design and management phases) that can be addressed to the major parts of the aspects contributing to a fire.

13.6 Refinery Process Unit Fire

13.6.1 Why It Happened

Following the results of the investigation performed by the company and the analysis of an amateur video, the company has formulated following assumptions concerning the accident.

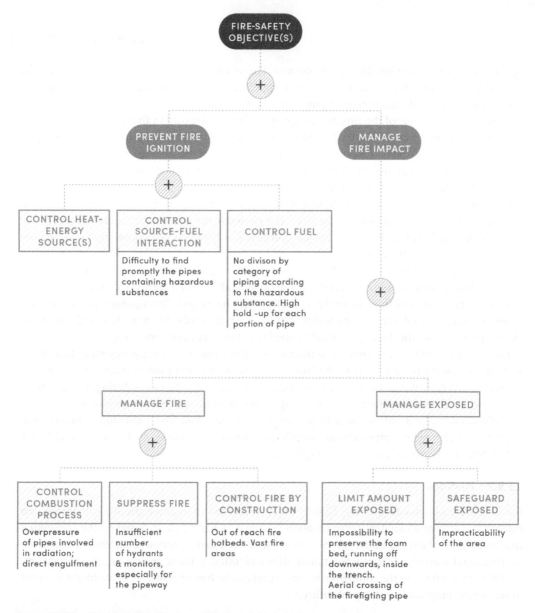

Figure 13.22 Graphical visualisation of the found shortcomings.

The accident, considering the products processed, could have originated by the failure of one of the following plant components:

- Pipes leading to the pressure gauges of reactor R-1702.
- Recycled gas pipe at the bottom of the reactor R-1702 having a quench function.
- Diathermic oil pipe (hot oil) entering or exiting the exchanger.
- Flanged joints exchanger E-1718, E-1709 and connection lines.

The company excludes a release from the hot oil circuit as triggering factor of the fire, based on the evidence gathered from the records on the pressure in the circuit which demonstrate the failure 30 minutes after fire start. Also the video confirms the pipe rupture 30 minutes after fire begin.

For the same reason, a release from the hydrogen pipes is not considered likely, the records demonstrate that the hydrogen pipe failed seven minutes after fire began.

Concerning the flange joints of exchangers E-1718, E-1719 experts requested the dismounting of the exchanger flanged joints, the joint gaskets resulted to be not damaged.

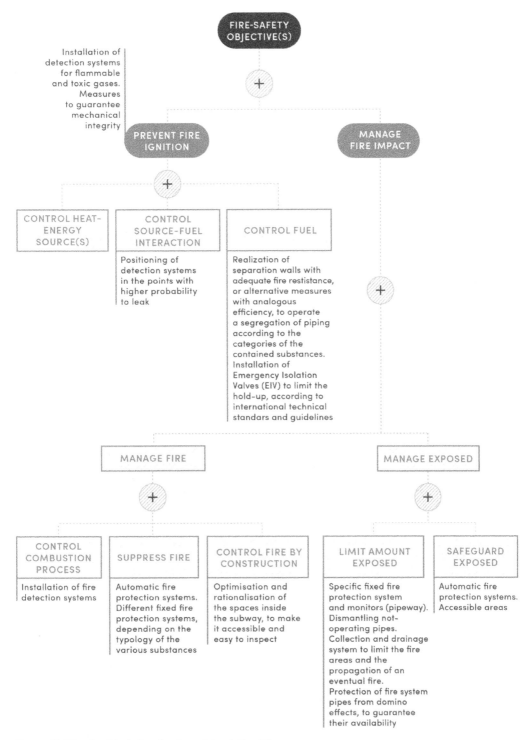

Figure 13.23 Graphical visualisation of the defined fire strategy.

For this reason, the company considers the failure of a pipe from the pressure measurement gauges of reactor R-1702 as the most likely accident triggering factor, and this assumption is supported by the following facts:

- This part is located in the area corresponding to the epicentre of the fire.
- The area corresponds to the area visually identified by the witnesses.
- The product release (hydrogen and fuel oil) from one of these pipes can cause a 6 m long jet flame as occurred.
- The product supposedly released would have had a high enough temperature and pressure to self-ignite or ignite against a hot spot of the plant like the hot oil circuit.
- The damages recorded are caused by overheating (flame exposition) and were not caused by overpressure or explosion.

The pressure measurement records confirm significant pressure changes at the beginning of the event. The company does not have any element allowing to identify the failure cause of that pipe.

Table 13.2 lists the events in a tabular timeline.

13.6.2   Findings

The onsite emergency response service of that shift was composed by six persons having following functions:

- Shift manager in charge with the emergency management.
- Gate guard responsible for the external communication.
- 1st product transfer operator in charge with firefighting.
- 2nd product transfer operator with firefighting.
- 3rd distillation plant worker in charge with firefighting.
- 5th processing plant worker in charge with firefighting.

All the teams assisted by the other shift personnel forming the operative team, participated from the beginning of the event in implementing the emergency response.

The operative team of the shift was constituted by eight workers with the following functions:

- Shift foreman coordinator responsible for securing the installations.
- Distillation plant Q1 control-room operator responsible for securing the plant from the control room.
- Processing plants Q2 control-room operator responsible for securing the plant from the control room.
- 1st distillation worker responsible for securing the installation equipment.
- 3rd services operator responsible for securing the installation equipment.
- Processing plant operator responsible for securing the installation equipment.
- Shift foreman in charge with product transfer responsible for the securing of the storage facilities.
- 3rd product transfer operator responsible for securing the Boccarda storage.

After approximately three minutes from beginning of the event, the fire brigade of Busalla (a team with six firefighters) arrived on site, and subsequently fire brigades from Bolzaneto, Genoa and Novi Ligure arrived with a total of 50 firefighters.

The offsite emergency plan in force has a temporary value and has a provisory character and was drawn up by the prefecture of Genoa in 1998. The updating activity of the offsite emergency plan has been requested by the prefecture and is still under elaboration. The new offsite emergency plan is under evaluation by the local authorities. The evaluation of the emergency response measures has to be considered preliminary awaiting the results of the technical assessment requested by the competent authorities, and the results of the investigation may help to identify the accident causes and indicate organisational measures to improve safety.

Table 13.2 Tabular timeline of the main events.

Progression	Time	Event
00.00″	21.40	Event takes place
00.30″	21.40	Shift supervisor declares the emergency
01.00″	21.41	Night porter phones National Fire Brigade and alerts refinery managers
01.00″	21.41	CTI operator activates foam monitor to protect E1701/E1702/E1709 heat exchangers
01.00″	21.41	4° field operator activates steam barriers on E1717/E1718 heat exchangers
01.00″	21.41	Shift supervisor activates foam pump to pressurise foam network
01.00″	21.41	5° field operator activates foam monitor located near the U1100 process unit in order to prepare a foam bed under the unit
02.00″	21.42	1° operator activates a monitor between conversion and distillation units
02.00″	21.42	1° and 2° operators from storage area activate foam monitors to prevent U1700 ground from hydrocarbon pool formation on the basis of the process unit
02.35″	21.42	From control-room panel operator Q2 activates the process shutdown of the main involved process unit (U1700B) together with the trip of C1701/2 compressors; furthermore, he or she activates the emergency depressurisation of the hydro desulphurisation process units (HDS). Depressurisation is completed in 15′.
03.00″	21.43	I vigili del fuoco di Busalla entrano in raffineria, prendono il comando delle operazioni e iniziano il raffreddamento dell'impianto in ciò coadiucati dalla squadra d'emergenza IPLOM Arrival of the National Fire Brigades; cooling actions directed to the plant structures with the support of the internal emergency team
05.00″	21.45	1° storage area field operator activates cooling rings on S48 and S49 storage tanks
06′.14″	21.46	Probable failure of 3″ diameter line with H2 quenching to R1702 reactor
08′.00″	21.48	2° storage area field operator activates cooling rings on S88, S89, S90 and S33 storage tanks
10′.00″	21.50	Town major arrives at the refinery
11′.00″	21.51	Refinery director arrives at the refinery
18′.00″	21.58	Bolzaneto National Fire Brigade team arrives at the refinery
25′.00″	22.05	Genova National Fire Brigade team arrives at the refinery
28′.00″	22.08	Novi Ligure National Fire Brigade team arrives at the refinery
30′.30″	22.10	8″ hot oil processing line catastrophic failure; BLEVE from 25 to 65 m
33′.00″	22.13	Topping unit general shutdown
39′.00″	22.19	Vacuum unit general shutdown
50′.00″	22.30	Fire becomes limited in extension
1h50′	23.30	Genova Airport National Fire Brigade team arrives at the refinery with Perlini trucks equipped with very high flow/rate – pressure monitors
3h 35′	01.15	Extremely reduced flames length due to nitrogen purge
3h 40′	01.20	Fire completely extinguished
	01.45	Emergency end declaration
	02.00	Plant purging
	02.00	National Fire Brigade team remains at the site for the night
	02.00	Fax communication to the Authorities Having Jurisdiction to notify the major accidents event according to the EU Seveso Directive

Examination of the effects of the fire

The high pressure of the circuit and the temperature of the leaked product (64 bar and 245 °C) determined initial flame lengths of approx. 6 m estimated by the witnesses.

Verifications carried out by TECSA S.p.A. using mathematical models confirm the possibility of flame lengths for similar events.

After inspections after the fire, damage to the structures and equipment from 14 up to 25 m was detected on an area of 10×10 m^2.

In particular, in a smaller area of approx. 5 m $\times$ 5 m (between 15 and 20 m) the temperatures reached resulted in the release of the product from the flanged couplings, the collapse of the hot oil and hydrogen pipes, the deformation of the support beams of the E1718 exchangers at an altitude of 18 m, the melting of the aluminium sheet and the roasting of the rock wool cushion of the thermal insulation of the equipment.

The main damage area coincides with the point of origin of the fire identified visually immediately by the CTP and other shift staff.

In this area, there are the reactor, the E1718 ABCDE exchangers, the hydrogen quench line and the measurement instruments of the Delta P of the R1702 reactor.

The fire after approx. 30 minutes extended to the hot oil circuit, determining the failure of the 8″pipe at a 25 m altitude for simultaneous yield and internal pressure with the release of the product at a pressure of about 11 bar and at a temperature of about 385 °C (data recorded at the distributed control system [DCS]) generating a flare ray of about 20 m with a progress from bottom to top in the direction of the topping column damaging the paint of the insulation and damaging some light points of the 1100 unit.

Water and foam consumption

During the emergency, approx. 10 000 m^3 of water with an average flow rate higher than 3000 m^3/h and 16 m^3 of foaming liquid were consumed.

The reduced consumption of foaming is due to the fact that the fire has developed in altitude without consequences on the ground. It was therefore sufficient to firstly create a foam mat at the base of the plants affected by the fire and keep it constant by the operation of a single foam dispenser. All the other firefighting equipment was used to supply only water in order to enhance the cooling of the plants resulting in this pre-eminent action to combat the fire.

After verification of the fire, it was verified that no damage was reported to the structures and equipment below 10 m.

Damages

The damages caused by the fire are limited to the unit of purification of the diesel fuel in the deck containing the light and heavy diesel fuel exchangers and in the R1702 reactor deck for an area of about 100 m^2 starting from a height of 14 m towards the high.

It was noted in particular

- damage to the support structure between 14 and 21 m (Figure 13.24);
- damage to the piping between 14 and 25 m;
- damage to electrical and instrumental equipment between 14 and 25 m;
- effects of overheating and sudden cooling (water jets) to heat exchangers between 14 and 25 m;
- damage to the insulation of the R1702 reactor;
- loss of the catalyst characteristics of R1702 and R1701A reactors due to prolonged plant shutdown;
- damage to light points and electrical instrumentation cables on adjacent systems;
- damage to paint and insulation of nearby systems.

Figure 13.24 Steel structure damaged (*Source:* Courtesy of IPLOM S.p.A.).

13.6.3 Lessons Learned and Recommendations

The operator decided to rebuild the plant with a new executive project, in consideration of the damage caused to the plant, maintaining the same production layout.

The new executive project foresees essentially

- the complete separation of the light fuel oil section and the heavy fuel section to avoid, for example, the possibility of domino effects;
- lowering the maximum height for the installation of exchangers from 25 to 15 m to facilitate fire extinguishing operations;
- reconstruction of the plant in compliance with the PED directive (CE no. 97/23);
- rationalising the piping system to minimise adjacencies, relocate valves on the hydrogen quench line in R1072 to maintain the line depressurised, reduce the number of measurement gauges and insertion of valves in a safe area for depressurising the hot oil circuit.

Figure 13.25   Forensic engineering highlighting the evidence collection, tagging and movement (*Source:* Courtesy of IPLOM S.p.A.).

The investigation required a preliminary onsite inspection, whose main step was the evidence collection. Figure 13.25 shows some forensic engineering highlights about evidence collection, tagging and movement. Forensic activities have been conducted both in order to discover the causes and the fire dynamics as well as to identify specific improvements. Fire dynamics assessment has been supported by both the use of digital data recorded by the DCS of the refinery (process control system), of amateurs' video, and the use of specific simulation carried out with quantitative risk assessment tools. In Figure 13.26, some screenshots of the evaluations carried out by TECSA S.r.l. with DNV Phast Professional are shown.

13.7   Fire in Historical Buildings

13.7.1   Introduction

The ancient Sabaudian Castle in Moncalieri, close to Turin, has been subjected to restoration works since 2005. All the wooden floor frames had to be reinforced. Three to five workers were involved in the building site, which occupied two floors (third and fourth). One of the workers made all the work on the wooden structures. The yard was open from 7 a.m. to 5 p.m. As often occurs during restoration works in ancient buildings in Europe, the environment was at high fire risk due to the widespread use of old wood. In this case, the floor and many of the wall frames were made of wood. The extensive use of false ceiling in bamboo was also encountered. A common problem in such conditions is the high fire risk, which is rarely accompanied by the adoption of fire prevention measures, such as fire detectors. This is a consequence of the high cost of the fire detection systems compared to the limited time of the work sites. This paper contains results that can be used in the future to discuss the fire detection methods that could be adopted in such structures.

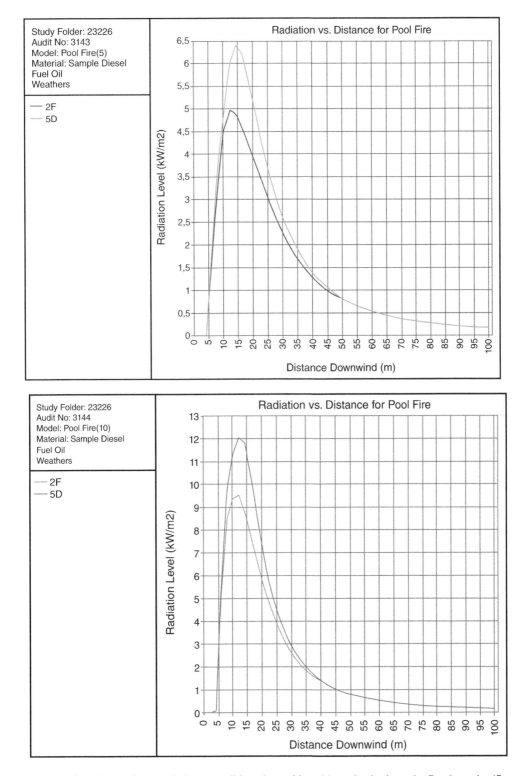

Figure 13.26 Simulations carried out to validate the accidental hypothesis about the fire dynamics (*Source:* Courtesy of TECSA S.r.l.). Radiation at 5 m (left) and 10 m (right) by pool fire in different weather conditions (2F and 5D).

13.7.1.1 Description of the Building and Works

The fire involved the southeast tower of the Moncalieri Castle. The tower has a rectangular base, which delimits the north and east wings of the castle. The tower is higher than the two adjacent wings: the former has four floors, an attic, a ground floor and a mezzanine. The two wings have three floors. A scheme of the tower and the north wing is shown in Figure 13.27.

The external tower walls are made of bricks with a thickness of roughly 1.6 m. The tower body is divided by walls that are similar to the external one. Three walls have an east–west direction, and one of these crosses the entire tower. Another internal wall is located in the north–south direction and is half the tower length.

Many internal walls divide the compartments on each floor. These walls were thin (roughly 10 cm thick) and made according to different techniques. A scheme of the third and fourth floors is shown in Figures 13.28 and 13.29.

If the vertical direction is considered, the Officers Club is located on the first floor and King Vittorio Emanuele II apartments are on the second floor. The first floor was not involved in the fire, whereas the second was damaged by the collapse of the ceiling.

The third floor is more complex. It is organised on two levels. There is the real third floor and a mezzanine (see Figure 13.28, rooms 1–6) that occupies the entire tower except for the so-called 'hall'. The 'hall' is therefore double the height of the other compartments on the third floor. The hall ceiling was in fact the false ceiling located below the fourth floor.

The fourth floor is depicted in Figure 13.29. It is composed of the compartments from 7 to 18. All the floors except those below compartments 7, 8, 17, and 18 were made entirely of a wooden frame structure, which included the main beams (28.5 × 60 cm cross section, spacing 250 cm), secondary beams (10 × 12 cm cross section, spacing 48 cm) and 35 × 3 cm wooden boards. Some compartments also had a false ceiling made of a wooden structure, bamboo and mortar.

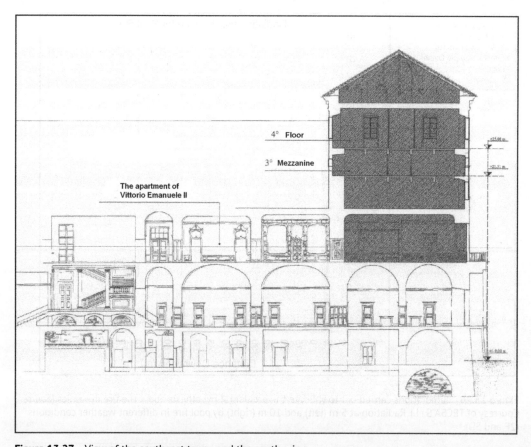

Figure 13.27 View of the southeast tower and the south wing.

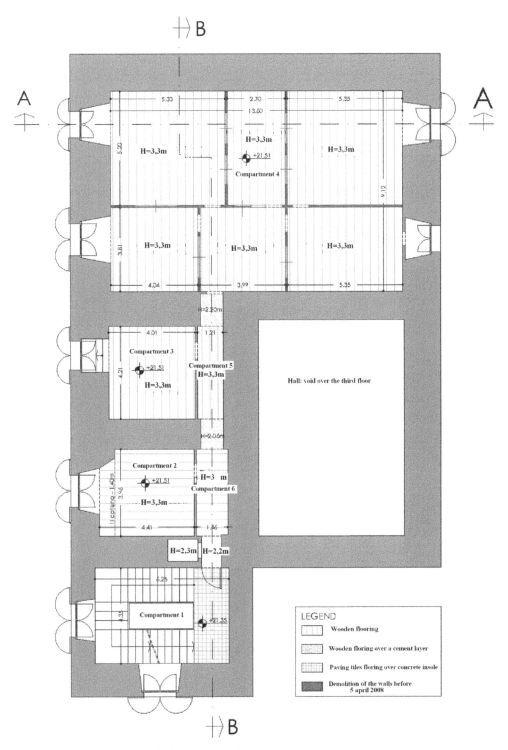

Figure 13.28 Section of the third floor mezzanine.

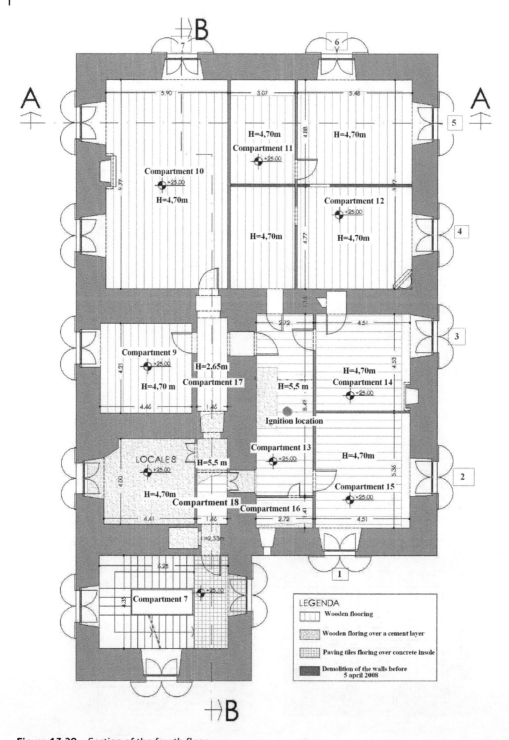

Figure 13.29 Section of the fourth floor.

13.7.2   The Fire

The fire was noticed at 5.45 a.m. by a person walking along a road at a distance of some 200 m from the tower. This person will be called witness 1. At that moment witness 1 saw flames through window No. 4 (see Figure 13.28), which was located in compartment No. 12 and also a smoke column rising from the northwest area of the tower roof. Witness 1 took a few minutes to reach the castle checkpoint. The person on duty in the castle (witness 2) was activated to inspect the tower. Witness 2 reached the second floor of the building at roughly 5.00 a.m., but could go no further up the stairs because of the smoke and the materials on fire falling from above.

The fire brigade reached the castle in a few minutes. They attacked the fire from the roof and the fourth floor windows, delivering foam from the south and east. At that moment, the flames had developed too much to be controlled. Due to the structure of the roof, the flames spread along the wooden frame and the firefighting activities were ineffective. At the same time, the flame spread to the third and second floors because the wooden floors collapsed.

In the fourth floor section, the numbering of the windows is also included (Figure 13.29).

13.7.2.1   The Fire Damage

Most of the damage occurred on the third and fourth floors in the tower. The floors on the third and fourth floors and the roof were completely destroyed. All the walls and the floors that were made of wood were completely destroyed. The second floor was also seriously damaged by the collapse of the ceiling.

In the first phase, the flames were visible through the window on the fourth floor, then spread to the north portion of the roof and then to the entire tower roof. Later the flames spread to the lower floors due to the collapse of the structure. At the end of the fire, some 40 hours later, only the brick structures remained at a level higher than the second floor. The stairs located in the northeast part of the tower did not suffer many damage as they were made of non-combustible materials.

In order to verify the assumptions and the fire hypothesis made on the basis of the investigation, a CFD fire dynamics simulation has been performed; therefore, a specific computational domain has been created (Figure 13.30).

13.7.3   Fire-Safety Lessons Learned

During the activities, and on the precise request by the Public Prosecutor, the authors investigated the fire risk reduction in consideration of the presence of eventual smoke/temperature detectors in

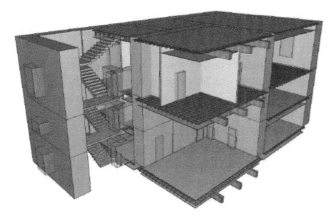

Figure 13.30   View of the simulation domain: third and fourth floors.

order to understand whether in this particular case such prevention measures would have changed the evolution of the fire.

This chapter shows, starting from a large fire, both the possibility of analysing fire dynamics taking advantage of calculation tools and the importance of evaluating the fire risk associated with particular buildings. In particular for the simulation part, the activities were planned and conducted in order to avoid unnecessary and insignificant calculations due to the great CFD simulation time requirements, by making assumptions and trials on a smaller scale and referring to literature data when the calculation tools were not suitable to represent some phenomena. Furthermore, the results of the investigation have been used to understand how this new fire case could be helpful to better deal with the fire risk in particular conditions: historical buildings, restoration works, prevention measures, etc.

This was done in order to support the development of performance-based fire engineering as an invaluable tool to ensure the best fire protection in all the particular cases where the simple application of codes/standards/and regulations is not enough or is not possible (historical buildings, very complex facility layouts, particular substances and the presence of equipment, ongoing activities that can greatly influence the fire risk, such as maintenance works, etc.).

The modelling activity led the authors to recognise that the cause of the fire was a smouldering combustion that lasted some tens of hours, between the false ceiling and the wooden floor on the fourth floor. The most probable ignition cause is connected to the restoration works that were being made on the wooden floor frame.

Modelling the fire allowed three phases to be recognised: (1) smouldering combustion, (2a) flaming combustion with oxygen limitation and (2b) full flaming combustion. The transition between phases (1) and (2a) could not be shown by the simulation, whereas the transition between phases (2a) and (2b) was due to the collapse of some of the windows.

13.8 Fire Safety Concepts Tree Applied to Real Events

In the previous chapters, several fire incidents have been presented. Lessons learnt from real events are fundamental to understand immediate and root causes that led to the fire. Root causes may belong to different fields and different stages of the project, from feasibility, to design, to construction, to operation and finally to decommissioning.

Fire safety, as pointed out in this book, is not a single activity at the beginning of a project while it is the certainty that a proper fire risk reduction is in place during the entire life cycle of the project.

The above-discussed incidents show that pitfalls are connected with several phases of the project as well as different aspects of the fire-safety management. The fire safety concepts tree, from the NFPA 550 standard, is a holistic approach that can be used for both qualitative fire risk assessment and deficiencies identification from real events. It focuses on both design preliminary activities and fire-safety management in operation.

Fire-safety management during operation guarantees a proper and acceptable fire risk level and detailed emergency operation requirements in the case of fire. This is generally achieved by a number of measures aimed to maintain the design requirements, to keep preventive and protective measures under control, to detail the emergency procedures and to define training and competency.

Application of the fire safety concepts tree to real incidents can highlight immediate and root causes that contributed to the fire event considering all the building/plant life-cycle phases and the effectiveness of fire risk reduction connected to those with a specific focus on inherent safety and engineered safety in design and commissioning phases and procedural safety in the operation phase (Figure 13.31).

Fire safety concepts tree structure allows to identify, with a reactive approach to real fire episodes, the fault elements considering the fire-safety objectives reached by the satisfaction of a series of elements interrelated.

Figure 13.31 Effectiveness in risk reduction vs. phases.

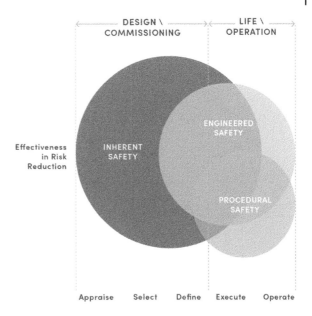

It is fundamental, in first instance, to understand if deficiencies are in the ignition prevention or in the fire impact management or in both. Given this first selection for each element, more specific faults should be discussed.

Fire ignition could have not been prevented due to a lack in heat/energy source control, in source–fuel interaction control or even in a lack of fuel control. On the other hand, fire impact management failure could be caused by a fault in fire management (suppression, combustion control, inherent safety and engineered safety) or a fault in exposure management (no safeguard of the exposed people/assets or no limitation of the exposure level).

Fire safety concepts tree offers a clear formal structure of the elements that, considered together, concur in achieving the fire-safety objectives. While intended to provide tools to assist the fire-safety practitioner (designer, engineer and code official) in communicating fire safety and protection concepts, its use can also assist with the analysis of codes and standards, with the development of performance-based designs and furthermore with, as shown below, a preliminary qualitative assessment of the faults that led to real fires for forensic purposes and/or for developing lessons learnt given a real negative episode (including near misses). During design, it can identify in a similar way any eventual gaps in the fire-safety proposed strategy and it should be intended to support fire-safety-related decisions together with the application of fire protection engineering principles. It can be applied to building design and to existing building management (and their changes), but the general principles can be applied to industrial assets too.

Assuming that it is not possible to achieve perfect prevention and perfect management (the first level of the tree), the structure helps in dealing with both the aspects and apply together in a balanced strategy. In fact, the likelihood of achieving defined fire-safety objectives is increased if both the aspects are considered, with a sort of redundancy.

Despite its flexibility, the fire safety concepts tree has some limitations such as the impossibility to consider time factors and the qualitative approach, but, if applied keeping them in mind, it could be a very important preliminary tool to support the fire-safety strategy initial design and the discussion of the gaps from real events.

Table 13.3 summarises the application of the fire safety concepts tree to the real fire events discussed in this chapter from a forensic perspective.

Table 13.3 Application of the fire safety concepts tree to the real fire events discussed in this chapter.

				"TORRE DEI MORO" BUILDING FIRE	"NORMAN ATLANTIC" FERRYBOAT FIRE
PREVENT FIRE IGNITION	CONTROL SOURCE – FUEL INTERACTIONS	CONTROL FUEL	ELINIMATE FUEL	It appears that find load was present on the balcony where fire started.	Total quantity of fire load was greater than expected to the a number of embarked trucks greater than allowed
			CONTROL FUEL IGNITABILITY	It appears that combustible façade as been selected during design phase.	--
		CONTROL SOURCE – FUEL INTERACTIONS	CONTROL SOURCE TRANSPORT	It appears that source has been a cigarette launched from a balcony. Several e episodes have been recorded in the past.	--
			CONTROL TRANSPER PROCESSES	It appears that convection and a radiation effects from the façade to the building and vice-versa resulted in a fast propagation. These effects have also been promoted by ambient temperature and wind velocity on the façade and in the cavity among the building and the cladding.	--
			CONTROL FUEL TRANSPORT	--	Trucks were not secured onboard. --
	CONTROL HEAT / ENERGY SOURCES		CONTROL ENERGY RELEASE RATE	It appears that head release rate from the combustible façade has been greatly influenced by corner effect, cavities and chimney effect.	Delay in the activation of drencher system resulted in an increased HRR that determined a severe fire
			ELIMINATE SORSES	It appears that, despite several episodes in the past, a cigarette launched from a balcony ignited fuel on the balcony where the fire started.	Reefer trucks, as occurred in the past, were kept with thermal engine in operation. Due to the gas emissions, the avoid) interruption, detection systems have been excluded.

ROOF MOUNTED PHOTOVOLTAIC PLANT FIRES	"THYSSEN KRUPP" STEEL PLANT FIRE	PIPEWAY TRENCH FIRE	REFINERY PROCESS PLANT FIRE	"MONCALIERI" HERITAGE CASTLE UNDER RESTORATION WORKS
Fire load in the compartments was greater than expected	No consideration about alternative non-mineral lubricants and hydraulic oils. Drainage system resulted not effective and fuel released accumulated under the plant originating with sparks multiple pool fires.	No division by hazardous category of the various pipes.	--	--
--	idem	--	--	Poor fire risk assessment did not consider the presence of a great quantity of wooden dust in false ceilings.
--	--	--	--	Poor fire risk assessment for restoration activities did not consider the use of portable sparks generating equipment on the wooden finitures of the compartment.
--	--	Difficulty to promptly find pipes containing hazardous chemicals during emergency response.	--	No provision of metal sheds or other incombustible separation to limit the probability of ignition of dust/wooden structures by sparks generated with equipment used by workers
--	Poor house Keeping resulted in a number of different fuels located in different areas.	--	Congested and complex area with no minimization of the flanges number.	Restoration works on wooden structures caused movement and deposition of large quantities of wooden dust in the compartment.
--	Thermal stress induced by relative small repeating pool fire caused the rupture off hydraulic oil pressurized hoses, resulting in a flash/jet fire with fast and severe effects	No separation of the pipes and no dismission of decommissioned pipes resulting in a congested and complex area.	--	--
--	Ignition sources were connected by sparks generated by friction, not eliminated, in several parts of the production line.	--	--	--

				"TORRE DEI MORO" BUILDING FIRE	"NORMAN ATLANTIC" FERRYBOAT FIRE
MANAGE FIRE IMPACT	MANAGE EXPOSED	SAFEGUARD EXPOSED		It appears that exposed have been granted a secure path along the building with separation and structural stability but the exit (entrance for the emergency services) gas been impacted by embers and materias from the façade.	Due to the greater number of trucks then allowed a number of emergency exits wen not available, Some exits were supposed to be used by the crew and there for they resulted unavailable during the emergency response. Safety means to escape from the ship wen lost at sea, resulting unavailable.
		LIMIT AMOUNT EXPOSED		--	Emergency rafts have been lost to the sea; passengers could not leave the ship. Detection has been delayed resulting in a non-tempestive communication. Poor communication and poor emergency signage have been discovered. Stability has been partially lost due to the severe effects of the fire, an incorrect load of the trucks and the loss of power
	MANAGE FIRE	CONTROL FIRE BY CONSTRUCTION		It appears that no considerations about fire propagation to/from the external façade have been done, while structural stability and fire compartments have been granted inside the building	Fire propagation on deck no. 4 where the fire started has been increased in velocity due to the windows openings, able to feed the fire. Furthermore, the propagation among different decks resulted from the missed closure of fire dampers in the ventilation system.
		SUPPRESS FIRE		It appears that no automatic suppression/mitigation protection systems where in place to limit the spreading on the façade, to limit the involvement of apartments. It appears that manual suppressions systems (hydrants and hoses) have resulted in limited performance/availability	Manual water deluge system ("drencher") has been actioned by the crew with a delay. Delay time has been caused by a late detection and acknowledgement of the fire, long travel distances to the drencher room, wrong signage of drencher valves for the different decks, Drencher has been ineffective also due to a reduced total flow-rate (problems with the water in-takes, 4 zones opened instead of 2, water pumps failure). Manual provisions (hydrants and hoses) resulted ineffective due to the severity of fire. Emergency management hasn't been effective due to poor communication and training of crew members.
		CONTROL COMBUSTION PROCESS		It appears that no considerations have been done about the possibility to limit fuel quantity in the façade elements.	Fuel quantity was greater than allowed and trucks were not secured to the ship

ROOF MOUNTED PHOTOVOLTAIC PLANT FIRES	"THYSSEN KRUPP" STEEL PLANT FIRE	PIPEWAY TRENCH FIRE	REFINERY PROCESS PLANT FIRE	"MONCALIERI" HERITAGE CASTLE UNDER RESTORATION WORKS
--	No proper instructions provided to the operators in order to electrically isolate the production line with the shutdown and the manage fires, resulting in operators directly exposed to the first fire and its eventual escalation.	Area not practicable by emergency services due to the severity (extension and HRR) of the trench fire, including congested areas of the pipe way under the road.	--	--
While this roof-mounted installation does not pose threats for occupants propagation of the fire from the roof to the storage internal compartments may lead to secondary fires	--	Impossibility to maintain the fire protection foam bed, running of downwards the entire length of the pipe way due to the terrain slop. Pool fire aerial crossing of the fire hydrants feeding line resetting in the impossibility to use part of the hydrants.	Congested area resulting in secondary effects and domino escalation from the initial jet-fire.	During inactivity period (from late afternoon to the morning) the construction site was kept closed; this resulted in the impossibility to reach the fire for a suppression attempt using portable means.
No consideration during design has been done about horizontal propagation on the roof, embers generation and propagation via vertical openings from the roof to the internals of the buildings.	No consideration about layout of the plant in terms of safety and a safety accessibility of the emergency team.	No fire zones defined during design phase, resulting in a vast area impacted by limited number of manual hydrants in the area.	No separation among the two main sections of the affected plant (light fuel oil and heavy fuel oil). Plant layout privileged extension in height.	--
--	No automatic systems to suppress fire were in place (neither a general fire protection systems nor a local-application system for congested areas). Fire emergency management relied on manual suppression even with a high demand proper emergency plan. Main event resulted from an unwanted and un managed pool fire. Application of suppression spreader the fire from the first fire zone to adjacent bunds, in creasing the total surface on fire. Detection of the fire was preformed by operators since detection systems were in place.	No fixed fire detection and fire protection systems installed. Availability of a limited number of manual hydrants in the area.	No fixed fire detection and fire protection systems in place, resulting in severe efforts of the emergency team due to the complexity to control the initial fire and its domino effects.	No fire/smoke detection systems were in place during restoration works, resulting a several hours delay in discovering the fire, that lasted shows in a smouldering invisible fire and then suddenly erupted in a full developed fire involving all the wooden structures and the entire roof.
--	--	Over pressure of pipes involved in radiation due to direct engulfment/impingement	--	During inactivity no surveillance have been put in place, resulting in the impossibility to notice any smoke coming from a smouldering fire originated and lasted for several hows in the false ceiling

From Table 13.3, it is possible to make some considerations that are related to the causation path and the fire strategy failures observed:

- fire-safety strategy should be based on preventive and mitigation controls;
- fire risk should be considered, with the same importance, in the design phase and in the operation phase;
- majority of severe fires show a causation path that include multiple failures and escalation factors;
- in several cases a proper fire risk assessment would have guaranteed a more robust strategy, whereas in the majority of the cases poor fire risk management in operation played a fundamental role in terms of ignition probability, speed of propagation, resulting severity and even fatalities;
- fire risk management during operation according to the requirements identified in the design phase is fundamental to avoid escalation factors that may lead to events that have been formally excluded from the initial risk assessment due to a very low probability of occurrence considering the effective controls in place;
- real fires often underline problems at the risk analysis stage, which are often not resolved during the life of the building or asset, even though it is actually foreseen that both in the event of changes and new facts, but also in the absence of such conditions at pre-established intervals, a review of the fire risk assessment carried out should be carried out;
- problems arising from poor fire risk analysis can be traced to a number of different cases: lack of a sufficient degree of detail in relation to complexity, use of incorrect methodologies given the context, deficiencies in input information and deficiencies in the quality of output data resulting in a lack of knowledge of the level of risk and an inability to make informed decisions;
- in some specific cases it is observed that the problems recorded in the fire risk assessment, when present, related to real aspects highlighted by the fire, are a function of a lack of precise knowledge of the phenomena that could actually develop (typical is the case where the risk assessment only considers the fire load, often forgetting the parameters that define the dynamics of the fire, such as HRR, specific process conditions, room layout, etc.).

14

Case Studies (Risk Assessment Examples)

The previous chapters of this volume first presented the concept of fire risk and the importance of its estimation in the design phases and its maintenance during the life cycle of assets at an acceptable level. This recognition and management of fire risk, but in some cases also of explosion risk, takes the form of a strategy based on a series of elements that contribute to preventing and mitigating the risk. The complex of multiple activities that lead to the definition of the strategy, including the definition of the requirements for maintaining the level of safety over time, depends on a series of factors specific to the reality to be examined, the purposes of the analysis, the methodology selected, any legal requirements, and the information available. This chapter provides several highly simplified case studies of fire risk assessment in different application areas, from the process industry to civil buildings of artistic and historical value. Each of the cases presented summarizes the peculiarities of the analysis, the objectives and the results achieved.

More examples are available as online-only resources in the companion website along with a presentation to be used during training courses that includes diagrams and pictures of the discussed fire cases.

Fire risk assessment for a company managing multiple railway stations:

Topic	Description
Scope	Scope of the fire risk assessment is a preliminary assessment of multiple assets (railway stations) distributed nationwide in Italy to evaluate the current fire-safety level and to identify any gaps (administrative, technological and organisational).
Context	Context is composed by several very large and complex railway stations under the same owner.
Criteria	Fire risk assessment should verify the current fire-safety level and the fire-safety management. The ultimate scope is to identify gaps and their solution priority also considering, given a single organisation, those that affects multiple buildings and areas in the same railway infrastructure and those that affects multiple assets of the organisation. Assessment should define priority considering cost and implementation time together with the risk reduction factor. Assessment is aimed to the identification of the correct budget to improve fire safety across the portfolio assets.
'Risk of what'	Fire risk to occupants including travellers, contractors, tenants, etc.
'Risk to what'	Risk is posed to all the people using the infrastructure for different reasons. Threats should be considered also from the relationship with other transportation systems and the fire risk posed to crowded areas and to heritage buildings.
'Risk from what'	Fire risk is posed by trains, technological systems, connected transportation hubs, stores and commercial facilities, including their storage areas in the undergrounds.

Fire Risk Management: Principles and Strategies for Buildings and Industrial Assets, First Edition.
Luca Fiorentini and Fabio Dattilo.
© 2023 John Wiley & Sons, Inc. Published 2023 by John Wiley & Sons, Inc.
Companion Website: www.wiley.com/go/Fiorentini/FireRiskManagement

Fire risk assessment is the key element of a Bow-Tie-based fire-safety management system (FSMS) to design the fire-safety strategy for railway stations.

Case studies in Italy

Railway stations are nowadays, among several other specific occupancies characterised by the presence of masses of people often in fast transit (as ports and airports), a clear example of infrastructure open to the public that are constantly subject to modifications. Changes take place to install temporary stores, new stores or facilities (even schools, healthcare facilities and hotels), advanced security measures due to recent episodes, new utility systems, and networks. Besides this, occupants' population in railway stations is particularly heterogeneous and much more different from the population in different and more controlled occupancies (ports and airports), also because railway station is more than an infrastructure hub. They indeed are centres open to the city where people can find commercial stores, services, restaurants, etc. This modification process, in Italy, is even 'dangerous' because the majority of the biggest railway stations are located in heritage buildings that pose both physical and permitting-related constraints. Stations are built around train tracks and composed by several buildings, with some underground floors (where usually connections to metro stations as well as utility occupancies are located, as warehouses for stores located on upper floors). Back in 2011, a specific decree requested railway stations to obtain their fire certificate for the entire entity and not limiting the fire certificates to single occupancies. Complexity should be considered in a uniform way across all the station with a single approach that should be properly maintained during the time. Furthermore, it is worth to mention the fact that until 2011 fire certificate application procedure in railway stations was supposed to cover only single selected occupancies on the basis of their specific activity and/or extension in surface. Since 2011 a specific fire-safety regulation (national level) enforced the requirement to extend an additional single fire certificate for the entire railway stations having a total amount of public open surface greater than 5000 m^2. This new requirement recognises the railway station being a single unique special occupancy and raised the need for a fire-safety general design strategy to overcome the limits of specific occupancies and to find a general approach able to consider all the additional and common areas that connect/serve tracks and commercial retail stores. The focus must also be on station buildings and associated utilities that complement the main areas open for the general public (the 'station' perceived by most of the travelling passengers in and out the main buildings). While these new to be considered spaces are mainly office occupancies rented to third parties, it is also true that real fire risks are associated with areas such as tunnels, utilities rooms (electrical substations), warehouses after located in the undergrounds (several floors) as well as connections (underground roods with significant traffic of motor and electricity powered vehicles), car parks, etc. Complexity becomes even worse considering that railway stations are always-on changing assets due to new commercial needs (growing number of passengers due to a specific competition among fares with air traffic especially at national level facilitated by the increase of high-speed train connections) and the stop operation for any revamping activities (even fire and security implementation of new measures). With this overview in mind, upper management has to face and win a specific game: in a given time (by law) and considering finite economic resources, without shutting down any part of the asset but during severe revamping projects of renovation and modernisation and with ongoing general and extraordinary maintenance (and inspection) activities, gain a proper fire-safety level and get the specific fire-safety certificate from authorities having jurisdiction. Besides this, a parallel workflow has put in place to increase security measures that collide with fire-safety requirements (in particular with the emergency evacuation of large masses of people). Both the modernisation processes should consider heritage buildings and specific conditions of railway stations in Italy: a frequent use of historical buildings for temporary exhibitions as well as specific periods (or days) of recording and extraordinary affluence of passengers due to touristic special dates especially in summer time in several cities should be added to the impressive movement of tourists during the entire year in Italian cities moving from city to city by trains.

On the basis of this very impressive complexity and considering a renovation for the periods of several years without any business interruption, a fire-safety management plan has been defined at the central level for the main 14 railway stations located in Italy: Bari Centrale, Bologna Centrale, Firenze S. Maria Novella, Genova Brignole, Genova Piazza Principe, Milano Centrale, Napoli Centrale, Palermo Centrale, Roma Termini, Roma Tiburtina, Torino Porta Nuova, Venezia Mestre, Venezia S. Lucia and Verona Porta Nuova.

Fire-safety plan started from the construction of an FSMS to deal both with the initial intervention phase and with the future use/modification of the railway stations. FSMS is composed by elements common to all the stations at a centralised level and customised elements specific to each single station. This plan enforces the use of a common approach, name assumptions, methods and tools with a shared dashboard with fire-safety-related key performance indicators (KPIs) to track the fire-safety level during time comparing different areas of the same station as well as different stations in a global benchmarking activity for the upper management. Dashboard serves as a decision tool to also allocate resources. FSMS encompasses several aspects: culture of safety, global policy, organisation and people, training, fire risk assessment, inspection/maintenance, emergency response, audit and feedback. The key element of the entire process is the fire risk assessment. This has been conducted with the Bow-Tie methodology, with a simple template Bow-Tie diagram (Figure 14.1), to identify the top events to be considered for subsequent more in-depth assessment also recurring to the use of simulation methods (fire and evacuation) in a performance-based environment to take into account all the specific aspects of the complexities, as described in the following sections.

The Bow-Tie method (from the oil and gas quantitative risk assessment) is the proper tool to

- represent the main cause and consequences in a logic notation;
- identify both prevention and protection barriers able to prevent the occurrence or mitigate the consequences (or reduce both) of the identified outcomes;
- consider both technical and management barriers.

Bow-Tie is characterised by a clear and simple methodology; it is very straightforward to use it to

- explain fire design considerations to both the upper management and the field operators;
- identify the need for new/different barriers;

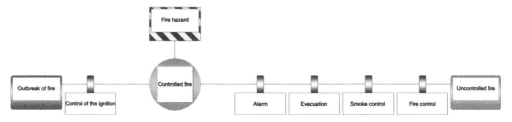

Figure 14.1 Bow-Ties developed to assess fire risk in multiple railway stations.

- make comparisons among areas, assets and stations at a global level;
- describe the current and intended level of fire safety;
- conduct technical audits using their results to visualise the current status of barriers against the minimum established level (e.g. using inspections result data on fire protection active systems to update the relevant barrier and the related key performance indicator/s).

Top event identified can be developed with engineering tools as well as can be used as the basis of the emergency and evacuation plan.

Given typical top events (each of them related to a single Bow-Tie diagram), those have been applied to a hierarchy. Hierarchy has been defined considering, the single building for each station, inside each building the various fire compartments located in floors (above and underground). Initially, each fire compartment has been defined in terms of fire-safety-related properties (surface, people, public presence, fire load (Figure 14.2), safety measures in place, etc.), and Bow-Ties have been applied to the compartments, intended as atomic units of the risk assessment, considering five standard barriers: ignition prevention, alarm, evacuation, smoke control and fire control. Each of them is composed of several elements to be judged:

- Ignition prevention
  - Permit to work
  - Electrical system safety level
  - Fire-safety quality and reliability of the building
  - Ignition source control
  - Inspections
  - Maintenance
- Alarm
  - Fire detection
  - Manual call points
  - Control room
  - Emergency plan
  - Alarming system
  - EVAC system
  - Fire drills
- Evacuation
  - Fire brigade
  - Emergency exit width
  - Emergency route length
  - Elevators
  - People with disability management
  - Railway station supervisor
- Smoke control
  - Separation
  - Compartments
  - Ventilation
  - Heat/smoke extraction
  - Material control
- Fire control
  - Sprinkler
  - Separation

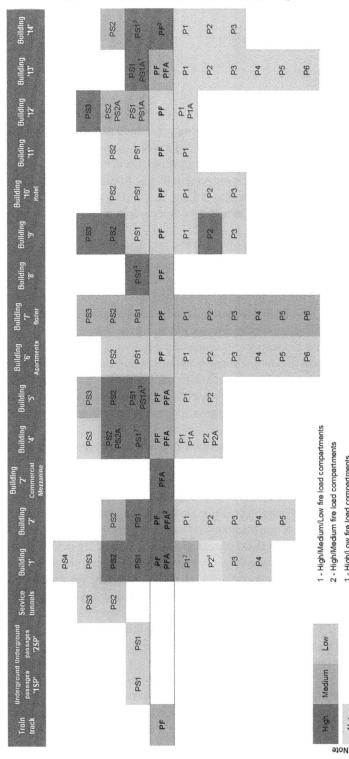

Data related to a case study developed as an example

Figure 14.2 Fire load.

- Compartments
- Fire load requirements/HRR
- Hydrants
- Fire brigade connections
- Fire brigade
- Portable fire protection means
- Intervention time

For each fire compartment, the specific conditions of the Bow-Tie identified barriers have been judged with brainstorming sessions of experts using predefined categories about the quality of the barrier in place (or the lack of the barrier if requested but not implemented). Due to the number of the fire compartments, the classification has been made using a tabular format (Figure 14.3). Weights employed in expert judgement during brainstorming sessions have been defined in a qualitative way as 'not necessary' (score 1), 'present and fully efficient' (score 1), 'compliant' (score 0.9), 'to be verified' (score 0.33), 'not compliant' (score 0.6) and 'absent/not working' (score 0).

From collected data, structured as the underlying Bow-Tie diagram, it is possible to have an over-view of the situation: specific problems of single barriers in compartments, quality problems in single buildings or even low performances across all the assets, affecting the entire station.

Taking the performance of the single barriers and having those represented in the logic diagram of the station (Figure 14.4), it is possible to visually identify absolute weakest points and even make combination of data to define general safety level (considering the weights of the barriers assigned in the Bow-Tie diagram) and to identify critic compartments/floors of single buildings or station areas given the relationship of the compartment calculated performance level or the most relevant single barrier performance level and specific characteristics of the fire hazard (quality of fixed fire protection systems with a significant fire load, week alarm systems/evacuation routes where public is located (Figure 14.5), quality of smoke detection and smoke extraction or ventilation system in undergrounds, etc.).

Activity allowed to have a risk-based picture of the fire-safety level across the assets of the orga-nisation in order to define priorities with an improvement plan. Plan has been defined with a cost–benefit analysis on the basis of the simulation of post-mitigation new conditions.

Methods such as Bow-Tie, LOPA and cost–benefit analysis have been selected as the approach to be included in an FSMS while improving existing situations. In fact, Bow-Tie can also be used to demonstrate/verify the impact of modifications (in organisations, systems, assets and compart-ments), even with the support of an HAZID method to verify temporary modification impact on the overall fire-safety level (closure of emergency exit due to temporary construction sites inside the station and impact on the evacuation strategy) in order to meet the requirements of a specific management of the change process. The FSM system has been defined using a standard SMS struc-ture, with a barrier-based approach, considering responsibility, operation, inspections, mainte-nance, MOC, emergency planning and response, KPIs and system review.

Conducted activity showed a lot of benefits and demonstrated how to manage all the require-ments coming from the application of several regulations in complex realities effectively.

The application of the FSMS, based on barrier assessment, also resulted in the possibility to col-lect information, data, documents, and performance indicator results to support the fire certificate request to the authorities, to take better informed decisions, to demonstrate to all the stakeholders the activities in place on real time and the design intent with the intended results and the path to achieve those.

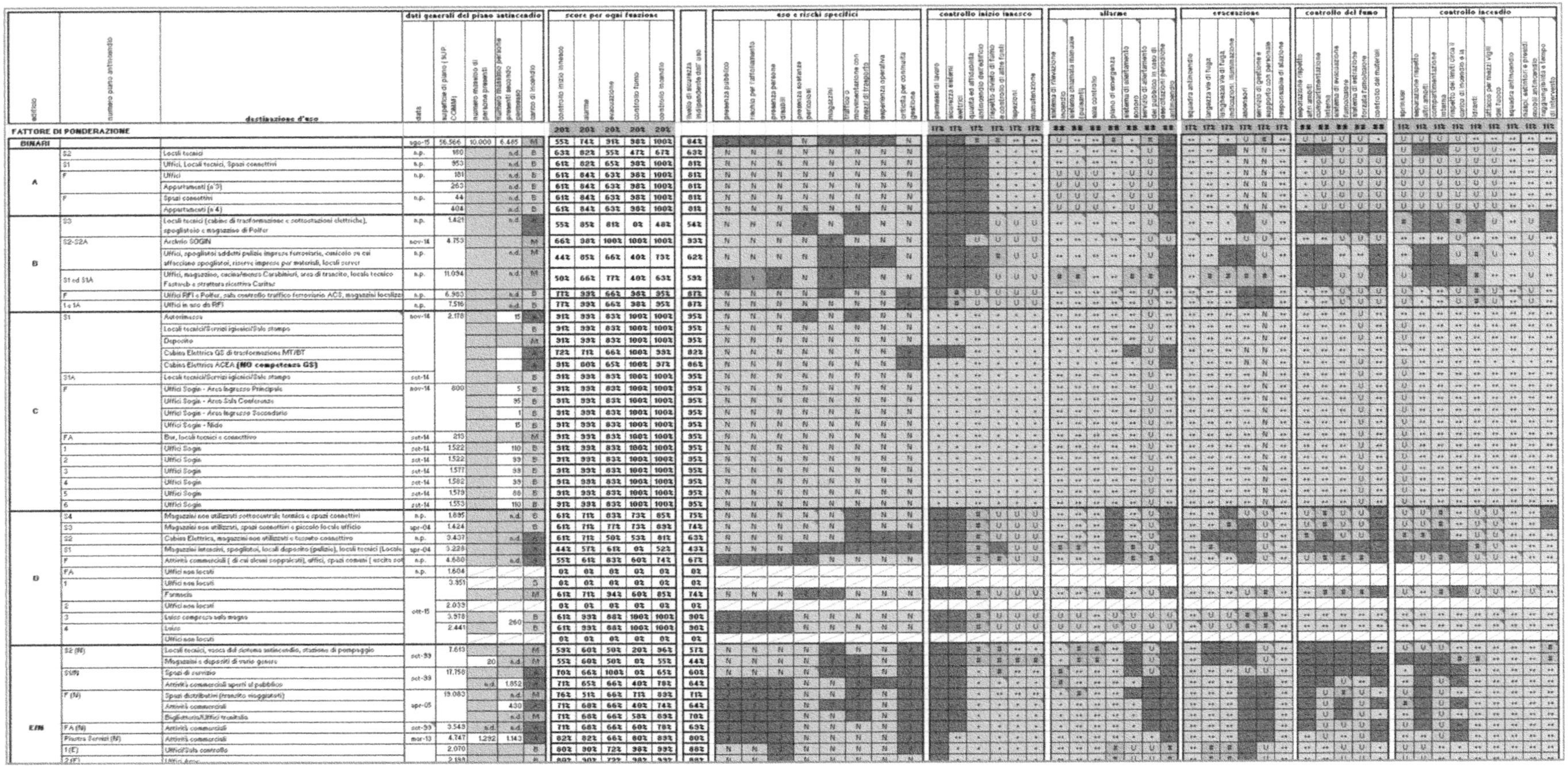

Figure 14.3 Tool.

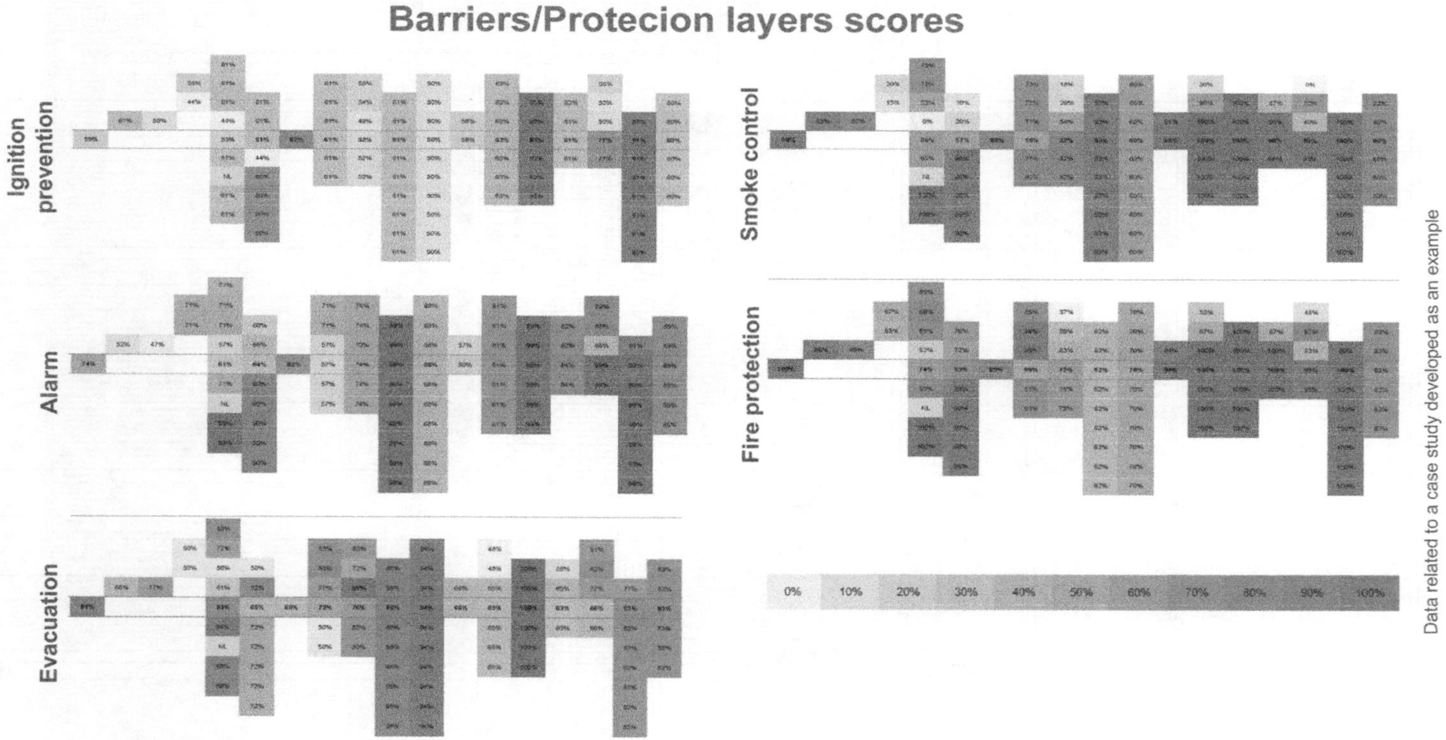

Barriers/Protecion layers scores

Data related to a case study developed as an example

Figure 14.4 Barriers/protection layer scores.

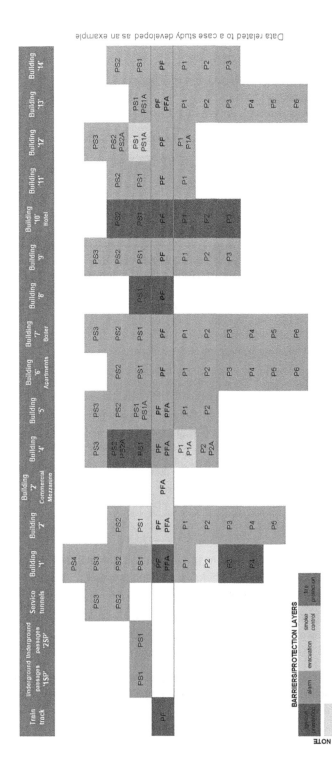

Figure 14.5 Weakest barriers and public.

Bow-Tie and barrier failure analysis, applied to an organisation, satisfy the requirements of an 'enhanced risk management':

- continual improvement;
- full accountability for risks;
- application of risk management in all decision-making cases;
- continual communications;
- full integration in the organisation's governance structure.

The fire risk assessment of the listed industrial chemical production facility:

14.1 Introduction

Topic	Description
Scope	Fire risk assessment related to a chemical industrial plant with different peculiarities due to the type of production and a multiple amount of combustible and flammable material placed in outdoor areas.
Context	Chemical industrial plant consisting of several buildings with different activities subject to fire prevention control. About 200 employees work inside the plant. Production takes place on a continuous cycle, seven days a week.
Criteria	Quantitative assessment of fire risk through the use of a simplified methodology divided into five phases capable of returning a quantitative assessment both for individual compartments and the complexity of the industrial complex.
'Risk of what'	Specialised design of verification of proper sizing and placement of flame detectors.
'Risk to what'	Insight into the peculiarities of the industrial complex and identification of applicable solutions according to current regulations and the state of the site.
'Risk from what'	Specialised in-depth study of a flammable and/or toxic substance detection system to prevent the escalation of a fire.

14.2 Facility Description

The case study considered a chemical plant in which there are facilities for the processing and transformation of raw materials through chemical reaction processes.

The activities carried out overall are as follows:

- Procurement activities and chemical transformation of raw materials into finished products.
- Storage and marketing of chemical products, bulk or packaged.
- Miscellaneous support activities.

The various production stages and the large quantity of products result in a significant need for storage, in fixed or mobile tanks, and significant handling, carried out with fixed lines or by unloading into containers.

The following homogeneous areas can be identified within the plant:

- Ethylene oxide storage and oxyethylation plant
- Production

- Storage of products in drums and small tanks
- Storage warehouses
- Service (offices)

Figure 14.6 shows a floor plan of the plant with the main uses of the warehouses identified. The fire risk assessment was extended to all compartments of the plant, with the objective of verifying compliance with the Fire Prevention Code (IFC 15) or specific regulations applicable to the activity.

14.3 Assessment

14.3.1 Selected Approach and Workflow

The methodology used in this case study for fire risk assessment is divided into five phases of analysis and insights as described below:

PHASE 1. Preliminary qualitative assessment of the macro-areas of the production system, with the aim of making an initial classification of the zones and identifying the areas of greatest potential risk.

PHASE 2. Quantitative assessment of individual areas in accordance with regulatory requirements.

PHASE 3. Cross-referencing of quantitative assessment data with fire risk classification (medium, high or very high) in order to define intervention priorities.

PHASE 4. Identification of measures to be introduced for fire risk mitigation.

PHASE 5. Specialised insights.

Figure 14.6 Ground floor plan of the industrial plant.

The previous steps are described in detail in Section 3.2.

14.3.2 Methods

As a preliminary step, a qualitative analysis of potential critical issues at the plant was carried out (PHASE 1). The methodology adopted involves schematising the production processes from the arrival of material at the plant to the finished product, highlighting their logical connections; an example diagram is shown in Figure 14.7. In the case under consideration, three main production lines were identified, which for the ease of reading have been simplified into lines A, B and C.

Appropriately developed checklists based on the peculiarities of the production process allowed obtaining an indicative assessment of potential critical issues. This assessment was then deepened with a quantitative assessment, capable of evaluating whether the potential criticalities highlighted by parametric indicators were confirmed, mitigated or increased according to the actual fire load or deviations from regulatory requirements.

The outcome of the first qualitative assessment aims to preliminarily highlight the potential risks related to the type of activity (production, storage, etc.) without going into the merits of fire-fighting apparatus, quantity of material present or other specific characteristics that may mitigate or burden the assessment of the specific fire risk.

The final evaluations are represented in a functional logic diagram (Figure 14.7) that sequences the production process, highlighting how a potential fire risk takes on different values (low, medium or high) both in relation to the type of process and its location in the production chain.

A quantitative analysis of the plant was then carried out to characterise the activities and determine the applicable methodology (PHASE 2) according to national regulation regarding fire prevention.

The result of the quantitative assessment of the activities is graphically represented in Figure 14.8, which punctually highlights the reference regulations to the individual compartments of the structure under consideration.

Once the applicable regulations were identified, a methodology was adopted that through the use of detailed tables and graphical diagrams is able to return the outcome of the fire risk assessment for both individual compartments and the facility as a whole. The point analysis was carried out by assigning a score ranging from 1 to 3 depending on whether the regulatory requirements are fully met (value 1), are compliant but with verification (value 2) or not compliant (value 3). The sum of the individual scores falls within a range to which three levels of evaluation were assigned: if the value obtained is 10, the requirement is met, so no further study and/or additional interventions are needed; if the score is between 11 and 30, implementation and/or verification interventions are needed; above the value of 30, the discrepancies are important and extensive.

The table shown in Figure 14.9 represents the final outcome of the fire risk assessment, as a summary of the specific detail report, in which all the actions required to comply with the regulatory requirements are highlighted punctually. The assessment reports for each regulatory requirement (reaction to fire, fire resistance, exodus, etc.), the level of performance required and the corresponding verification of compliance.

Upon completion of the individual detailed analyses, the methodology adopted involves the preparation of a concluding summary table as shown in Figure 14.10, relating to all zones/departments of the establishment under consideration. As can be seen in the case under review (Figure 14.10), there are some conformities to be verified (yellow colour) and some non-conformities (red colour) that need further investigation and/or adjustment work. A reading of the overall document for all compartments highlights that there are non-conformity points while other non-conformities affect

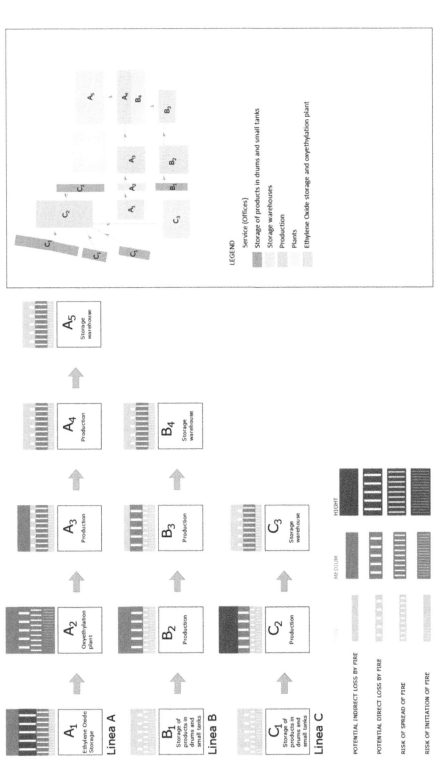

Figure 14.7 Qualitative classification of production processes.

Figure 14.8 Classification of activities.

multiple compartments across the board. Therefore, the targeted action on individual strategies (e.g. fire operations, fire control, fire-safety management, etc.) can mitigate fire risk in the complexity of the plant.

However, the result in Figure 14.10 is not sufficient by itself to completely quantify the fire risk, so the methodology adopted the use of matrices and data were cross-referenced with the fire load in order to establish priorities for action (PHASE 3).

In particular, matrices have been used that show in abscissa a score ranging from 1 to 3 depending on whether the requirement is compliant (1), compliant with verification (2) or not compliant (3) and in ordinate three levels depending on the fire load (medium, high or very high). The matrix shown in Figure 14.6 is a representative of the method applied with respect to the warehouse named A5.

The methodology adopted requires that at the end of the individual point assessments, the results are graphically represented in easy-to-understand floor plans or graphic diagrams in order to prioritise interventions and plan the necessary actions to reduce the fire risk. The graphical representation of the result of the quantitative fire risk assessment is depicted in Figure 14.12. As can be seen in the present case, there are compartments with compliance to be verified (M2, H2 and V2), which need further investigation and/or retrofitting works.

The analysis of the final fire risk assessment (Figure 14.11), together with the detailed tables (Figure 14.10) for the individual compartment, made it possible to identify any non-conformities in a timely manner and at the same time plan interventions according to the assigned priority.

PHASE 4 involves determining the necessary measures for fire risk mitigation. The specific reference standards provide guidance regarding the identification of interventions to be implemented

WAREHOUSE A5

QUALITATIVE FIRE RISK ASSESSMENT DM 18/10/2019

Risk profiles
Rvita — A4
Rbeni — 1
Rambiente — NOT SIGNIFICANT
Number of people — 4

		COMPLIANCE	COMPLIANCE WITH VERIFICATION	NOT CONFORMING		
STRATEGY S1 REACTION TO FIRE	LEVEL 1 PERFORMANCE				COMPLIANCE	1,00
	LEVEL 2 PERFORMANCE	X				
	LEVEL 3 PERFORMANCE					
	LEVEL 4 PERFORMANCE					
STRATEGY S2 FIRE RESISTANCE	LEVEL 1 PERFORMANCE				COMPLIANCE WITH VERIFICATION	2,00
	LEVEL 2 PERFORMANCE					
	LEVEL 3 PERFORMANCE		X			
	LEVEL 4 PERFORMANCE					
	LEVEL 5 PERFORMANCE					
STRATEGY S3 COMPARTMENTALIZATION	LEVEL 1 PERFORMANCE				COMPLIANCE WITH VERIFICATION	2,00
	LEVEL 2 PERFORMANCE		X			
	LEVEL 3 PERFORMANCE					
STRATEGY S4 EXODUS	LEVEL 1 PERFORMANCE				COMPLIANCE WITH VERIFICATION	2,00
	LEVEL 2 PERFORMANCE		X			
STRATEGY S5 FIRE SAFETY MANAGEMENT	LEVEL 1 PERFORMANCE				COMPLIANCE WITH VERIFICATION	2,00
	LEVEL 2 PERFORMANCE		X			
	LEVEL 3 PERFORMANCE					
STRATEGY S6 FIRE CONTROL	LEVEL 1 PERFORMANCE				COMPLIANCE	
	LEVEL 2 PERFORMANCE	X				
	LEVEL 3 PERFORMANCE					3,00
	LEVEL 4 PERFORMANCE			X	NOT CONFOMING	
	LEVEL 5 PERFORMANCE					
STRATEGY S7 DETECTION AND ALARM	LEVEL 1 PERFORMANCE					3,00
	LEVEL 2 PERFORMANCE				NOT CONFORMING	
	LEVEL 3 PERFORMANCE			X		3,00
	LEVEL 4 PERFORMANCE					3,00
STRATEGY S8 SMOKE AND HEAT CONTROL	LEVEL 1 PERFORMANCE				NOT CONFORMING	3,00
	LEVEL 2 PERFORMANCE			X		
	LEVEL 3 PERFORMANCE					
STRATEGY S9 FIREFIGHTING OPERATIONS	LEVEL 1 PERFORMANCE				NOT CONFORMING	3,00
	LEVEL 2 PERFORMANCE			X		
	LEVEL 3 PERFORMANCE					
	LEVEL 4 PERFORMANCE					
STRATEGY S10 PLANT SAFETY	LEVEL 1 PERFORMANCE		X		COMPLIANCE	1,00

BENCHMARKS

COMPLIANCE	COMPLIANCE WITH VERIFICATION	NOT CONFORMING

SCORE AWARDED ACCORDING TO COMPLIANCE AS PER DM 18/10/2019

1	2	3
P = 10	11 ≤ P ≤ 30	P ≥ 31

28,00

1	COMPLIANCE
2	COMPLIANCE WITH VERIFICATION
3	NOT CONFORMING

Figure 14.9 Sample table compliance with regulatory requirements – Warehouse A5.

Figure 14.10 Quantitative analysis of regulatory compliance.

WAREHOUSE A5

RISK QUANTIFICATION MATRIX					

FIRE LOAD qf (MJ/m²)	RATING				
FIRE LOAD qf <900 MJ/m² Fire load value derived from Min sterial Decree 03/09/2021. Below this limit, workplaces are defined as low fire risk	**MEDIUM (M)**	M1	M2	M3	
FIRE LOAD 900 < qf < 1,200 MJ/m² fire load value derived from DM 18/10/2019 Strategy S 3 fire load value derived from DM 18/10/2019 Strategy S 5	**HIGH (H)**	H1	H2	H3	
FIRE LOAD qf > 1,200 MJ/m² fire load value derived from DM 18/10/2019 Strategy S 3 fire load value derived from DM 18/10/2019 Strategy S 5	**VERY HIGH (V)**	V1	V2	V3	5000 MJ/m²

PRIORITY OF INTERVENTION	Score of 1 Compliance with regultory requirements	1			
	Score between 11 ≤ P ≤ 30 Compliance with verification to regulatory requirements		2		
	Score including P > 30 Non-compliance with regulatory requirements			3	28

PRIORITY RISK LEGEND	
V3	LACK OF SECURITY REQUIREMENTS, URGENT SCHEDULING
H3	LACK OF SECURITY REQUIREMENTS, PROGRAMMING IN THE VERY SHORT TERM
M3	LACK OF SECURITY REQUIREMENTS, PROGRAMMING IN THE SHORT TERM
V2	INADEQUACY OF SECURITY REQUIREMENTS, PLANNING IN THE VERY SHORT TERM
H2	INADEQUACY OF SECURITY REQUIREMENTS, PLANNING IN THE SHORT TERM
M2	INADEQUACY OF SECURITY REQUIREMENTS, PLANNING IN THE MEDIUM TERM
V1	MAINTENANCE OF RISK MANAGEMENT MEASURES IN VIEW OF THE HIGH FIRE LOAD
H1	MAINTENANCE OF RISK MANAGEMENT MEASURES IN VIEW OF THE MEDIUM FIRE LOAD
M1	MAINTENANCE OF RISK MANAGEMENT MEASURES IN VIEW OF THE LOW FIRE LOAD

OUTCOME OF THE RISK ASSESSMENT

Warehouse A5 has a very high fire load 5000 MJ/m² and nonconformities as per DM 18/10/2019 regarding several startegies.
CORRECTIVE ACTIONS NEEDED TO BE PLANNED IN THE VERY SHORT TERM

Figure 14.11 An illustrative example of the matrix adopted with regard to Warehouse A5.

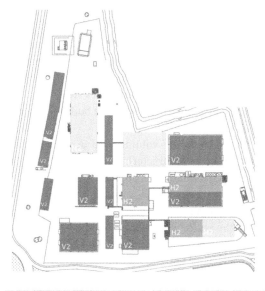

PRIORITY RISK LEGEND	
V3	LACK OF SECURITY REQUIREMENTS, URGENT SCHEDULING
H3	LACK OF SECURITY REQUIREMENTS, PROGRAMMING IN THE VERY SHORT TERM
M3	LACK OF SECURITY REQUIREMENTS, PROGRAMMING IN THE SHORT TERM
V2	INADEQUACY OF SECURITY REQUIREMENTS, PLANNING IN THE VERY SHORT TERM
H2	INADEQUACY OF SECURITY REQUIREMENTS, PLANNING IN THE SHORT TERM
M2	INADEQUACY OF SECURITY REQUIREMENTS, PLANNING IN THE MEDIUM TERM
V1	MAINTENANCE OF RISK MANAGEMENT MEASURES IN VIEW OF THE HIGH FIRE LOAD
H1	MAINTENANCE OF RISK MANAGEMENT MEASURES IN VIEW OF THE MEDIUM FIRE LOAD
M1	MAINTENANCE OF RISK MANAGEMENT MEASURES IN VIEW OF THE LOW FIRE LOAD

Figure 14.12 Plant floor plan with the results of the risk assessment highlighted.

according to the requirements to be met. In some cases, the adaptation interventions are easy to implement such as the implementation of management measures, whereas in other cases they may present critical issues related to economic aspects or difficult to implement; therefore, through the application of performance methodologies, in-depth investigations are possible that can ascertain the actual performance being analysed (PHASE 5).

Specialised in-depth investigations can cover all requirements such as fire resistance of structures, with the aim of ascertaining whether existing structures guarantee the required performance or upgrades are needed. In the case under consideration, in addition to the macro-zones highlighted in Figure 14.6, there are also several point stockpiles of combustible material serving the activity. Therefore, the following is a summary of the in-depth study carried out.

14.3.3 Fire Risk Assessment

In the process of fire risk assessment, careful analyses of workplaces were preliminarily carried out in the consideration of

- risk profiles R_L, R_p and R_e (they are respectively the risk for occupants' life, properties and the environment);
- fire hazard;
- context and environment in which the hazards are embedded;
- number and type of occupants exposed to the hazard.

The R_L risk profile can be identified through the combination of the prevailing characteristics of the occupants located in the fire compartment (δ_{occ} factor) and the prevailing characteristic fire growth rate (δ_α factor).

A non-exhaustive indication of the R_L risk profile for the most common types of use (occupancy) is given in Table 14.1.

Table 14.2 shows the criterion for assigning the factor (δ_α factor) according to the prevailing fire velocity. In the case study under consideration, the prevailing characteristic fire speed was assumed to be 300 seconds (average) for all compartments.

Table 14.3 highlights the R_L value identified for each of the compartments under study.

In all compartments, the occupants are in an awake state and familiar with the building (prevalent characteristics of the occupants ($\delta_{occ} = A$)).

Based on quantity, type and distribution of material, building volume, and accessibility to rescue operations, it is possible to define the parameter $\delta_a = 4$ (characteristic fire growth rate: 75 ultra-rapid). Therefore, the allocation of the R_L parameter conducted in application of the indications of national regulation and taking into account the values of δ_{occ} and δ_a defined in the previous sections for the compartments under consideration is equal to $R_L = A4$.

The attribution of the R_P risk profile was made for the entire activity according to the strategic nature of the building and the assets it contains (see the Italian Fire Prevention Code – IFC15). In the present case, the building is not artistically and historically listed and is not strategic for the purposes of public rescue and civil defence planning, so the R_P risk profile attributed is $R_P = 1$ (Table 14.3).

The environmental risk can be considered mitigated by the application of all firefighting measures related to the R_L and R_P risk profiles, which generally allow this risk to be considered not significant.

Following the identification of the risk profiles R_L, R_P and R_E, the fire risk was determined, according to different criteria and methods depending on the classification of the compartments.

In the specific case under consideration, the risk profile assigned is R4.

The determination of the risk profile is the main input data, together with the specific fire load of the individual compartments, to determine the performance requirements to be met.

Table 14.1 Prevalent characteristics of occupants for R_L determination.

Prevalent characteristics of occupants δ_{occ}		Examples
A	Occupants are in a waking state and are familiar with the building	Offices not open to the public, schools, private garages, private sports centres, general manufacturing activities, warehouses, industrial sheds
B	Occupants are in an awake state and are unfamiliar with the building	Commercial businesses, public garages, exhibition and public entertainment businesses, convention centres, offices open to the public, restaurants, medical offices, medical clinics and public sports centres
C	Occupants may be asleep:	
Ci	In individual activity of long duration	Civilian homes
Cii	In managed activity of long duration	Dormitories, residence halls, student residences and residences for self-sufficient people
Ciii	In managed activity of short duration	Hotels and mountain lodges
D	Occupants receive medical care	Inpatient hospital stays, intensive care, operating rooms, residences for dependent and nursing caregivers
E	Occupants in transit	Train stations, airports and subway stations

[1] When C is used in this document, the relevant indication applies to Ci, Cii and Ciii.

Table 14.2 Prevailing characteristic rate of fire growth for RL determination.

δ_α	t_α [1]	Criteria
1	600 s slow	Activity areas with specific fire load $q_f \leq 200$ MJ/m^2 or where there are predominant materials or other combustibles that contribute negligibly to the fire.
2	300 s medium	Areas of activity where materials or other combustibles that contribute moderately to fire are predominantly present.
3	150 s rapid	Environments where significant amounts of stacked plastics, synthetic textiles, electrical and electronic equipment and combustible materials not classified by reaction to fire are present. Areas where vertical stacking of significant quantities of combustible materials occurs with 3.0 m $< h \leq 5.0$ m [2]. Storage areas classified HHS3 or activities classified HHP1, according to UNI EN 12845. Environments with technological or process facilities using significant amounts of combustible materials. Environments with the simultaneous presence of combustible materials and hazardous workings for fire purposes.
4	75 s ultra-rapid	Areas where vertical stacking of significant amounts of combustible materials with $h > 5.0$ m [2] occurs. Storage areas classified HHS4 or activities classified HHP2, HHP3 or HHP4, according to EN 12845. Areas where significant quantities of substances or mixtures hazardous to fire are present or in process, or cellular/expanded plastic materials or combustible foams not classified for reaction to fire.

Unless more in-depth evaluations are made by the designer (e.g. literature data, direct measurements, ...), at least the quantities of materials in compartments with specific fire load $q_f \leq 200$ MJ/m^2 are considered not significant for this classification.

[1] Prevailing characteristic fire growth rate.

[2] With h stacking height.

Table 14.3 R_{beni} profile determination.

		Constrained activity or scope	
22		No	Yes
Activity or strategic area	No	$R_P = 1$	$R_P = 2$
	Yes	$R_P = 3$	$R_P = 4$

In the case study under consideration, relating to an industrial chemical plant, the performance levels to be guaranteed are very high as there are significant amounts of combustible material within the individual raw material storage, production and finished goods warehouse areas.

For example, the qualitative analysis (PHASE 1) in Figure 14.7 preliminarily shows that areas where flammable or combustible materials are stored and processed present very high potential risks, compared to the finished material storage activity. This expeditious analysis of fire risk allowed a specific level of attention to be placed on these areas.

However, a qualitative assessment is not meaningful unless the level of fire risk is somehow determined based on the actual fire load and the condition of the premises. For example, the outcome of the quantitative analysis carried out in the areas where acetylene is present preliminarily showed a high overall risk, which on further investigation appears to be partially mitigated by existing measures. The adoption of management strategies, such as the implementation of fire-fighting operations, ensures that regulatory requirements are met.

The quantitative analysis also revealed that storage warehouses, in view of the high fire load, are inadequate for the amount of material present. In these specific cases, it is possible to achieve the required performance levels either through passive protection interventions (fire proofing) or through the use of performance methodologies capable of demonstrating that in the event of a fire the measures present such as active protection of sprinkler systems are able to guarantee the safe exodus of the occupants. Through fire simulations combined with structural analysis (deformations) and performance evaluations of the fixed sprinkler systems, it is possible to determine the actual level of resistance of the structures and demonstrate the fulfilment of the required performance levels.

Specific insights into the design of gas detection systems, to be installed in hazardous areas, are carried out in this case as there are several storage tanks in the chemical plant under consideration.

14.3.4 Specific Insights

In the event of a release of a flammable and/or toxic substance, it is crucial in order to avoid escalation to a major accident that the release or fire be detected as early as possible.

This can only be achieved with a sensor system that constantly monitors critical plant areas and, as appropriate, notifies the control room of the incident or triggers safety systems automatically.

The purpose of this semi-quantitative study is to define the characteristics of the fire and gas detection system for outdoor areas.

This type of analysis is carried out in two phases:

1) Definition of site-specific performance targets (performance goals);
2) Semi-quantitative analysis to define the positioning of fire and gas detectors in order to achieve the goals defined in PHASE 1.

The result of this process is a fire and gas detection mapping report that includes

a) description of the analysis process and main results;
b) graphical representation of the fire and gas detection coverage.

There are no flame detection systems in the areas of the plant covered by this study, so a specialised in-depth analysis was carried out in order to bring the level of coverage to an acceptable level with respect to the hazards present. The level of coverage was quantified taking into account the characteristics of a specific multi-spectrum detector suitable to detecting hydrogen and methanol fires, both of which are present at the site.

Regarding the quantitative parameters for the coverage levels of the detection systems, the following categories (Grades B and C) were adopted, applying the company's best practices based on experience, considering the hazards and the possibility of fire extension.

- *Standard risk – Grade B*: Assigned to the area around equipment in process areas with the presence of flammables where a fire could have significant consequences. Fire starts and fires developed may be due to one or a combination of the following factors:
  - Fuel.
  - Established fire risk.
  - Potential for escalation to more severe consequences.

Grade B areas are extended at least 2 m around a piece of equipment or the corresponding containment basin/boundary and should be surrounded by Grade C areas or be confined through fire-resistant structures.

- *Low risk – Grade C*: Assigned to process areas with the presence of flammables where all of the following conditions are met:
  - The flashpoint of the flammable substance is above 60 °C.
  - The release pressure is less than 1 barg.
  - Any puddles of flammable substance would have an area of less than 50 m^2.

Grade C is assigned to areas, including those around equipment, within areas where hydrocarbons are processed, that are less susceptible to fire than Grade B areas, where it is still desirable to have some fire protection.

Similar to Grade B, Grade C areas also extend at least 2 m around equipment or the corresponding containment basin/diked area.

Grades B and C refer to the maximum severity of a fire in terms of the radiant heat output (RHO) that must be detected by a flame detector in less than 10 seconds. These severity levels translate into distances depending on the actual location of the detector, which is given in terms of 'D' and discussed in detail below.

Table 14.4 shows the thresholds corresponding to Grades B and C, whereas for Grade S areas the thresholds must be defined in relation to the specific hazards.

The alarm detection time for areas that fall into Grades B and C must be less than 10 seconds. The coverage requirements for fire detection in the different areas are expressed in percentages and are defined as modelling ideal events.

This definition does not take into account the frequency or probability of a given event, but rather is a quantification of the performance of the detection system should a fire occur at a given point within the area of interest (categorised area).

Table 14.5 shows the minimum coverage percentages required of the flame or gas detector, which take into account shaded areas due to obstructions and loss of coverage for any other reason.

Table 14.4 Typical categorisation for areas with hydrocarbon hazards and associated fire size.

Grade	Severity of fire (RHO) for	Severity of fire (RHO) for
Standard risk (B)	Single alarm (1ooN)	Confirmed alarm (2ooN)
Low risk (C)	90 kW (1.5D)	360 kW (3.0D)
Special risk (S)	640 kW (4.0D)	640 kW (4.0D)

It should be noted that these coverage factors are to be used as a general rule, and in each case an in-depth engineering assessment must be made in the light of the knowledge/interpretation of individual cases. Depending on where the obstructions are located in some cases, a Grade B area with 78% coverage might be more reliable than an area with 90% coverage, and flame detection coverage assessment (FDA) detectors are depicted with two-dimensional images that reproduce the field of view of each detector. The ground footprint is divided into detectable fire dimensions, as shown in Figure 14.13. The use of a multi-spectrum infrared (IR) detector, characterised by an effective viewing distance of 20.25 m referring to the fire of a standard 30 × 30 cm *n*-heptane basin, was considered for the evaluation of the coverage levels reported in the present study.

As determined during the performance objective specification procedure, each area was categorised according to its local hazards and escalation risks. Specifically in the mapping, the colour orange corresponds to Grade B, green to Grade C whereas the colours light blue (standard risk) and blue (low risk) are assigned to Grade S (special risk). Grade S in the present study was used in reference to the presence of methanol and hydrogen in the processes given the transparent flame characteristic of these substances.

A special software overlays the detector footprint with the Grademap and, through a crossover table, builds a graphical representation of the coverage provided by detectors in the area (Figure 14.14).

The final graphic file returns an objective estimate of the level of fire detection coverage in the area. The level of coverage calculated for each area, expressed in percentage terms, can then be used to determine the adequacy of the existing detection system. The fire detection coverage

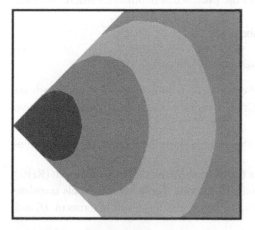

Figure 14.13 Graphic representation of a detector (footprint on the ground).

Table 14.5 Criteria for minimum coverage factors.

Grade	Coverage factor target
B	80%
C	70%

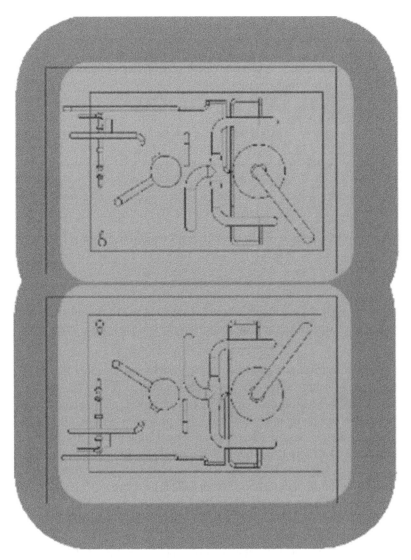

Figure 14.14 Graphic representation of the grading (Grade B surrounded by Grade C).

assessment (FDA) tool is a two-dimensional representation of a three-dimensional space, and the results of the FDA must be interpreted in the light of the skills of the analyst who must assess the reliability of the results. Table 14.6 shows the colour scale with the purpose of illustrating the meaning of the different categories and associated colours.

In addition, a third Grade S, special risk, has been defined to apply to areas where the risk, although having similarities with Grade B or Grade C, is characterised by differences that do not make it directly comparable to Grade B or Grade C:

- *Special risk – Grade S*: Assigned to areas that cannot be classified as either Grade B or Grade C. In these cases, the closest risk grade is identified and changes from the identified grade are defined.

Table 14.6 Risk level categorisation.

Grade		FDA colour map	Fire size (RHO) for single alarm (1ooN)	Fire size (RHO) for confirmed alarm(2ooN)
Standard risk (B)		Orange	Orange	Orange
Low risk (C)		Green	Green	Green
Special risk (S)	Light blue	Light blue	Light blue	360 kW (3.0D)
	Blue	Blue	Blue	640 kW (4.0D)

To specify the detection performance required for a Grade S area, it is necessary to determine the following factors:

- The nearest applicable grade and any exceptions or modifications needed (e.g. Grade B without automatic actions).

Or

- The type of fire and fuel (the hazard), the risk that needs to be mitigated and the planned actions, including the time of intervention (the role), the required sensitivity and the system outputs (the function).

Flame detectors should be positioned to allow a good view of the local hazard (as identified by the categorisation and mapping process) and to allow easy access for maintenance.

In the present in-depth analysis, the outdoor areas of the chemical plant identified in Figure 14.10 in which various mixed flammable fluids are present were examined. It is to be believed, depending on the pressure and nature of a potential release, that puddle fires and jets of ignited liquid phase substance could form if an ignition is present. No fire detection systems are present in the area under investigation. Therefore, a specialised study was carried out in order to determine the type and location of detectors to be installed.

Table 14.7 summarises the case under consideration.

At the end of data processing, the programme returns the outcome of the evaluation with the location of the detectors as listed in Table 14.8.

The outcome of the assessment with the identification of the detectors to be installed is included in Annex A.

14.4 Results

Regarding the chemical plant in question, the fire risk assessment was carried out through the use of a five-step methodology with the objective of characterising and verifying compliance with regulatory requirements and highlighting any non-conformities.

The analysis of the fire risk assessment allowed for the early identification of possible non-conformities and the planning of interventions according to the assigned priority.

At the end of the assessment, all interventions and/or specialised investigations to be carried out to bring the plant into compliance with the regulatory standard were identified on time. In the present case study, a specialised in-depth assessment was carried out for outdoor areas

Table 14.7 Summary of relevant information for fire detection.

Description of the hazard

The area contains several mixed flammable fluids. It is to be believed, depending on the pressure and nature of a potential release, that puddle fires and jets of ignited liquid phase substance could form if an ignition is present.

Voting strategy/control measures

The cause/effect diagram is not available; however, it is assumed that

a) the intervention of a single detector triggers the alarm in the control room;
b) in areas where the number of detectors is more than one, the intervention of at least two detectors triggers the confirmed alarm in the control room and the activation of the automatic extinguishing system;
c) audible and visual alarms are present in the control room;
d) a plant-wide acoustic alerting system is present.

Grading

Grade	Description	Substance
C	IBC bulk containers approximately $1 \times 1 \times 1$ m in size located in outdoor areas in different parts of the plant.	Miscellaneous

Note: The above grade has been applied conservatively in view of the fact that different substances having different flammability characteristics and in varying quantities may be present in the areas indicated. The probability of a significant accidental event occurring is low because the tanks are at atmospheric pressure; however, the areas will be equipped with a fixed extinguishing system.

Table 14.8 Summary of technologies planned for fire detection.

Area	Detector	Height (m)	Comments
External areas southeast EL: 0000			
External plant areas (1)	Multi-spectrum IR	4.5	
		4.5	
External areas northwest EL: 0000			
External plant areas (2)	Multi-spectrum IR	3.2	
		3.2	
		4.5	
		2.4	
External areas northeast EL: 0000			
External plant areas (3)	Multi-spectrum IR	5.0	Height determined by taking into account the presence of the REI120 walls of height 3.3 m and the two rows of IBC bulk containers about 4 m high.
		5.0	
		5.0	
		5.0	

where substances with different flammability characteristics and in varying quantities are present and stored in IBC bulk containers with the aim of identifying the measures necessary for risk mitigation.

The outcome of the assessment represented graphically in Annex A made it possible to identify the number, type and location of the fire detector to be installed to bring the acceptable level of risk with respect to the potential hazards present. In particular, the conducted activities highlighted in-depth areas (Figure 14.15), with a specified grade map (Figure 14.16), including external areas (Figures 14.17–14.19). Grading is based on the location, typology and number of the detectors selected for the various areas, as summarised in Tables 14.9, 14.10 and 14.11.

Fire risk assessment for companies managing multiple PV plants:

Topic	Description
Scope	Given the large number of fires affecting PV plants (originated by the PV installations or escalated by the PV installations), the scope of the activity verifies the fire risk level of the plants in a portfolio that lists plants in different configurations, locations and remotely operated.
Context	Context includes different configurations: PV plant installed in urban areas (mainly on rooftops) and in rural areas.
Criteria	Due to the large number of the installations, template configurations have been selected. Each template configuration has been assessed with Bow-Tie methodology, and each specific plant, given the template Bow-Tie, has been individually assessed with a LOPA quantitative assessment to verify the specific fire risk level and impacts.
'Risk of what'	Fire risk is posed by fires originating in PV plant installations and by fires originating from outside the PV plants (including wildfire propagating to the PV plant areas and compartment fires propagating to PV plants installed on building roofs).
'Risk to what'	Risk is posed to occupants, assets (including third-party assets affected by the PV fires), environment and reputation (PV plants may be installed on or nearby strategic infrastructure).
'Risk from what'	Fire threats may originate by PV plant component failures or PV plants may determine the escalation of external fires.

The fire risk analysis for photovoltaic systems represents a novelty in the panorama of fire risk analysis for which recognised and established working methods are not available, both for the technical-plant peculiarities and the recent affirmation of this industrial sector.

This methodological proposal is based on the Bow-Tie method and consisted of the following seven steps:

1) Creation of some Bow-Tie models, conceptually applicable to all the systems under examination.
2) Extraction of critical barriers.
3) For each plant, assign the performance standards to the critical barriers and define their level of effectiveness.
4) Frequency quantification of Bow-Tie models (frequency of causes and PFD of barriers) with a qualitative or a semi-quantitative approach (LOPA).
5) For each system, the definition of parametric values such as hazard, vulnerability and exposure.
6) For each system, and for each consequence, the definition of the risk level.
7) Definition of a programme of recommendations.

Figure 14.15 Floor plan of the chemical plant with in-depth areas highlighted (fire–gas mapping).

Grade Map Colour Key

Grade	Grademap Colour	Alarm Coverage	Ctl Action Coverage	Votes
A	Light Cyan	1.5D	3.0D	2
B	Orange	1.5D	3.0D	2
C	Green	4.0D	4.0D	2
S	Blue	4.0D	4.0D	2

Assessment Colour Key

Alarm Target Met	Control Action Target Met	Assessment Colour	Notes
Yes	Yes	Green	1
No	Yes	Yellow	5
Yes	No	Orange	2
Late	Late	Brown	3
No	No	Red	4

Notes:

1. Control and alarm action voting requirements are both met.
2. At least one detector can 'see' the fire, but not enough to obtain the votes required for control action.
3. This case occurs when detectors are too far from the hazard to 'see' the specfied target fire. However, if the fire incident escalates then the fire detection system will respond.
4. No detector can see this hazard and there will be no response.
5. This case arises when the voting requirement can be met for the larger control action fire size, but not for the smaller alarm action fire size.

Figure 14.16 Interpretation of FDA Grademap and assessment colours.

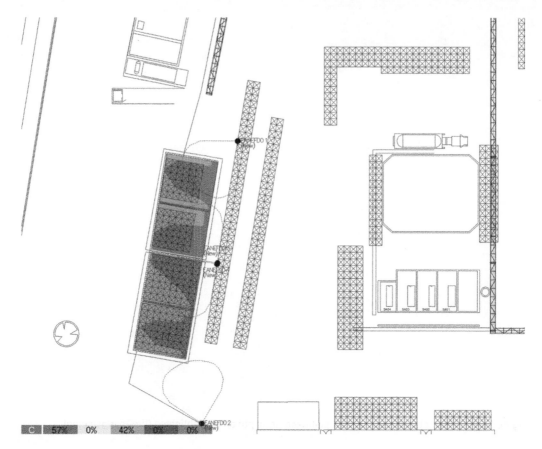

Figure 14.17 External areas in southeast – coverage map.

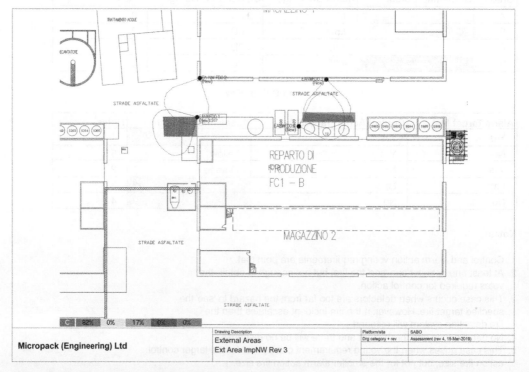

Figure 14.18 External areas in northeast – coverage map.

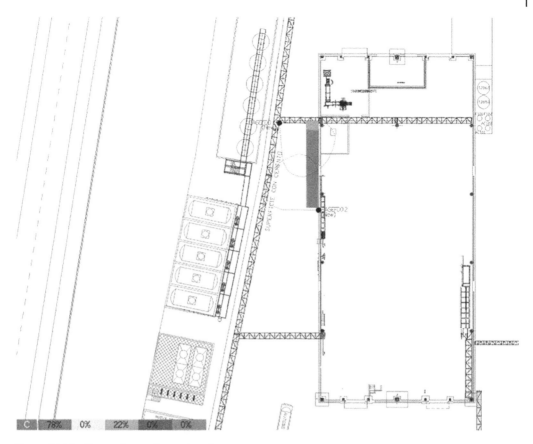

Figure 14.19 External areas in northeast – coverage map.

Table 14.9 External areas in southeast – coverage map.

Area	Detector	Height (m)	Comments
External areas southeast EL: 0000			
External plant areas (1)	Multi-spectrum IR	4.5	
		4.5	

Table 14.10 External areas in northeast – coverage map.

Area	Detector	Height (m)	Comments
External areas northwest EL: 0000			
External plant areas (2)	Multi-spectrum IR	3.2	
		3.2	
		4.5	
		2.4	
		5.0	

Table 14.11 External areas in northeast – coverage map.

Area	Detector	Height (m)	Comments
External areas northeast EL: 0000			
External plant areas (3)	Multi-spectrum IR	5.0 5.0 5.0 5.0	Height determined by taking into account the presence of the REI120 walls of height 3.3 m and the two rows of cisterns about 4 m high.

STEP 1

There are two 'type' Bow-Ties identified and conceptually applicable to all systems:

- One is relative to the entire photovoltaic system, excluding the inverter cabin.
- The other considers the inverter cabin only.

This distinction between the two plant areas has become necessary due to the characteristics of the inverter cabin, so it is necessary to diversify the causes of fire in principle with respect to outdoor plant areas.

During the Bow-Tie workshop sessions, relevant threats, consequences and barriers have been identified for the specific top events being analysed.

An example of the Bow-Tie model is shown in Figure 14.20.

STEP 2

We proceed to identify which barriers, among those identified during the previous phase, are considered 'critical'. The following barriers are defined as critical:

- Whose failure causes a fire.
- The failure of which contributes to the outbreak of fire.
- The purpose of which is to prevent a fire.
- The purpose of which is to mitigate the consequences of a fire.

STEP 3

For each system, the effectiveness of each individual barrier identified in the Bow-Tie is assessed, following what already discussed in Chapter 9, in the paragraph 'Bow-Tie'.

STEP 4

Bow-Ties produced in STEP 1 are quantified in frequency through a semi-quantitative analysis (by orders of magnitude) using a LOPA analysis.

The frequency of cases is obtained from internationally recognised and proven databases, as well as the PFD of barriers, trying to make the most of organisation's operational experience. Given the absence of recognised fire risk analysis methodologies in the sector, a quantification by orders of magnitude is considered sufficient, making any attempt to improve the numerical accuracy of the frequency and PFD data difficult and of little benefit.

The frequency of causes and the PFD of barriers are then combined according to a LOPA analysis to determine the frequency of occurrence of the consequences.

At the end, then, the frequency values of the following scenarios (both for the two Bow-Ties) are known:

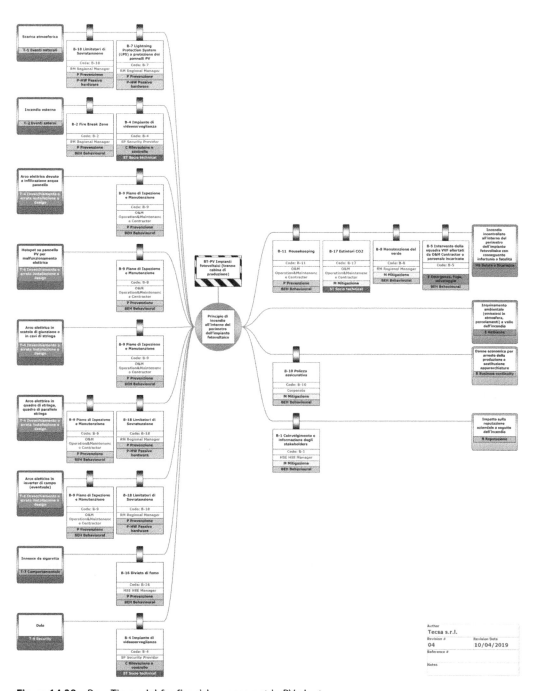

Figure 14.20 Bow-Tie model for fire risk assessment in PV plants.

- *Health and safety*: Uncontrolled fire within the perimeter of the photovoltaic system resulting in an accident or fatality or occupational disease.
- *Environment*: Environmental pollution (atmospheric emissions and percolation) after the fire.
- *Business continuity*: Economic damage due to production stop and equipment replacement.
- *Reputation*: Impact on the company's reputation following the fire.

In this way, Bow-Tie models have been preliminarily quantified.

STEP 5

At this point, it is necessary to decline the information obtained in PHASE 4 with the peculiarities of each system.

This option is permitted thanks to the adoption of certain 'weight' factors: dangerousness (P), vulnerability (V) and exposure (E).

Dangerousness (P)

Hazard is the intrinsic 'propensity' of the site where the system is located to suffer a fire. It is therefore an expression of the greater or lesser probability of occurrence of the initiating causes of a fire.

Analysing the causes defined in the Bow-Tie type, only the causes connected to natural external events can be weighed through a site-dependent factor since the other causes (electrical malfunctions of various kinds and cigarette triggers) cannot be correlated to the choice of the site where the system is installed: the following are therefore the only causes:

- Atmospheric download.
- External fire.

With reference to atmospheric discharges, the map of lightning density in Italy according to the CEI 81–3 standard shows different values in relation to the different areas of Italy (Figure 14.21).

It is therefore proposed to adopt the values shown in this map as an estimate of the frequency of occurrence of atmospheric discharge system by system.

With reference to the external fire, being able to attribute it to phenomena correlated to the environmental temperature, it is of interest to analyse the average annual temperature in Italy. Figure 14.22 (Source: Wikipedia) shows how there is a clear difference in this value between the different regions of Italy.

It is therefore proposed to adopt a correction factor for the frequency of occurrence of an external fire depending on the PV plant location.

Vulnerability (V)

Vulnerability is the inherent characteristic of the system (regardless of where it is located) of whether or not it can withstand a fire. It is related both to the type of system and the way in which the control measures installed on it are managed.

First of all, the effectiveness of the barriers, evaluated in the previous phase of work, is taken into account and then the peculiarities of the barriers on a plant-by-plant basis, using a 'weight' factor (*w*) in the PFD assignment. The factor '*w*' is evaluated in relation to the effectiveness of the barrier.

Exposure (E)

Exposure is a qualitative measure of the importance of the target in the event of a fire. It takes into account the following parameters:

- Organisational assets involved in possible fire.
- Personnel involved (number of people).

Figure 14.21 Map of lightning density in Italy.

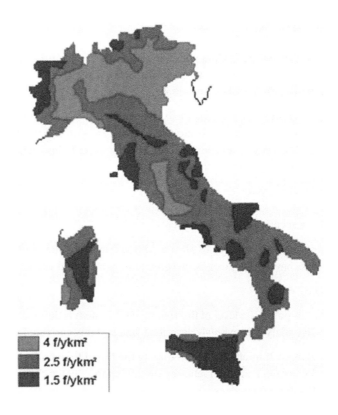

Figure 14.22 Annual average temperature in Italy.

- Severity of the business interruption in the case of production failure from that specific plant.
- Severity of reputational risk in the case of fire.
- Impact on neighbouring areas.

The exposure therefore contributes to increasing the level of severity defined a priori in STEP 4 taking into account the peculiarities of the system, its crowding and its surroundings.

The severity of all the consequences for installations whose consequences (identified in Bow-Tie) have a significant impact on

- Public service.
- Strategic infrastructure (e.g. airport operations and high-voltage lines).
- Local communities.

Similarly, the severity of all the consequences for the systems is further increased for those PV plants having a big extension (greater than a defined value) or a big installed power.

STEP 6

At this point, the level of fire risk for each system is defined, by means of the SFPE matrix and the LOPA analysis previously described, and suitably corrected the following weight factors:

- *Danger*: It intervenes by modifying the initial frequency of the cause 'external fire'.
- *Vulnerability*: It intervenes by modifying the PFD of the barriers and changing the frequency of the top event.
- *Exposure*: It intervenes by modifying the severity associated with the consequences.

STEP 7

In conclusion, having defined the acceptability risk criteria, the plants are listed in order of decreasing risk (high, medium and low), and for each plant an action plan is defined to improve the effectiveness of the existing barriers (by first intervening on very low, then low and finally good barriers).

Priority of intervention has been given to installations with high, then medium and finally low fire risk.

15

Conclusions

As stated in ISO 31000 prominent standard on risk management, *organizations of all types and sizes face internal and external factors and influences that make it uncertain whether and when they will achieve their objectives. The effect this uncertainty has on organization's objectives is a risk.*

All activities of an organization involve risk.

Risk management is therefore an integral part of all organisational processes and of decision-making. It should be systematic, structured and timely. It also should be based on the best available information and tailored.

It should consider human and cultural factors ('soft' factors) together with technical and organisational factors ('hard' factors).

Intuitively and understandably, fire risk is one of the risks that requires the most attention and the development of a robust strategy, maintained over time to ensure the safety of occupants, the environment, assets and business continuity.

Despite all the efforts made over the years for fire risk prevention and mitigation in various sectors, from civil construction to the most complex industrial plants, a number of severe fires still happen. Some of the most recent fires, which have unfortunately and negatively affected various sectors, including civil construction, rail and maritime transportation, the steel and chemical process industries, forestry (Grenfell Tower, Viareggio, Norman Atlantic, Thyssenkrupp and Pedrogao Grande) and the world of alternative renewable energies (with fires involving storage systems, photovoltaic modules, wind turbines, etc.) as some real and interesting incidents discussed in this book, underline that continuous technological and scientific progress can only be accompanied by a continuous improvement in the approach to fire prevention, which can take into account both new risks and the lessons learnt from occurred fires.

Therefore, fire risk assessment becomes the best tool to deal with complexity and emerging threats that determine unexpected fire scenarios.

Fire risks and associated fire scenarios should be countered with a proper and solid fire-safety strategy that allows owners to maintain an acceptable risk level over time.

The aforementioned events showed several shortcomings in a number of protection layers; therefore, fire-safety strategy selection should carefully take into account several key factors, including the efforts to maintain it for the building or the industrial plant life cycle. While design stages require a specific fire risk assessment, with a level of detail proportional to the complexity and to the fire hazards, the outcomes of this analysis should be considered to define fire-safety criteria as the basis of the strategy to be maintained during subsequent operation phases. Several methods are available for risk identification and risk analysis, with qualitative, quantitative or

Fire Risk Management: Principles and Strategies for Buildings and Industrial Assets, First Edition.
Luca Fiorentini and Fabio Dattilo.
© 2023 John Wiley & Sons, Inc. Published 2023 by John Wiley & Sons, Inc.
Companion Website: www.wiley.com/go/Fiorentini/FireRiskManagement

mixed approaches, and their use is supported by internationally available standards and guidelines. Their selection, and eventually their combination, should be based on the adoption of a coherent framework that will allow risk-informed and risk-based decisions.

This is also extremely important when a performance-based approach is selected instead of adopting the compliance with prescriptive requirements. Within a performance-based approach, performance criteria should be defined for each vulnerability to be considered (people, environment, property and business continuity), and performance level should be set, verified and maintained over time.

Fire-safety management during operation (including emergency situations) becomes even more important since all the elements of the fire strategy should work together to guarantee the global risk reduction factor.

Alternative fire strategies can be evaluated against their cost and maintainability over time, considering all the modifications that may affect the fire strategy and the fire-safety elements over time. It clearly appears that fire-safety design should lead to a holistic fire strategy whose responsibility shifts from designers to owners and, during the initial design phase, should become part of the overall design phase of the building performance, where several different pools of expertise and interests are present or represented.

While fire-safety experts are familiar with the many documents available in the specialised technical literature (codes, standards, handbooks, papers and articles), other actors involved in the design phases often have a generally limited view of fire-safety issues. In addition, fire-safety planners themselves, due to the complexity of projects, methods and tools, often work in multidisciplinary collaboration groups and need a common basis for information exchange.

The basis can be the fire-safety strategy derived from the fire risk assessment, and different fire strategies may be effective in achieving the global fire risk reduction to reach an acceptable/tolerable level with potential different requirements for the operation phase that should be known and managed, including any need of fire risk assessment and fire strategy review in the case of modifications that may impact the fire-safety level.

This comprehensive risk-based approach results in a fire-safety strategy that is the holistic composition of preventive and protection measures (each one having its risk reduction factor) aimed at maintaining an acceptable risk level. Its proper management over time in large and complex environments could give a number of benefits even in case some prescriptive requirements are not met or could not be met (heritage buildings, assets under modifications, etc.) or in case multiple asset owners should take risk-informed decisions to deploy limited resources to their asset portfolio and they are requested by their stakeholders to justify the fire-safety-related investments with the achieved risk reduction.

A similar approach can be followed both in buildings and industrial installations. The differences can only be attributed to the type of fire (and associated dynamics as well as direct and indirect or domino effects), the type of occupants and the methodologies generally used to conduct assessments to estimate the probability of occurrence and the expected consequences.

It has been pointed out, with the support of the discussion of real severe fire events used as lessons learnt (a fundamental step of a complete risk management framework), that same principles of formulating a strategy from the key elements affecting fire safety and the criteria for maintaining requirements over time are valid in different areas and across different sectors.

The holistic approach to fire-safety design (considering specific insights regarding human behaviour in fire, environmental impact of fires, availability on demand of preventive and protection measures, etc.) requires competences shared among different experts led by a common fire-safety culture, a common fire-safety goals and a strategic view. This view can help in integrating fire-safety considerations into organisation management systems, in order to achieve a proactive and a generative approach during the operation phase to managing fire safety with the combination of fire risk assessment practices and fire risk management activities under the satisfaction of the fire principles aimed to the solution of complexity and new or emerging threats. These activities can be accomplished considering the abundant technical literature available including prominent technical standards (especially those by SFPE and NFPA) that, starting from the fire principles, lead towards the definition of a balanced fire safety strategy given the specific socio-technical system under consideration, being it a civil building, an industrial asset or a special premise posing peculiar fire risks.

Bibliography

Ahmad, S., Hashim, H., and Hassim, M. (2016). A graphical method for assessing inherent safety during research and development phase of process design. *Journal of Loss Prevention in the Process Industries* 42: 59–69.

AIChE-CCPS (2018). *Bow Ties in Risk Management – A Concept Book for Process Safety*, 1e. Hoboken, NJ, USA: The American Institute of Chemical Engineers.

American Institute of Chemical Engineers (2008). *Inherently Safer Chemical Processes: A Life Cycle Approach*. New York, NY: AIChE.

American Institute of Chemical Engineers (2018). *Bow Tie Risk Management*, 1e. New York, NY: Wiley.

Amyotte, P. (2020). The role of inherently safer design in process safety. *The Canadian Journal of Chemical Engineering*, 99. 10.1002/cjce.23987.

Amyotte, P.R., Khan, F.I. (2021). The role of inherently safer design in process safety. *The Canadian Journal of Chemical Engineering*, 99: 853– 871.

ANAS S.p.A. Direzione Centrale Progettazione (2006). Linee Guida per la progettazione della sicurezza nelle Gallerie Stradali.

Assael, M. and Kakosimos, K. (2010). *Fires, Explosions, and Toxic Gas Dispersions. Effects Calculation and Risk Analysis*. Boca Raton, FL: CRC Press/Taylor & Francis.

Aven, T. (2012). *Foundations of Risk Analysis*, 2e. Chichester, UK: John Wiley.

Aven, T. and Zio, E. (2018). *Knowledge in Risk Assessment and Management*. New York, NY: Wiley.

Bansal, A., Kauffmann, R., Mark, R., and Peters, E. (1991). *Financial Risk and Financial Risk Management Technology (RTM): Issues and Advances*. New York, NY: Leonard N. Stern School of Business, New York University.

Baron, H. (2015). *The Oil & Gas Engineering Guide*, 2e. Paris: Editions Technip.

Barry, T. (2002). *Risk-Informed, Performance-Based Industrial Fire Protection*, 1e. Knoxville, TN: Tennessee Valley Pub.

Baybutt, P. (2015). Calibration of risk matrices for process safety. *Journal of Loss Prevention in the Process Industries* 38: 163–168.

Benintendi, R. (2021). *Process Safety Calculations*, 2e. Amsterdam: Elsevier.

Bi, J., Guo, X., Liu, M., and Wang, X. (2010). High effective dehydration of bio-ethanol into ethylene over nanoscale HZSM-5 zeolite catalysts. *Catalysis Today* 149 (1–2): 143–147.

Bigi, S. (2018). Gli strumenti ingegneristici a supporto dell'indagine forense in caso di incendio. Caso studio: La propagazione dell'incendio a bordo della M/N Norman Atlantic, dallo studio della dinamica occorsa alle "lessons learnt".

Bragg, S. (2015). *Enterprise Risk Management*. Accounting Tools.

Fire Risk Management: Principles and Strategies for Buildings and Industrial Assets, First Edition.
Luca Fiorentini and Fabio Dattilo.
© 2023 John Wiley & Sons, Inc. Published 2023 by John Wiley & Sons, Inc.
Companion Website: www.wiley.com/go/Fiorentini/FireRiskManagement

British Standard Institution (2014). *BS ISO 55001:2014 Asset Management Systems – Requirements.* British Standard Institution.

Britton, L. (1994). Loss case histories in pressurized ethylene systems. *Process Safety Progress* 13 (3): 128–138.

BS 9999:2008. Code of practice for fire safety in the design, management and use of buildings. British Standards Institution (BSI). http://www.bsigroup.com.

Cancelliere, P.G., Manzini, G., and Mazzaro, M. (2017). A review of the photovoltaic module and panel fire tests. IFireSS 2017 – 2nd International Fire Safety Symposium, Naples, Italy.

CCPS (1999). *Guidelines for Chemical Process Quantitative Risk Analysis*, 2e. New York, NY: Wiley.

CCPS (2008). *Guidelines for Hazard Evaluation Procedures*, 3e. New York, NY: Wiley.

CCPS (2009). *Inherently Safer Chemical Processes: A Life Cycle Approach*, 2e. New York, NY: Wiley.

CCPS (2011). *Layer of Protection Analysis*, 1e. New York, NY: Wiley.

CCPS (2013). *Guidelines for Enabling Conditions and Conditional Modifiers in Layer of Protection Analysis*, 1e. New York, NY: Wiley.

CCPS (2015). *Guidelines for Initiating Events and Independent Protection Layers in Layer of Protection Analysis* [Electronic Resource], 1e. New York, NY: Wiley.

CCPS (2018). *Bow Ties in Risk Management: A Concept Book for Process Safety*, 1e. New York, NY: Wiley.

CCPS (Center for Chemical Process Safety) (2003). *Guidelines for Investigating Chemical Process Incidents*, 2e. New York, NY: American Institute of Chemical Engineers.

CERMAP (2021). Cryogenic studies section physical explosion: LNG rapid phase transitions (RPT). (G. d. France, A cura di) Tratto il giorno January 22, 2021 da https://www.youtube.com/watch?v=h-EY82cVKuA.

Chiaia, B., Marmo, L., Fiorentini, L. et al. (2017). Incendio della motonave Norman Atlantic: Indagini multidisciplinari in incidente probatorio. IF CRASC 17. Milano: Dario Flaccovio Editore, 129–140.

COSO (2017). *Enterprise Risk Management – Integrating with Strategy and Performance*. COSO.

Cote, A. (2008). *Fire Protection Handbook*. Quincy, MA: National Fire Protection Association.

Cox, A., Ang, M., and Lees, F. (1998). *Classification of Hazardous Locations*. Rugby: Institute of Chemical Engineers.

Cox, L. (2008). What's wrong whit risk matrices? *Risk Analysis* 28 (2): 497–512.

Cozzani, V., Tugnoli, A., and Salzano, E. (2009). The development of an inherent safety approach to the prevention of domino accidents. *Accident Analysis & Prevention* 41 (6): 1216–1227.

Crawley, F. (2020). *A Guide to Hazard Identification Methods*, 2e. San Diego, CA: Elsevier.

Daèid, N.N. (2004). *Fire Investigation*. Boca Raton, FL: CRC Press.

Danzi, E., Marmo, L., and Fiorentini, L. (2021). FLAME: A parametric fire risk assessment method supporting performance based approaches. *Fire Technology* 57: 721–765.

Dattilo, F., Puccia, V., Fiorentini, L. et al. (2008). *L'applicazione alla "Fire Safety Engineering" di strumenti dell'analisi di rischio per aumentare l'efficienza dello studio e l'ottimizzazione del livello degli interventi sul progetto antincendio*. Pisa, Italy: Atti Convegno Valutazione Gestione Rischio.

David, R. and Wilkinson, G. (s.d.). Back to basics: Risk matrices and ALARP.

Decree of the Minister of the Interior. (s.d.). Approval of fire prevention technical standards, pursuant to Article 15 of Legislative Decree 139 of 8 March 2006.

Demichela, M., Piccinini, N., Ciarambino, I., and Contini, S. (2004). How to avoid the generation of logic loops in the construction of fault trees. *Reliability Engineering & System Safety* 84 (2): 197–207.

Dipartimento dei Vigili del fuoco, del Soccorso pubblico e della Difesa Civile (2022). *Codice di prevenzione incendi*. Roma: Ministero dell'Interno della Repubblica Italiana.

Directive 2012/18/EU of the European Parliament and of the Council of 4 July 2012 on the control of major-accident hazards involving dangerous substances, amending and subsequently repealing Council Directive 96/82/EC (2012, July 4).

DNV (2002). Application of QRA in operational safety issues. (HSE, A cura di) Tratto il giorno April 27, 2021 da https://www.hse.gov.uk/research/rrpdf/rr025.pdf.

DNV GL (2014). *Recommended Development and Operation of Liquefied Natural Gas Bunkering Facilities*. London: DNV GL AS.

Dow Chemical Company (1987). *Fire & Explosion Index Dow (F&I)*. American Institute of Chemical Engineers.

ECHA (2021a). Fuels, Diesel – Substance Information – ECHA. Tratto il giorno January 22, 2021 da https://echa.europa.eu/substance-information/-/substanceinfo/100.063.455.

ECHA (2021b). Natural Gas, Sweetened, Liquefied – Substance Information. Tratto il giorno January 22, 2021 da Echa.Europa.eu: https://echa.europa.eu/substance-information/-/substanceinfo/100.096.166.

Energy Institute (2021). How to win hearts and minds: The theory behind the program. Tratto il giorno January 22, 2021 da Energy Institute: https://heartsandminds.energyinst.org/__data/assets/pdf_file/0016/31291/Poster-The-theory-behind-the-programme.pdf.

England, D. and Painting, A. (2022). *An Effective Strategy for Safe Design in Engineering and Construction*. Oxford: Jhon Wiley & Sons Ltd.

EUROPA – eMARS Accident Details – European Commission [Internet]. Minerva.jrc.ec.europa.eu (cited 15 November 2017). Available at https://minerva.jrc.ec.europa.eu/en/emars/accident/view/19158d8a-2bb2-4ea5-dd8a-b0e5a9b7cb0a.

EUROPA – eMARS Accident Details – European Commission [Internet]. Minerva.jrc.ec.europa.eu (cited 15 November 2017). Available at https://minerva.jrc.ec.europa.eu/en/emars/accident/view/5a988875-96e2-2e88-5700-099e5b52f5bc.

European Commission (2020). *Compressor Station Facility Failure Modes: Causes, Taxonomy and Effects*. Luxembourg: European Union.

European Committee for Standardization (2016). *EN 1473: Installation and Equipment for Liquefied Natural Gas – Design of Onshore Installations*. European Committee for Standardization.

European Parliament (2014, February 26). Directive 2014/34/EU of the European Parliament and of the Council of 26 February 2014 on the harmonization of the laws of the Member States relating to equipment and protective systems intended for use in potentially explosive atmospheres.

Fiorentini, L. and Marmo, L. (2011). *La valutazione dei rischi di incendio*. Roma: EPC.

Fiorentini, L. and Marmo, L. (2019). *Principles of Forensic Engineering Applied to Industrial Accidents*. Hoboken, NJ: Wiley.

Fiorentini, L., Marmo, L., Danzi, E., and Puccia, V. (2015). Fires in photovoltaic systems: Lessons learned from fire investigations in Italy. SFPE Magazine, 99.

Fiorentini, L., Marmo, L., Danzi, E., and Puccia, V. (2016a). Fire risk assessment of photovoltaic plants. A case study moving from two large fires: From accident investigation and forensic engineering to fire risk assessment for reconstruction and permitting purposes. *Chemical Engineering Transactions* 48: 427–432.

Fiorentini, L., Marmo, L., Danzi, E., and Puccia, V. (2016b). Fire risk assessment of photovoltaic plants. A case study moving from two large fires: From accident investigation and forensic engineering to fire risk assessment for reconstruction and permitting purposes. Konzerthaus Freiburg (Germany): ICHEME.

Fiorentini, L., Marmo, L., Danzi, E. et al. (2014). Fire risk analysis of photovoltaic plants. A case study moving from two large fires: From accident investigation and forensic engineering to fire risk assessment for reconstruction and permitting purposes. Warsaw, Poland: 47th ESReDa Seminar on "Fire Risk Analysis".

Fiorentini, L., Rossini, V., and Tafaro, S. (2012). L'incendio di una pipe-way di raffineria: L'indagine di un incidente industriale rilevante per il miglioramento della sicurezza. IF CRASC 12. Pisa, Italy: Dario Flaccovio Editore.

Fiorentini, L. and Sicari, R. (2020). *Analisi, Valutazione E Gestione Operativa Del Rischio*, 1e. Roma: EPC editore.

Forck, F. and Fry, K. (2016). *Cause Analysis Manual*, 1e. Brookfield, WI: Rothstein Publishing.

Fraser, J. and Simkins, B. (2010). *Enterprise Risk Management, Today's Leading Research and Best Practise for Tomorrow's Executives*. New York, NY: Wiley.

Gadd, S., Keeley, D., and Balmforth, H. (2003). Research Report RR151 – Good practice and pitfalls in risk assessment. Tratto il giorno May 11, 2021 da London: Health and Safety Executive: https://www.hse.gov.uk/research/rrpdf/rr151.pdf.

Gould, J., Glossop, M., and Ioannides, A. (2005). *Review of Hazard Identification Techniques*. Sheffield, UK: Health & Safety Laboratory.

Guarascio, M., Lombardi, M., Rossi, G., and Sciarra, G. (s.d.). Risk analysis and acceptability criteria. *WIT Transactions on the Built Environment* 94.

Guariniello, R. (2022). Prevenzione incendi: Luci e ombre. *Diritto & Pratica del Lavoro* 12.

Haddon, W. (1980). Advances in the epidemiology of injuries as a basis for public policy. *Public Health Rep.* 95 (5), 411–421.

Hoover, J., Bailey, J., Willauer, H., and Williams, F. (2005). *Evaluation of Submarine Hydraulic System Explosion and Fire Hazards*. Ft. Belvoir: Defense Technical Information Center.

Hoover, J., Bailey, J., Willauer, H., and Williams, F. (2008). Preliminary investigations into methods of mitigating hydraulic fluid mist explosions. *Fire Safety Journal* 43 (3): 237–240.

Hopkin, P. (2018). *Fundamentals of Risk Management*, 5e. London: Kogan Page.

HSE (2001). *Reducing Risks, Protecting People*. Norwich, UK: HSE Books.

HSE (2021). Risk management: Expert guidance. Tratto il giorno May 13, 2021 da Hse.gov.uk: https://www.hse.gov.uk/managing/theory/index.htm.

HSE (n.d.). *A Simplified Approach to Estimating Individual Risk*. London: The Health and Safety Executive.

HSE (s.d.). COMAH Guidance for the Surface Engineering Sector. Tratto il giorno May 11, 2021 da Health and Safety Executive: https://www.hse.gov.uk/surfaceengineering/comahguidance.pdf.

Hse.gov.uk (2020). Guidance on ALARP Decisions in COMAH – SPC/Permissioning/37. (s.d.). Tratto il giorno October 5, 2020 da https://www.hse.gov.uk/foi/internalops/hid_circs/permissioning/spc_perm_37.

Hunziker, S. (2019). *Enterprise Risk Management – Modern Approaches to Balancing Risk and Reward*. Rotkreuz: Springer Gabler.

Idaho National Laboratory (2005). *The SPAR-H Human Reliability Analysis Method*. Washington, U.S.: Nuclear Regulatory Commission Office of Nuclear Regulatory Research.

IEC (2016a). *61611-1 Functional Safety – Safety Instrumented Systems for the Process Industry Sector – Part 1: Framework, Definitions, System, Hardware and Application Programming Requirements*. Geneva, Switzerland : IEC.

IEC (2016b). *61882 Hazard and Operability Studies (HAZOP Studies) – Application Guide*. Geneva, Switzerland: IEC.

IEC (2019). *IEC 31010, Risk Management – Risk Assessment Techniques*. Geneva, Switzerland: International Electrotechnical Commission.

IEC (2020). *IEC 61511 – Functional Safety – Safety Instrumented Systems for the Process Industry Sector*, 2e. International Electrotechnical Commission.

Imperial Chemical Industries (1985). *The Mond Index*. London: ICI.

International Fire Code (2009). International Code Council http://www.iccsafe.org.

IOS (2012). *ISO 39001 – Road Traffic Safety Management Systems*. International Organization for Standardization.

IOS (2014). *ISO 19600 – Compliance Management Systems*. International Organization for Standardization.

ISO (2009). *ISO Guide 73, Risk Management – Vocabulary*. International Standard Organization.

ISO (2011). *ISO 19011, Guidelines for Auditing Management Systems*. International Standard Organization.

ISO (2012a). *ISO 17732: 2020-1 – Fire Safety Engineering – Fire Risk Assessment – Part 1: General*. Geneva, Switzerland: ISO.

ISO (2012b). *ISO 22301, Societal Security – Business Continuity Management Systems – Requirements*. International Standard Organization.

ISO (2013). *ISO/TR 31004, Risk Management – Guidance for the Implementation of ISO 31000*. International Standard Organization.

ISO (2014). *ISO 55000, Asset Management: Overview, Principles, and Terminology*. International Standard Organization.

ISO (2015a). *ISO 14001, Environmental Management Systems – Requirements with Guidance for Use*. International Standard Organization.

ISO (2015b). *ISO 31000 Risk Management – A Practical Guide for SMES*. International Standard Organization.

ISO (2015c). *ISO 9000: Quality Management Systems – Fundamentals and Vocabulary*. International Standard Organization.

ISO (2016a). *ISO 16294: Natural Gas Fuelling Stations – LGN Stations for Fuelling Vehicles*. Geneva, Switzerland: ISO.

ISO (2016b). *ISO 17776, Petroleum and Natural Gas Industries – Offshore Production Installations – Major Accident Hazard Management during the Design of New Installations*. International Standard Organization.

ISO (2018a). *ISO 31000, Risk Management – Guidelines*. Geneva, Switzerland: International Standard Organization.

ISO (2018b). *ISO 45000, Occupational Health and Safety*. International Standard Organization.

ISO (2019). *ISO Standard 10418: Petroleum and Natural Gas Industries — Offshore Production Installations — Process safety systems*. Geneve, Switzerland: International Standard Organization.

ISO (2020). *ISO 31022: Risk Management – Guidelines for the Management of Legal Risk*, 1e. Geneva, Switzerland: International Standard Organization.

ISO (s.d.). *The Implementation of ISO 31000*. International Standard Organization.

ISO/IEC (2012). *ISO/IEC 17024: Conformity Assessment – General Requirements for Bodies Operating Certification of Persons*. International Standard Organization/International Electrotechnical Commission.

ISO/IEC (2013a). *ISO/IEC 27001: Information Technology – Security Techniques – Information Security Management Systems – Requirements*. International Standard Organization/International Electrotechnical Commission.

ISO/IEC (2013b). *ISO/IEC 27002: Information Technology – Security Techniques – Code of Practice for Information Security Controls*. International Standard Organization/International Electrotechnical Commission.

ISO/IEC (2018a). *ISO/IEC 27000: Information Technology – Security Techniques – Information Security Management Systems – Overview and Vocabulary*. International Standard Organization/International Electrotechnical Commission.

ISO/IEC (2018b). *ISO/IEC 27005, Information Technology. Security Techniques*. Information Security Risk Management.

IWA (2020). *IWA 31: Risk Management – Guidelines on Using ISO 31000 in Management Systems*. Geneva, Switzerland: International Workshop Agreement.

Joglar, F. and Ontiveros, V. (2022). What's new: SFPE engineering guide to fire risk assessment. *Fire Protection Engineering* Q2.

Khan, F. and Abbasi, S. (1998). Multivariate hazard identification and ranking system. *Process Safety Progress* 17 (3): 157–170.

Khan, F., Husain, T., and Abbasi, S. (2001). Safety Weighted Hazard index (SWeHI). *Process Safety and Environmental Protection* 79 (2): 65–80.

Khoo, B. and Rein, G. (2022). Building fire protection layers: How they failed in the 1973 Summerland fire. *Fire Protection Engineering* Q1.

Kidam, K., Sahak, H., Hassim, M. et al. (2016). Inherently safer design review and their timing during chemical process development and design. *Journal of Loss Prevention in the Process Industries* 42: 47–58.

Kletz, T. and Amyotte, P. (2010). *Process Plants: A Handbook for Inherently Safer Design*, 2e. Boca Raton, FL: CRC Press/Taylor & Francis.

Lentini, J.J. (2006). *Scientific Protocols for Fire Investigation*. Boca Raton, FL: CRC Press.

London, H.B. (A cura di). (1996). Health Safety Executive Improving Inherent Safety. Offshore Technology Reports. Tratto il giorno January 22, 2021 da https://www.hse.gov.uk/research/othpdf/500-599/oth521.pdf.

Louisot, J. and Ketcham, C. (2014). *ERM – Enterprise Risk Management*. Hoboken, NJ: Wiley.

Lumbe Aas, A. (2008). The human factor assessment and classification system (HFACS) for the oil & gas industry.

Mannan, S. and Lees, F. (2012). *Lee's Loss Prevention in the Process Industries*, 4e. Boston, MA: Butterworth-Heinemann.

Mansfield, D., Turney, R., Rogers, R. et al. How to integrate inherent SHE in process development and plant design. ICHEME Symposium Series No. 139.

Manton, M. and Eccles, B. (2014). A simple end-to-end guide to Comah report writing and management. Symposium series No. 159. London: IChemE. (IChemE, A cura di) Tratto il giorno April 22, 2021 da https://www.icheme.org/media/8906/xxiv-paper-10.pdf.

Maritan, F., Masiero, C., and Rossato, M. (2020). Un metodo di valutazione dei rischi basato su riferimenti normativi: Dalla BS 18004 alla ISO/TR 14121-2. *Igiene & Sicurezza del Lavoro* 8-9: 455–460.

Marmo, L., Piccinini, N., and Fiorentini, L. (2013). Missing safety measures led to the jet fire and seven deaths at a steel plant in Turin. Dynamics and lessons learned. *Journal of Loss Prevention in the Process Industries* 26 (1): 215–224.

Marmo, L., Piccinini, N., Russo, G. et al. (2013). Multiple tank explosions in an edible-oil refinery plant: A case study. *Chemical Engineering & Technology* 36 (7): 1131–1137.

McGrattan, K., Baum, H., and Rehm, R. (1998). Large eddy simulations of smoke movement. *Fire Safety Journal* 30 (2): 161–178.

McGrattan, K., Hamins, A., and Stroup, D. (1998). Sprinkler, smoke & heat vent, draft curtain interaction. Large scale experiments and model development. Gaithersburg, MD: National Institute of Standards and Technology, NISTIR 6196-1.

McGrattan, K., Hostikka, S., Floyd, J. et al. In cooperation with VTT Technical Research Centre of Finland (2008). NIST Special Publication 1018-5. Fire dynamics simulator (Version 5) technical reference guide. NIST – National Institute of Standards and Technology, US Department of Commerce.

McGrattan, K., Klein, B., Hostikka, S., and Floyd, J., NIST special publication 1019-05 fire dynamics simulator (version 5), user's guide, NIST National Institute of Standards and Technology, U.S. Department of Commerce, USA.

McLeod, R. (2019). *Human Factors in Barrier Thinking*. Presentation to 21st National Biomedical & Clinical Engineering conference, Nottingham.

McSween, T. (2003). *The Values-Based Safety Process: Improving Your Safety Culture with Behavior-Based Safety*, 2e. Hoboken, NJ: Wiley.

Middleton, M. and Franks, A. (2001). Using risk matrices. *The Chemical Engineer*, 723 34–37.

Ministero dell'Interno (2022). Technical fire prevention standards. (F. A. Ponziani, A cura di) Rome: Direzione centrale per la prevenzione e la sicurezza tecnica.

National Fire Protection Association (2019). *NFPA 59A: Standard for the Production, Storage, and Handling of Liquefied Natural Gas (LGN)*. Quincy, MA: NFPA.

National Fire Protection Association (2022a). *NFPA 550 – Guide to the Fire Safety Concepts Tree*. Quincy, MA: NFPA.

National Fire Protection Association (2022b). *NFPA 551 – Guide for the Evaluation of Fire Risk Assessments*. Quincy, MA: NFPA.

NFPA (National Fire Protection Association) (2017a). *NFPA 550: Guide to the Fire Safety Concepts Tree*. NFPA (National Fire Protection Association).

NFPA (National Fire Protection Association) (2017b). *NFPA 921: Guide for Fire and Explosion Investigations*. NFPA (National Fire Protection Association).

NFPA (National Fire Protection Association) (2017c). *NFPA 921: Guide for Fire and Explosion Investigations*. NFPA (National Fire Protection Association).

NFPA 101. (2021).Life safety code. National Fire Protection Association. http://www.nfpa.org.

NFPA. (2021). *NFPA 921 Guide for Fire and Explosion Investigations*, 2008e. Quincy, MA: National Fire Protection Association.

Olivo, J. (1994). Loss prevention in a modern ethylene plant. *Journal of Loss Prevention in the Process Industries* 7 (5): 403–412.

Olson, D. and Dash, D. (2017). *Enterprise Risk Management Models*. New York, NY: Springer.

Olson, D. and Wu, D. (2017). *Enterprise Risk Management Models*. New York, NY: Springer.

Pasquini, M. (2020). *Tecnica della prevenzione incendi – Teoria dei fenomeni di combustione e pratiche per la prevenzione*. Palermo, Italy: Dario Flaccovio Editore.

Perry, M. (2019). Why reputation could be your biggest future risk. Racounter (0583), 1–4.

Philley, J. (2020). Collar Hazards with a Bow-Tie. Tratto il giorno September 28, 2020 da Chemicalprocessing.com: http://www.chemicalprocessing.com/articles/2005/612.html?page=1.

Piccinini, N. and Ciarambino, I. (1997). Operability analysis devoted to the development of logic trees. *Reliability Engineering & System Safety* 55 (3): 227–241.

Popov, G., Lyon, B., and Hollcroft, B. (2016). *Risk Assessment. A Practical Guide to Assessing Operational Risks*. New York, NY: Wiley.

Posada, J., Patel, A., Roes, A. et al. (2013). Potential of bioethanol as a chemical building block for biorefineries: Preliminary sustainability assessment of 12 bioethanol-based products. *Bioresource Technology* 135: 490–499.

Quintiere, J.G. (1998). *Principles of Fire Behavior*. Albany, NY: Delmar Publishers.

Quintiere, J.G. (2006). *Fundamentals of Fire Phenomena*. Chichester, UK: John Wiley & Sons Ltd.

Quintiere, J.G. and Karlsson, B. (2000). *Enclosure Fire Dynamics*. Boca Raton, FL: CRC Press.

Rahman, M., Heikkilä, A., and Hurme, M. (2005). Comparison of inherent safety indices in process concept evaluation. *Journal of Loss Prevention in the Process Industries* 18 (4–6): 327–334.

Rasmussen, J. (1983). Skills, rules, and knowledge: Signals, signs, and symbols, and other distinctions in human performance models. *IEEE Transactions on Systems, Man, and Cybernetics* 13 (3): 257–266.

Rathnayaka, S., Khan, F., and Amyotte, P. (2014). Risk-based process plant design considering inherent safety. *Safety Science* 70: 438–464.

Rausand, M. and Haugen, S. (2013). *Risk Assessment*. New York, NY: Wiley.

Reason, J. (1990a). *Human Error*. Cambridge: Cambridge University Press.

Reason, J. (1990b). The contribution of latent human failures to the breakdown of complex systems. *Philosophical Transactions of the Royal Society of London B, Biological Sciences* 327 (1241): 475–484.

Rees, M. (2016). *Business Risk and Simulation Modelling in Practice*. New York, NY: Wiley.

Rivière, C. and Marlair, G. (2009). BIOSAFUEL®, a pre-diagnosis tool of risks pertaining to biofuels chains. *Journal of Loss Prevention in the Process Industries* 22 (2): 228–236.

Russo, P., Coccorullo, I., and Russo, G. (2017). Incendio di un capannone durante l'incendio dell'impianto fotovoltaico. IF CRASC 17. Milano: Dario Flaccovio Editore, 179–188.

Saraf, S. (2010). Biodiesel accident trend continues in 2010. [Blog] Risk Safety, Available at: http://risk-safety.com/biodiesel-accident-trend-continues-in-2010 (Accessed 15 May 2021).

Schmidt, M. (2016). Making sense of risk tolerance criteria. *Journal of Loss Prevention in the Process Industries* 41: 344–354.

Shariff, A. and Leong, C. (2009). Inherent risk assessment – a new concept to evaluate risk in preliminary design stage. *Process Safety and Environmental Protection* 87 (6): 371–376.

Shariff, A., Leong, C., and Zaini, D. (2012). Using process stream index (PSI) to assess inherent safety level during preliminary design stage. *Safety Science* 50 (4): 1098–1103.

Shariff, A., Wahab, N., and Rusli, R. (2016). Assessing the hazards from a BLEVE and minimizing its impacts using the inherent safety concept. *Journal of Loss Prevention in the Process Industries* 41: 303–314.

Siu, N., Herring, S., Cadwallader, L. et al. (1997). *Interim Qualitative Risk Assessment for an LGN Refueling Station and Review of Relevant Safety Issues*. Washington, DC: United States Department of Energy, Office of Energy Efficiency and Renewable Energy.

Siu, N., Herring, S., Cadwallader, L. et al. (1999). *Qualitative Risk Assessment for an LNG Refueling Station and Review of Relevant Safety Issues*, Revision 2. Washington, DC: Department of Energy, Office of Environmental.

Sklet, S. (2006). Safety barriers: Definition, classification, and performance. *Journal of Loss Prevention in the Process Industries* 5 (19): 494–506.

Spinardi, G. (2016). Fire safety regulation: Prescription, performance and professionalism. *Fire Safety Journal* 80: 83–88.

Standards Council of Canada (2006). *CAN/CSA Z276-01 Natural Gas (LGN) – Production, Storage, and Handling*. Standards Council of Canada.

Strobhar, D. (2013). *Human Factors in Process Plant Operation*. Highland Park: Momentum Press.

Tafaro, S. (2011). *Un caso studio per la valutazione del rischio industriale – Incendio in una pipe-way di raffineria*. Roma: Corpo nazionale dei Vigili del Fuoco – Direzione Centrale per la Formazione.

Taylor, J. (2016). *Human Error in Process Plant Design and Operations*, 1e. Boca Raton, FL: Taylor & Francis.

The British Standard Institution (2007). *PAS 911:2007 – Fire Strategies – Guidance and Framework for Their Formulation*. London: BSI.

The British Standard Institution (2017). *BS 9999: 2017 – Fire Safety Design, Management and Use of Buildings – Code of Practice*. London: BSI.

The British Standard Institution (2019). *BS 7974: 2019 – Application of Fire Safety Engineering Principles to the Design of Buildings – Code of Practice*. London: BSI.

The Keil Centre and Edmonds, J. (2016). *Human Factors in the Chemical and Process Industries: Making It Work in Practice*, 1e. Amsterdam: Elsevier.

Thiruvenkataswamy, P., Eljack, F., Roy, N. et al. (2016). Safety and techno-economic analysis of ethylene technologies. *Journal of Loss Prevention in the Process Industries* 39: 74–84.

TNO (1992) Methods for the determination of possible damage to people and objects resulting from releases of hazardous materials.

TNO (1997) Methods for determining and processing probabilities.

TNO (1999) Guidelines for quantitative risk assessment.

TNO (2005) Methods for the calculation of physical effects due to releases of hazardous materials (liquids and gases).

Tranchard, S. (2018). The new ISO 31000 keeps risk management simple. Tratto il giorno September 21, 2020 da ISO.org: https://www.iso.org/news/ref2263.html.

Tschurtz, H. and Schedl, G. (2010). An integrated project management life cycle supporting system safety. In: *Making Systems Safer* (C. Dale and T. Anderson). London: Springer.

U.S. Bureau of Labor Statistics (2013). Occupations with high fatal work injury rates, 2013. Injuries, Illnesses, and Fatalities, 2013 Chart Package. Washington, DC.

UNI – Ente Italiano di Normazione (2020). *UNI ISO/TR 16732 – Ingegneria della sicurezza contro l'incendio – Valutazione del rischio d'incendio*. Roma: UNI.

Walton, W., Thomas, P., and Ohmiya, Y. (2016). Estimating temperatures in compartment fires. In: *SFPE Handbook of Fire Protection Engineering* (M. Hurley), 196–1023. New York, NY: Springer. doi:10.1007978-1-4939-2565-0_30.

Ward, E., Lee, G., Botelho, D. et al. (2000). Consequence-based criteria for the Gulf of Mexico: Philosophy & results. Offshore Technology Conference.

Weyenbergea, V., Crielc, P., Deckersa, X. et al. (2017). Response surface modelling in quantitative risk analysis for life safety in case of fire. *Fire Safety Journal*, 91, 1007–1015.

Wolski, A., Dembsey, N.A., and Meacham, B. (2000). Accommodating perceptions of risk in performance-based building fire safety code development. *Fire Safety Journal* 34: 297–309.

Woods, D., Dekker, S., Cook, R. et al. (2010). *Behind Human Error*, 2e. Farnham, UK: Ashgate.

World Economic Forum (2020). The global risks report 2020.

Yeoh, G.H. and Yuen, K.K. (2008). *Computational Fluid Dynamics in Fire Engineering – Theory, Modeling & Practice*. London: Butterworth-Heinemann.

Index

Note: page numbers in *italics* refer to figures; those in **bold** to tables.

Fire Risk Management: Principles and Strategies for Buildings and Industrial Assets, First Edition.
Luca Fiorentini and Fabio Dattilo.
© 2023 John Wiley & Sons, Inc. Published 2023 by John Wiley & Sons, Inc.
Companion Website: www.wiley.com/go/Fiorentini/FireRiskManagement

Printed and bound by CPI Group (UK) Ltd, Croydon, CR0 4YY

16/04/2025

14658473-0005